Renewable Energy Integration

Renewable Energy Integration

Practical Management of Variability, Uncertainty, and Flexibility in Power Grids

Lawrence E. Jones, Ph.D.

AMSTERDAM • BOSTON • HEIDELBERG • LONDON
NEW YORK • OXFORD • PARIS • SAN DIEGO
SAN FRANCISCO • SINGAPORE • SYDNEY • TOKYO
Academic Press is an Imprint of Elsevier

Academic Press is an imprint of Elsevier
32 Jamestown Road, London NW1 7BY, UK
225 Wyman Street, Waltham, MA 02451, USA
525 B Street, Suite 1800, San Diego, CA 92101-4495, USA

Notice

No responsibility is assumed by the publisher for any injury and/or damage to persons or property as a matter of products liability, negligence or otherwise, or from any use or operation of any methods, products, instructions or ideas contained in the material herein. Because of rapid advances in the medical sciences, in particular, independent verification of diagnoses and drug dosages should be made.

Library of Congress Cataloging-in-Publication Data
Application Submitted

British Library Cataloguing-in-Publication Data
A catalogue record for this book is available from the British Library

ISBN: 978-0-12-407910-6

For information on all Academic Press publications
visit our website at elsevierdirect.com

Praise for Renewable Energy Integration

In order to double the share of renewable energy in the global energy mix – one of the three goals of the UN Sustainable Energy for All initiative - there will need to be tools and methods for integrating high levels of variable renewable electricity into power systems and markets worldwide. This book makes an important contribution to the regulatory, operations, economic and technical aspects of that challenge. By bringing together cutting edge approaches, Dr. Jones has done much of the hard work for us. It is an extraordinary snapshot of the state-of-the-art, and I am very glad to recommend it to decision-makers in both industrialized and emerging economies alike.

Dr. Kandeh Yumkella, Under Secretary of the United Nations, Special Representative to the United Nations Secretary General, and CEO for UN Sustainable Energy for All (SE4All) Initiative General of the United Nations, Special Representative of the United Nations Secretary Sustainable Energy for All (SE4All)

With the demand for water, food and energy growing beyond all measure and with the supply of these inextricably linked 'resource spheres' under increasing threat, we are facing what many experts predict will be a 'perfect storm'. The threat to human life, as well as to whole sectors of the economy, is very real. Renewable energy can be a vital part of the solution and if this comprehensive and authoritative set of essays can help to accelerate both the generation and integration of renewable energy supplies then it will have served an invaluable purpose.

Paul Polman, Chief Executive Officer of Unilever, and Chairman, World Business Council for Sustainable Development

A typically outstanding effort by Dr. Jones and his assembled expert authors. A timely, "must read" for managing the energy trifecta of addressing climate concerns and energy poverty while maintaining economic viability and promoting more secure, reliable and sustainable fuel choices. The chapters deal head on with the key issues of the day (VER, storage, distributed energy, etc.) and suggest that while we should enjoy the success of the unconventionals revolution, we need to use the breathing space this moment provides to seriously move on to more sustainable energy forms.

Frank Verrastro, Senior Vice President and James Schlesinger Chair for Energy & Geopolitics, Center for Strategic and International Studies

Bravo! This book is an important resource. As renewable energy plays an increasingly important role in electric grids in the years ahead, this rich volume will help policymakers, utility executives, technology providers and many more.

David Sandalow, Inaugural Fellow, Center on Global Energy Policy, Columbia University

The efficient integration of renewable energy is one of the most important challenges posed by the move towards sustainable energy systems. Renewable energy challenges the norms and traditions accumulated over the last century, and it requires new dynamic approaches that match the needs and requirements of a modern, sustainable power system. Many of these issues are considered in this publication, which gives new insights into how power systems can move forward and provide society with clean, reliable and affordable electricity.

Christian Pilgaard Zinglersen, Deputy Permanent Secretary, Danish Ministry of Climate, Energy and Building

The use of renewable energy in modern power systems has accelerated rapidly in recent years — beyond what some skeptics thought possible. There could not be a more timely topic than the practical integration of these resources into large-scale grids. This collection of expert guidance is not only valuable now, but surely will need a fresh edition on an annual basis for the foreseeable future as technology continues to evolve.

Reid Detchon, Vice President, United Nations Foundation, and Executive Director, Energy Futures Coalition

Dr. Lawrence Jones has assembled an exceptional team of experts to provide deep insights into the challenges of fully leveraging renewable generation across the globe. This book will serve as a great reference source for interested readers from all levels of knowledge regardless of their area of interest. From policy to engineering to operations, it has insights for all. Innovation in the electric energy sector offers great promise for clean, reliable, resilient and affordable power across the globe, however this same innovation is increasing the complexity of an already complex system. This book gives the reader an introduction into this promise as well as into the complexity that it will bring.

Becky Harrison, Chief Executive Officer, GridWise Alliance

Transitioning our power system to clean, renewable energy is one of the most important challenges of our lifetime. In many ways the task is familiar, as since the days of Edison and Westinghouse grid operators have accommodated fluctuating electricity demand and abrupt power plant failures to keep electricity supply and demand in balance. From remote Pacific islands to mainland Europe, Jones insightfully spans the globe to distill the success stories of grid operators who now reliably obtain more than a quarter of their electricity from wind and solar energy. The path forward for integrating even higher levels of renewable energy is clear, and we have the technology to do it today.

Rob Gramlich, Senior Vice President, American Wind Energy Association

Electrical systems around the world are undergoing radical change due to the rapid growth of solar and wind energy. We must modernize the grid to make it compatible with these critically important energy sources. This collection provides real-world examples of how the power sector, and society's leaders generally, can achieve this goal, which is key to energy security, environmental protection, and economic progress.

Andrew L. Shapiro, Founder & Partner, Broadscale Group

As the world searches for pathways towards a sustainable and inclusive energy future, one of the fundamental opportunities lies in ensuring that renewable energy technologies meet their vast potential. To that end, it has become evident that we need to urgently address the tools, regulations, and operational and institutional issues that will serve to elegantly integrate renewable energy generation into the wider power system. Through rigorous analysis and sensitively designed contributions, Dr. Jones has brought us a book on just the right topic at just the right time. It clearly and coherently presents the state-of-the-art on this complex set of issues, and provides us with the confidence that these challenges can be addressed.

Dr. Morgan Bazilian, Adjunct Professor, Sustainable Engineering Lab, Columbia University

To simultaneously address climate change and meet the needs of the global poor for clean energy, renewable energy on a very large scale will have to play a central role. This book provides a detailed response to the central challenge in making this dream a reality: how to integrate clean but intermittent energy sources within utility systems that require a high degree of central planning and coordination.

Alan Miller, Principal Climate Change Specialist, International Finance Corporation (retired)

Solar and wind power is growing around the globe. Merits are obvious; fuel free electricity production is advantageous in terms of climate footprint and absence of other pollutants. However, integration of these variable power sources is challenging. This book is a comprehensive collection of contributions ranging from very technical challenges to market models and policies for this new era of electricity. Read and you will broaden and deepen your expertise in how to best integrate renewables in our power systems.

Dr. Magnus Olofsson, President, Elforsk—Swedish Electrical Utilities' Research & Development Company

Great book! Lawrence Jones has managed to capture the most important renewable energy topics in a single volume, and he has done so through the contributions of working experts in each topic. If you are interested in renewable energy integration, this book captures the current state-of-the-art for the entire field.

Mark Ahlstrom, CEO WindLogics

Renewable generation is becoming ever more prolific. The timing of this book is perfect. It combines practical examples with theory and will guide decision makers dealing with today's issues as well as those seeking ways to deal with tomorrow's challenges. The lessons learned will help avoid pitfalls and provide insight and inspiration. The topics covered are relevant to both developed and developing countries, those countries starting from a low renewables base as well as those with high proportions of renewables.

Eric Pyle, Chief Executive New Zealand Wind Energy Association

The timing of the publication is just perfect. Renewable energy has gone mainstream globally i.e. 45 GW of new wind installations in 2013. The content and focus of this remarkable book is both unique and demanding. It's all about integration: of markets, physical infrastructure, policies. This integrated approach is as often lacking in current debates as it is needed for progress. And the design both of the modern electricity markets and a modern grid are crucial for a transition to safer, cleaner energy world of the future. No transition without transmission, and no communication without electrification. Reading this book you might learn how integration can accelerate the transition.

Dr. Klaus Rave, Chairman Global Wind Energy Council

With wind and solar energy expanding at an ever-quickening pace, the time is right for a thorough and cross-disciplinary assessment of the integration challenge. This book hits the mark, with the industry's leading experts addressing a wide assortment of topics that are central to managing higher shares of variable generation.

Dr. Ryan H. Wiser, Staff Scientist, Lawrence Berkeley National Laboratory

Renewable Energy Integration is a critically needed and wonderfully comprehensive book that highlights the next frontier; not how much renewable energy potential exists, but how to most effectively and seamlessly merge this new power system with the old one.

Daniel Kammen, Class of 1935 Distinguished Professor of Energy, University of California, Berkeley

Understanding the intricacies discussed in Renewable Energy Integration is a predicate for achieving universal access to affordable, sustainable, reliable energy across a diverse portfolio of fuel sources. Towards this end, we must be able to maintain the balance and resilience of the power grid using technology, regulatory, and market forces. Dr. Lawrence Jones' outstanding compendium, based on an in-depth array of insights from an unique cast of renowned thought leaders, demonstrates that he clearly understands how critical this subject is for quality of life, continued economic growth and prosperity around the globe.

Hon. Vicky A. Bailey, former Assistant Secretary, International Affairs and Domestic Policy, Department of Energy and Former Commissioner, Federal Energy Regulatory Commission

There are many that have made a convincing case that we could move to 80% renewable electricity generation by 2030. As we unlock the greatest wealth creation opportunity since the mobile phone revolution, I am sure this resource from Dr. Jones and his assembled dreamteam will find its way onto the desks of every major grid operator and electricity policymaker in the World.

Jigar Shah, Founder SunEdison and Author of Creating Climate Wealth

The future of the energy landscape cannot be envisioned without taking into account renewable energy. It is a secret for no one however that the integration of renewable energy into the grid is an important challenge that will need to be overcome if we want to ensure its deployment to full capacity. Dr. Lawrence Jones brings together critical contributions from experts across the globe to address precisely these issues in a must-read, unique publication. It is an invaluable resource for anyone in the industry who wants a comprehensive overview of one of today and tomorrow's hottest topics.

Pierre Bernard, Founder and Managing Partner, Bernard Energy Advocacy

This book is dedicated to my parents, Emmanuel E. W. Jones, Jr. and Comfort H. Jones, who taught me many valuable lessons in life, two of which guided me especially on this journey: to always value and respect humanity and nature; and to work with people from different backgrounds toward a higher purpose.

This book honors the operators of power grids around the world. They are the unsung heroes and heroines who work around the clock, ensuring that we have electricity to light up our nights and fuel our lives.

Contents

PART 4   FORECASTING RENEWABLES

PART 7   DEMAND RESPONSE AND DISTRIBUTED ENERGY RESOURCES

PART 8 VARIABLE ENERGY RESOURCES IN ISLAND POWER SYSTEMS

PART 9 SOLAR, TIDAL AND WAVE ENERGY INTEGRATION

PART 10 ENABLING AND DISRUPTIVE TECHNOLOGIES FOR RENEWABLE INTEGRATION

About the Contributors

V.K. Agrawal is Executive Director at Power System Operation Corporation Ltd (POSOCO), in charge of the National Load Dispatch Centre of India. He has an M. Tech in 'Power Apparatus and Systems' from Indian Institute of Technology, Delhi and more than 33 years of experience in the power sector. Mr Agrawal has worked on the expansion and synchronization of large regional grids and development of power markets in India.

Stefano Alessandrini earned his Ph.D. in Environmental Science at the University of Piemonte Orientale, Italy. He works as a scientist at NCAR, USA. His main areas of interest are wind power forecast and mesoscale and air pollution modeling. He has published several articles in peer-reviewed journals.

Anders N. Andersen holds a master's degree in Mathematics and Physics, and a bachelor's degree in Economics from Aarhus University. He is Head of the Energy Systems Department at EMD International A/S and is Extension Associate Professor in energy planning at Aalborg University. He has more than 25 years of experience in the energy industry.

Göran Andersson obtained his M.S. (1975) and Ph.D. (1980) degrees from the University of Lund, Sweden. Since 2000 he is full professor in electric power systems at ETH Zürich, where he also heads the powers system laboratory. His research interests include power systems dynamics and control, power markets, and future energy systems.

Reza Arghandeh joined the University of California, Berkeley, California Institute of Energy and Environment in 2013. Dr Arghandeh is currently Vice-Chair of ASME Advanced Energy Systems committee. He received his M.S. in Industrial Engineering and Ph.D. in Electrical Engineering from Virginia Tech, USA. He holds a M.S. in Mechanical Engineering from the University of Manchester, UK.

George W. Arnold has over 40 years of experience in the telecommunications, information technology, and energy sectors in both industry and government. Dr Arnold was a Vice President at Bell Labs and served as National Coordinator for Smart Grid Interoperability at the National Institute of Standards and Technology. He received his Eng.Sc.D. degree in EE and CS from Columbia University.

Tatiana M.L. Assis received her D.Sc. degree in Electrical Engineering from the Federal University of Rio de Janeiro, Brazil. Over the last 15 years, Dr Assis has worked with power systems dynamics and her experience includes collaborations with the Brazilian Electric Power Research Center and the Brazilian ISO. She is currently a Professor at Federal University of Rio de Janeiro.

Chaitanya A. Baone received the B.Tech. degree in Electrical Engineering from Visvesvaraya National Institute of Technology, Nagpur, India, in 2006, and the M.S. and Ph.D. degrees in Electrical Engineering from the University of Wisconsin–Madison, Madison in 2009 and 2012, respectively. Dr Baone is currently with the Electric Power Systems Laboratory, GE Global Research Center, Niskayuna, New York, where his research interests include power system dynamics, control, estimation, and optimization.

Carl Barker holds a B.Eng from Staffordshire University and an M.Sc. from Bath University in the UK. He is presently the Chief Engineer at Alstom Grid, and is responsible for HVDC Grids, within the

Power Electronic Activities business. Carl is a Chartered Engineer in the UK and a member of the IET (UK), a Senior Member of the IEEE, and a member of CIGRE

Diane Broad has 18 years experience in the energy industry, working as a consulting engineer with utilities and project developers to increase the proportion of renewable energy in the electric system. Ms Broad is currently a Managing Consultant with Ecofys. She holds a BSEE from Colorado State University and is a registered Professional Engineer.

Maxime Baudette is a Ph.D. student in the Electric Power Systems department at the KTH Royal Institute of Technology, Stockholm, Sweden. He received the M.Sc. degree in Electrical Engineering from KTH (Sweden) and Supélec (France) in 2013.

Audun Botterud has 15 years of experience with electricity markets and renewable energy in the United States and Europe. Dr Botterud is a research scientist at Argonne National Laboratory. He received his M.Sc. in Industrial Engineering (1997) and his Ph.D. in Electrical Power Engineering (2004) from the Norwegian University of Science and Technology in Trondheim, Norway.

Richard Candy has worked for the South African power utility Eskom since 1972. Dr Candy's expertise lies within the system operations department where he has focused his attention on the control room, Man Machine Interface where he has provided considerable advances in information visualization, advanced alarm processing, and situational awareness facilities for the control staff.

Spyros Chatzivasileiadis received his Diploma in Electrical and Computer Engineering from the National Technical University of Athens, Greece (2007) and his Ph.D. in Power Systems from ETH Zürich (2013). Spyros is currently a research fellow at Lawrence Berkeley National Laboratory, California. His research interests include power system planning and operation, VSC-HVDC lines, and load management.

Puneet Chitkara has a Ph.D. in energy economics from IGIDR, Mumbai, India and is a Director with AF-Mercados EMI. He works on electricity market design, transmission, power trading, and renewable energy. Dr Chitkara's areas of concentration include power system economics, application of game theoretic, and optimization techniques to power systems.

Erik Connors is a senior research associate at SA Technologies in Marietta, GA, involved in the development of advanced user-centric system designs for the power system industry. Dr Connors received his Ph.D. from the College of Information Science and Technology at Pennsylvania State University.

Anish De is an engineer and has an MBA from XIM, India, and is the CEO of AF-Mercados EMI in India since 2008. He specializes in the fields of energy market design, pricing, trading, renewable energy, fuels, and utility regulation. He has been actively involved in energy policy formulation in India and South Asia.

Luca Delle Monache is a recognized expert in developing methods for uncertainty quantification, and he has published 25 peer-reviewed articles and two book chapters. Dr Delle Monache has received an M.S. in Math from the University of Rome, Italy, an M.S. in Meteorology from the San Jose State University, USA, and a Ph.D. in Atmospheric Science from the University of British Columbia, Canada. Currently he is a scientist at NCAR, USA.

Christopher L. DeMarco has been a member of the faculty at the University of Wisconsin–Madison, Madison, since 1985. Dr DeMarco has served as Electrical and Computer Engineering Department Chair, and is currently Grainger Professor of Power Engineering and Site Director for the Power Systems Engineering Research Center (PSERC). His research and teaching interests center on dynamics and control of energy systems.

Daniel Dobbeni has over 40 years of experience in the energy industry. He was CEO of the Elia Group from 2003 to 2012 and President of the European Network of Transmission System Operators from incorporation till June 2013. He is currently Chairman of 50Hertz GmbH, and Coreso S.A. Mr. Dobbeni received his MSc in Industrial Engineering in Brussels and Business Management from Vlerick, Belgium and CEDEP, France.

Ken Dragoon has more than 30 years in the power industry with responsibilities including power system planning, renewable resource acquisition and integration, and risk analysis. He is currently a Managing Consultant for Ecofys, a sustainable energy consulting firm. He authored a wind integration book: Valuing Wind Energy on Integrated Power Systems published by Elsevier in 2010.

John Dumas has 27 years of experience in the electric power industry. He joined Electric Reliability Council of Texas in 2004 and is currently the Director of Wholesale Market Operations where he is responsible for all Day-Ahead, Real-Time, and Congestion Revenue Rights market activities. Mr Dumas earned his bachelor's degree in Electrical Engineering from the University of Texas at Arlington.

Erik Ela is a senior engineer at National Renewable Energy Laboratory and a lead in steady-state power system operations, wholesale electricity market design, and other topics related to bulk power integration of renewable resources. He has the B.S. and M.S. degrees in Electrical Engineering and previously worked for the New York Independent System Operator implementing improvements to markets and operations.

Mica R. Endsley is currently serving as the Chief Scientist of the United States Air Force. Prior to assuming this position she served as President and CEO of SA Technologies, and as the faculty at Texas Tech University and MIT. Dr Endsley received her Ph.D. from the University of Southern California in Industrial and Systems Engineering.

Pavel V. Etingov graduated with honors from Irkutsk State Technical University specializing in electrical engineering in 1997. P.V. Etingov received his Ph.D. degree in 2003 from the Energy Systems Institute of the Russian Academy of Sciences, Irkutsk, Russia. He is currently a senior research engineer at Pacific Northwest National Laboratory, Richland, WA.

Steven Fine is an expert on environmental markets and has led numerous engagements, including work for the Edison Electric Institute, the American Wind Energy Association, and America's Natural Gas Alliance. He was an invited panelist to a U.S. Senate discussion on the future of air regulatory legislation conducted by Senators Carper and Alexander.

Jarett Goldsmith is an expert in wave and tidal energy at DNV GL - Energy, joining in 2011 to expand the renewables advisory group's service offerings within North America. He received his B.Sc. in Mechanical Engineering from the University of California, Santa Barbara as the valedictorian of the College of Engineering, and holds joint M.Sc. degrees in Management and Engineering of Environment and Energy from three leading European universities.

Santiago Grijalva is the Director of the Advanced Computational Electricity Systems Laboratory and Associate Director of the Strategic Energy Institute at the Georgia Institute of Technology. Dr Grijalva is a leading researcher on decentralized control and management architectures for sustainable electricity systems. His graduate degrees in Electrical and Computer Engineering are from the University of Illinois at Urbana–Champaign.

Udi Helman has worked on analysis and design of electric power markets since the mid-1990s, including with the U.S. Federal Energy Regulatory Commission and the California ISO. His current work is focused on renewable energy and storage. He has a Ph.D. in applied economics and systems analysis from The Johns Hopkins University.

Anders Plejdrup Houmøller has more than 20 years of experience in the electricity supply industry and seven years of experience from the IT and software industry. Mr Houmøller holds a Master degree in Physics, a Bachelor degree in Mathematics, a Bachelor degree in Computer Science, and a Bachelor degree in Commerce.

Mark Howells is professor at the Division of Energy System Analysis at KTH. His work focuses on developing methodologies to support decision making and understand systems interactions from the local to the global level. Previously, Mark worked as an economist and energy planner at the International Atomic Energy Agency.

Brendan Kirby is a private consultant with 39 years of experience in the energy industry, retired from the Oak Ridge National Laboratory. Brendan is a licensed Professional Engineer with an M.S degree in Electrical Engineering from Carnegie-Mellon University and a B.S. in Electrical Engineering from Lehigh University. Publications are available at www.consultkirby.com

Kiran Kumaraswamy is an expert on transmission markets and has performed numerous transmission and power market studies. His expertise includes the areas of transmission asset valuation, due diligence, Locational Marginal Price (LMP) forecasting, merchant transmission investment assessment and power systems modeling. He also specializes in distributed generation modeling, generation interconnection and NERC Reliability Standards Compliance.

Helena Lindquist is the CEO and founder of LightSwitch, a company devoted to accelerating international knowledge and technology transfer in the fields of renewable energy and sustainability. She has an academic and professional background in international relations and European affairs, and holds a Masters Degree in European Studies from the University of Bath, UK.

Clyde Loutan has over 25 years of experience in the energy industry. He started his career at Pacific Gas and Electric Company and is currently a Senior Advisor at the California Independent System Operator Corporation focusing on renewable integration and control performance. Mr. Loutan holds B.S. and M.S. degrees in Electrical Engineering from Howard University, Washington D.C.

Jian Ma has a Ph.D. degree in Electrical Engineering from The University of Queensland, Brisbane, Australia, in 2008. He is currently a Senior Electrical Engineer with Burns & McDonnell, Kansas City, MO. Mr. Ma is a Senior Member of IEEE, a licensed Professional Engineer in the State of Washington and a certified Project Management Professional.

Phillipe Mack holds an MsC in Electromechanical Engineering. He founded Pepite 10 years ago, with the perspective of commercializing university research on machine learning in the industrial world. His main focus is on smart grid applications and energy efficiency in process industries.

David Maggio joined the ERCOT ISO in 2007 and is currently manager of Congestion Revenue Rights. His group is primarily responsible for the facilitation of Congestion Revenue Rights auctions and promoting consistency between the various ERCOT Markets. He received his B.S. and M.S. degrees in Electrical Engineering from the University of Illinois at Urbana–Champaign.

Yuri V. Makarov has over 34 years of experience in power systems engineering. He joined Pacific Northwest National Laboratory in 2005, and is currently Chief Scientists in Power Systems. Dr. Makarov received his M.Sc. and Ph.D. degrees from the Saint Petersburg State Technical University, Russia.

Dimitris Mentis works as a researcher in the Division of Energy Systems Analysis at KTH, focusing on assessing renewable energy potentials through complex geographic information system analyses. Further, he is working on CLEWS (Climate, Land-use, Energy and Water Systems) modeling. Previously, he worked at the Hellenic Ministry of Energy, Environment and Climate Change on desalination projects.

Michael Milligan came to the Wind Energy Program at the National Renewable Energy Laboratory in 1992, and is now principal analyst in the Transmission and Grid Integration Group. Dr Milligan has published more than 175 papers, reports, and book chapters on power system planning and operation with large amounts of renewable energy.

David Mohler, Vice President of Emerging Technology, is responsible for the development and application of technologies for Duke Energy. David led the establishment of the company's technology office in 2006. He received a Master of Arts degree from Xavier University of Cincinnati and a M. Sc degree from the University of Pennsylvania.

Matthias Müller-Mienack received his Ph.D. in Electrical Engineering from the Brandenburg Technical University in Germany. He worked for 50Hertz (TSO) in Strategic grid planning, offshore project manager, Head of European Grid Concepts, Head of Corporate Strategy. In parallel, he was Convenor of the ENTSO-E WG "2050 Electricity Highways". Currently Dr Müller-Mienack is Technical Head of the Elia Group institute "GridLab".

Tim Mundon has more than 10 years of experience working on the development of wave energy and is currently the Director of Marine Operations for Oscilla Power. Dr Mundon received his Ph.D. from the University of Edinburgh in 2005 where he focused on the use of active control to optimize wave energy devices.

Ijeoma Onyeji is a research analyst at New Energy Insights (UK). She has extensive experience within the field of sustainable energy and has consulted with a variety of organizations, including the International Atomic Energy Agency (IAEA), United Nations Industrial Development Organization (UNIDO), and the former African Institute for Applied Economics. She holds an M.Sc. in Economics and Econometrics.

Andrew L. Ott is Executive Vice President of Markets for PJM Interconnection. Mr Ott has been with PJM for more than 15 years and is responsible for PJM market operations, external affairs, market design, and settlements divisions. He received a B.Sc. degree in Electrical Engineering from Pennsylvania State University and a M.Sc. in Applied Statistics from Villanova University.

Mark Rothleder is Vice President, Market Quality and Renewable Integration at the California Independent System Operator Corporation. Employed at the ISO since 1997, Mr Rothleder is the longest serving employee. Mr Rothleder holds a B.S. degree in Electrical Engineering from the CSU, Sacramento and an M.S. in Information Systems from the University of Phoenix.

Peter Schell has over 15 years of experience in the energy sector. He played a pivotal role in the liberalization of the electricity and gas markets in Belgium before becoming the General Manager of Ampacimon, a spin-off of the University of Liège, Belgium and a leading company in the Dynamic Line Rating market. Mr Schell holds a M.Sc. in Aeronautical and Space Engineering from the University of Stuttgart, Germany.

Fereidoon P. Sioshansi is President of Menlo Energy Economics and Editor and publisher of EEnergy Informer newsletter. His professional experience includes working at SCE, EPRI, NERA, and Henwood Energy. Since 2006, he has edited eight books published by Academic Press. Dr Sioshansi has degrees in engineering & economics including a Ph.D. in Economics from Purdue University.

J. Charles Smith is the Executive Director of the Utility Variable-Generation Integration Group and a Fellow of the IEEE. He is a guest editor for the IEEE Power and Energy magazine. He received his BSME and M.S. degrees from MIT in 1970, and has over 40 years of experience in the electric power industry.

Sushil K. Soonee, a graduate from Indian Institute of Technology Kharagpur, is currently the Chief Executive Officer of Power System Operation Corporation Ltd (POSOCO). He has over three decades of experience in Power System Operation of the Eastern, Southern, and Northern Grids of India. Mr Soonee has worked extensively on Integration of State Grids to form Regional Grids and subsequently the formation of the National Grid of India.

Daniel Sowder has led numerous technology development and demonstration projects related to distributed energy technologies. After serving as a nuclear submarine officer in the U.S. Navy, he joined Duke Energy's Emerging Technology Office in 2010. He holds degrees from the U.S. Naval Academy, University of North Carolina, and Old Dominion University.

Glauco N. Taranto is associate professor of Electrical Engineering at the Federal University of Rio de Janeiro/COPPE, Brazil, since 1995. In 2006, he was a visiting fellow at CESI, Milan, Italy. Dr Taranto received his Ph.D. degree in Electrical Power Engineering from Rensselaer Polytechnic Institute, Troy, New York, USA in 1994.

T. Bruce Tsuchida, Principal at The Brattle Group, has over 20 years of experience in power generation development, utility operation, and power market analysis. He received his M.S. in Technology and Policy, and M.S. in Electrical Engineering and Computer Science from the Massachusetts Institute of Technology, and B.Eng. in Mechanical Engineering from Waseda University.

Andreas Ulbig is with the Power Systems Laboratory of ETH Zurich, Switzerland since 2008. Prior to this he worked for RTE, the French transmission system operator, and the International Energy

Agency in Paris. He received his M.Sc. from Supélec, Paris and his Dipl.-Ing. degree in Engineering Cybernetics from the University of Stuttgart, Germany.

Luigi Vanfretti is an associate professor at KTH Royal Institute of Technology, Stockholm, Sweden. Since 2011 he has served as scientific advisor for the R&D division of Statnett SF, the Norwegian transmission system operator, where he is currently employed as Special Advisor in Strategy and Public Affairs. Dr Vanfretti obtained his M.Sc. and Ph.D. Degrees from Rensselaer Polytechnic Institute in Troy, New York. He is a senior member of the IEEE.

Alexandra von Meier has studied electric power systems and renewable energy since the late 1980s. Dr von Meier currently directs electric grid research at the California Institute for Energy and Environment and is an adjunct associate professor in EECS at University of California, Berkeley. She holds a Ph.D. in energy and resources from University of California, Berkeley.

Xing Wang joined Alstom Grid in 2001, and is currently R&D Director for Transmission and Pipelines of Network Management Systems business. He received his Ph.D. degree in Electrical Engineering from Brunel University, UK. Dr Wang has over 20 years' experiences in power industry. Dr Wang was involved in many electricity markets implementation project.

Manuel Welsch works as lead researcher in the Division of Energy Systems Analysis at KTH, focusing on the development and application of energy models. Before joining KTH, Manuel worked at the UNIDO on UN-Energy, the Energy Facility of the European Commission, and at Bernard Engineers on the design of hydro power projects.

Austin D. White is a senior engineer at Oklahoma Gas & Electric Co. He is currently responsible for transmission/substation protective system settings and coordination, disturbance event analysis and system modeling/simulation. White earned a BSEE degree from Oklahoma Christian University in 2001, followed by a master's degree in engineering and technology management from Oklahoma State University in 2008. He is a licensed professional engineer in Oklahoma.

About the Editor

Dr. Lawrence E. Jones is a thought leader and practitioner with over twenty years of experience in the energy industry. His expertise includes renewable energy integration and the application of smarter technologies in the engineering and operations of electric power grids and other critical infrastructures. He also focuses on system resiliency, disruptive and innovative business models, and strategies for addressing challenges at the food-energy-water nexus.

He joined Alstom Grid Inc. in 2000, and is currently North America Vice President for Utility Innovations and Infrastructure Resilience, and serves on the company's global Business Development team for Smart Grids and Smart Cities. He was previously Vice President for Regulatory Affairs, Policy and Industry Relations. He also served as Director of Strategy and Special Projects, Worldwide, in the Network Management Systems business and led its global Renewable Energy Integration activities.

Dr. Jones is an advocate for the use of smart, clean, and renewable energy technologies enabled by resilient infrastructures around the world. He was the principal investigator of the 2010/11 Global Survey on Renewable Energy Integration in Power Grids funded by the Office of Energy Efficiency and Renewable Energy (EERE) at U.S. Department of Energy. For this work, he received the Renewable Energy World Network 2012 Excellence in Renewable Award for Leadership in Technology, and the Utility Variable Generation Integration Group 2012 Achievement Award. In 2000, he co-founded the International Workshop on Large-Scale Integration of Wind Power and Transmission Networks for Off Shore Wind Farms.

In September 2010, Dr. Jones was appointed by the United States Department of Commerce's National Institute of Standards and Technology (NIST) to a three-year term on the 15-member Federal Smart Grid Advisory Committee, and was reappointed to serve another three-year term which ends in 2016. He also serves on the Advisory Boards of several industry conferences and smart grid research programs within the Americas, Europe and Africa.

Dr. Jones is frequently an invited speaker at industry conferences and academic symposia for diverse audiences including: utility executives, investors and policy makers. He has published and been cited in scholarly journals, trade magazines and newspapers, and has also appeared on TV & Radio around the world.

Dr. Jones is involved with several philanthropic and social entrepreneurial initiatives. He is co-founder and President of the Board of Directors of the Center for Sustainable Development in Africa (CSDA), and a member of the Board of Chess Challenge in DC. He also serves on the Advisory Committee of The Energy Action Project (EnAct) - a multimedia platform which seeks to help a global audience understand the scope of the energy poverty challenge and the collaborative action needed to address it.

Born in Liberia, Dr. Jones received his MSc, Licentiate and PhD degrees in Electrical Engineering from the Royal Institute of Technology in Stockholm, Sweden.

Acknowledgments

One of the most valuable experiences of serving as a book editor is having the chance to collaborate and cocreate with many wonderful people around the world.

The idea of this book was born from a conversation I had in 2011 with Tiffany Gasbarrini, then a Senior Acquisitions Editor at Elsevier. For over a year, she persistently asked if I would serve as editor for a book on renewable integration for which I repeatedly demurred. But after months of careful consideration and discussion with my family, Tiffany's persistence thankfully paid off. Hence, you hold this book in your hands.

Special thanks to the team at Elsevier; in particular Laura Colantoni, Joe Hayton, Poulouse Joseph, Rajakumar Murthy, Sruthi Satheesh and Kattie Washington, for their commitment and dedication to seeing the idea come to fruition.

This book is underpinned by the knowledge, insights and experiences of leading experts and practitioners who graciously devoted their time and talent to writing the chapters. Thanks to: V.K. Agrawal, Stefano Alessandrini, Anders N. Andersen, Göran Andersson, Reza Arghandeh , Tatiana M. L. Assis, Chaitanya A. Baone, Carl Barker, Diane Broad, Maxime Baudette, Audun Botterud, Richard Candy, Spyros Chatzivasileiadis, Puneet Chitkara, Erik Connors, Anish De, Luca Delle Monache, Christopher L. DeMarco, Ken Dragoon, John Dumas, Erik Ela, Mica Endsley, Pavel V. Etingov, Steven Fine, Jarett Goldsmith, Santiago Grijalva, Udi Helman, Anders Plejdrup Houmøller, Mark Howells, Brendan Kirby, Kiran Kumaraswamy, Helena Lindquist, Clyde Loutan, Jian Ma, Phillipe Mack, David Maggio, Yuri V. Makarov, Dimitris Mentis, Michael Milligan, David Mohler, Matthias Müller-Mienack, Tim Mundon, Ijeoma Onyeji, Andrew L. Ott, Mark Rothleder, Peter Schell, Fereidoon P. Sioshansi, Sushil K. Soonee, Daniel Sowder, Glauco N. Taranto, T. Bruce Tsuchida, Andreas Ulbig, Luigi Vanfretti, Alexandra von Meier, Xing Wang, Manuel Welsch, and Austin D. White.

Thanks are also due to two forward-thinking leaders, George Arnold and Daniel Dobbeni, for their eloquent forewords, which highlight the drivers and opportunities for integrating renewable energy from their unique vantage points in the U.S. and Europe, respectively.

Making predictions is never easy. However, when it comes to integrating renewables, I can, with a high degree of confidence, place my bets on projections made by J. Charles Smith. As the Executive Director of the Utility Variable Generation Integration Group, and an eminent thought leader in the field, Charlie's impressive prescience made him the unquestionable choice to write the epilogue for this book. I am grateful to him.

I was fortunate to have extraordinary teachers and role models who inspired me and shared their wisdom and expertise, which prepared me for this task. In this context, there are four individuals I would like to express my thanks to: Barbro Hellermark, my Swedish teacher at the University of Stockholm, whose brilliant pedagogical skills helped me to quickly become fluent in the language, and also gave me the foundation to function in today's multilingual world; Professor Göran Andersson, my PhD supervisor at the Royal Institute of Technology in Sweden, who, as head of the Power Systems Group, demonstrated the importance of being able to work on a team with individuals from different cultures; Professor Emeritus Gustaf Olsson at Lund University for our stimulating in-depth discussions, which led to interesting 'A-ha!' moments, helping me appreciate the complexity and enormity of the water-energy nexus problem, as well as the need for holistic approaches to tackle mega challenges like climate change; and finally, to Margaretha Andolf, the former head of Language and Didactics at the Royal Institute of Technology, for assigning me a class project that taught me a lot about inter-cultural communications, and eventually led me to write the book, *Visiting Students in Stockholm, Encountering and Adjusting to Swedish Culture.*

I would like to acknowledge and offer thanks to the following individuals, who provided early praise for this book: Mark Ahlstrom, Vicky Bailey, Morgan Bazilian, Pierre Bernard, Reid Detchon, Rob Gramlich, Becky Harrison, Daniel Kammen, Alan Miller, Magnus Olofsson, Paul Polman, Eric Pyle, Klaus Rave, David Sandalow, Andrew L. Shapiro, Jigar Shah, Frank Verrastro, Ryan H. Wiser, Kandeh Yumkella, and Christian Pilgaard Zinglersen.

I am also thankful to the intellectual sparring partners with whom I have worked, brainstormed, coauthored, and debated over the years, always with the mutual goal of advancing solutions for integrating renewables. They include: Thomas Ackermann, Charlton Clark, David Elzinga, Russell Philbrick, Olof Samuelsson, Jon O'Sullivan, Eric Goutard, Ali Sadjadpour, and Robert Zavadil.

I am always mindful of the village that it took to get me to where I am, so it is in the spirit of gratefulness that I acknowledge the support over the years of my family and friends around the world.

I owe an infinite debt of gratitude to my mother, Comfort Hadoo Jones, and my departed father, Emmanuel E. W. Jones, Jr. They made countless sacrifices to ensure my siblings and I received a well-rounded education. I will be forever grateful to them for instilling in us the importance of faith, courage, and determination. They taught us to always strive for excellence in everything we do, and to remember that "No Man Is an Island" - we will always need one another in our interconnected and interdependent world.

To serve as editor for a book with so many contributors requires that one patiently listens to a myriad of disparate views, before synthesizing and integrating them into a scholarly roadmap that tells a cohesive story. In this regard, I am extraordinarily grateful to my uncle, Charles Gyude Bryant, II, who not only taught me about the importance of listening, but exemplified this estimable skill when he successfully led the transition of Liberia from war, to peace and democracy. Sadly, he died a few months before this book was completed. I will always strive to be as good a listener as he was.

I am thankful for the relationships with my siblings, Jerome, Vivien, and Jestina. Vivien had the foresight to persuade me to take a leap of faith, and make the transition from Sweden to the USA, in order to broaden my horizon. Without that decision, the opportunity to work on this book would probably not have come about.

Writing a book always takes the biggest toll on the immediate family. Therefore, my immeasurable gratitude goes to my wife, Facia, and our daughter, Nohealani. They are my endless source of love, joy, inspiration, enthusiasm, and optimism. Although this endeavor meant less time for us to spend together over the many weekends this effort demanded, they never complained. Instead, Facia was steadfast in encouraging me to remain focused on completing this journey. Nohealani's innocent inquisitiveness helped sharpen my explanations of wind and solar energy.

I hope that this book will catalyze greater investments in, and integration of, renewable energy resources. Thus, in the near future, Nohealani and her contemporaries will continue to reap from, enjoy, and maintain a more sustainable planet universally gifted to all of us by divine providence.

Lawrence E. Jones
Washington DC, May, 2014

Foreword from Europe

From the early days of Thomas Edison till the middle of last decade, the growth of the power industry was based on four main pillars. In first place, in order to meet the instantaneous power consumption, generation plants were instructed to deliver electricity starting from the lowest marginal cost (nuclear) up to the most expensive (fuel). This so-called merit order model combined with local energy resources and predictable demand curves determined the portfolio of generation technologies. Secondly, economies of scale were readily available with increasingly larger centrally dispatched power plants. Thirdly, grids connecting demand areas with different demand curves improved reliability and lowered peak demand, reducing overall cost. Finally, when combined these features allowed for lower tariffs, encouraging an ever-growing demand. They also contributed to appearance of increasingly larger vertically integrated utilities. These national champions and their predecessors successfully developed one of the most important industries in mankind; an industry that is on the very basis of today's economies, health, and welfare.

In those years, the future of a power systems could be planned with the near certainty that it would materialize in due time. Return on investment was (nearly) guaranteed as customers had often only one national supplier that, in return for this privilege, would ensure long-term security of supply and day-to-day reliability. Rising issues were permits and rights of ways, while the inroad of new technologies, such as combined heat and power generation and combined cycle gas plants, was perceived not as a threat but as an opportunity.

The European power industry is upside down after 17 years, witnessed by the decreasing share value of what remains from the national champions!

At a joint press conference in Brussels, October 11, the CEOs of 10 leading European companies, representing half of Europe's power capacity, painted a bleak future and raised concerns about security of supply, as they close loss-making (fossil) plants.

What happened?

As with each black swan event, several elements concurred to disrupt the four pillars.

In the first place, the generation mix changed fast and drastically. Attractive support mechanisms for combined heat and power generation and renewable energy sources led new investors, such as private capital, municipalities, energy intensive industries, small and medium enterprises, as well as millions of residential customers, to enter the power industry. Two support mechanisms are the foundations of this revolution. One gives priority access to the grid for CO_2-friendly generation while the second offers a fixed and attractive return on investment, which often shields the owner from potential disruption of the wholesale market during 20 years.

Europe's ambitious target of 20% renewable energy sources in 2020 (or 33% of renewable generation for electricity) prompted several member states to propose highly attractive support mechanisms. Denmark, Germany, Spain, Italy, Ireland, and Belgium for example have seen their share of renewable energy sources, manly wind and solar, increase drastically in less than 5 years.

While previous (and current) support mechanisms for nuclear and fossil fuel generation are funded by the Europe Union and the concerned states, the cost of the support for renewable energy sources appears as a surcharge added on top of the transmission and/or distribution tariffs. This cost is much more visible and measurable.

An unprecedented wave of investments induced major side effects.

In the first place, the contemplated growth in wind and solar power generation attracted worldwide manufacturers, leading for example to price plummeting for photovoltaic panels. Soon, these technologies will become competitive without support mechanisms; challenging the second pillar. The operational impact of large shares of variable generation grew quickly, whether in terms of large power flows impacting neighboring countries or reduced number of operating hours for fossil plants (less than 10% of the time during sunny and/or windy months in some countries).

Secondly, exploitation of shale gas in the United States and a very low CO_2 price led to a trend reversal in less than 3 years. Coal plants became the cheapest to operate at the expense of more efficient and modern gas plants while fast ramping gas plants are disappearing at the very moment that system operators hardly need them.

Thirdly, the tremendous growth of renewable energy sources with nearly zero-marginal cost pulled down the wholesale price from around 80 €/MWh in 2008 to 45 €/MWh in 2013. Together with the decreasing number of generation hours, fossil plants are facing a bleak future; challenging the merit order model and first pillar.

Finally, the financial and economic crisis in Europe with its toll on employment amplified the impact of the fast rising cost of support mechanisms for renewable energy sources. Large industries started to complain heavily about worldwide competitiveness, attracting media and political attention. And although, residential customers are not (yet) reacting to this situation, some policy markers are now moving from "saving the planet first" toward "saving us first and then eventually the planet".

Renewable energy sources are more and more depicted as guilty for higher electricity bills, future blackouts, decreasing reliability, insufficient security of supply, inefficient electricity markets, etc.

Could these assertions be true? Should all support to CO_2-friendly generation be stopped?

As usual, the picture is neither white nor black and the initiative of the editor, Lawrence Jones, comes at the right moment. There is actually an urgent need to clarify the real challenges induced by larger shares of (variable) renewable energy sources and to put forward practical and efficient solutions.

The various authors provide state-of-the-art contributions from a research and development perspective as well as case studies. The major role of grids and interconnections between power systems is highlighted as well as the need to invest in advanced forecasting of wind and solar generation, especially for the design of energy management system. Several contributions also demonstrate that smart tools are readily available to maximize the efficient use of the existing assets, such as dynamic line rating, phasor measurements, data mining, demand response and flexibility. As former CEO of the Elia Group, a fully unbundled transmission system operator listed on the stock exchange with branches in Belgium and Germany, I have witnessed the added value brought by dynamic line rating as a cost efficient and reliable solution to increase transmission capacity. Another example is phasor measurements that were recognized by the members of the GO15, (i.e. the association of the world's 16 largest power system operators with a total of 3.5 billion customers), as a major tool to ensure reliability in a cost-effective way. The need for flexible generation and demand, an issue faced by all power systems with an increasing share of wind and solar energy, is a common theme of several contributions. Improved visualization in control centers, intentional islanding of distribution network operation, energy storage, demand response, high-performance computing, integration of renewable energy sources in ancillary services are all examples of the creativity unleashed by the energy system mutation toward a carbon-free power generation. From the many challenges humanity will face in the future, there is one this generation has to tackle now! Path the way toward an affordable, climate friendly, and secure supply of electricity for all inhabitants of this planet, whether my two grandsons Ethan and James or any present and future children's and I am grateful to Lawrence Jones and all authors for having taken the initiative to share such wealth of information at this critical moment.

Daniel Dobbeni
Belgium

Foreword from the USA

In his second Inaugural Address in January 2013, President Obama stated, "We will respond to the threat of climate change, knowing that the failure to do so would betray our children and future generations."

Reducing carbon pollution is a key element of the United States climate action plan. Today's electric grid represents the largest single source—accounting for over one-third of carbon emissions into the atmosphere. It is imperative that we reduce the carbon footprint of the power grid. At the federal level, the Environmental Protection Administration has proposed carbon pollution standards for new power plants, and 35 states have introduced renewable portfolio standards. Penetration of renewable energy sources is increasing rapidly. Since 2009, the United States more than doubled the generation of electricity from wind and solar energy, and renewable energy sources accounted for about half of the new generating capacity installed in 2012.

The power grid has benefited from many technological innovations over the past century. However in some respects, the basic architecture of the grid is little changed. The traditional operating paradigm assumes controllable generation and variable demand. Today's grid is designed to ensure near-instantaneous balance between generation and consumption by controlling generation to match stochastically variable demand. The widespread use of renewable generation necessitates the reverse paradigm: variable generation and controllable demand. In other words, renewable generation is stochastically variable but not controllable (we cannot control when the sun shines or when the wind blows), and thus demand must become much more controllable in order to maintain balance. Energy storage technologies, not widely used today, will be needed to provide additional flexibility by buffering short-term mismatches between generation and demand. Reversing the grid's operating paradigm and incorporating new technologies such as energy storage requires fundamental reengineering of the grid as it exists today.

Reengineering the grid to accommodate widespread use of renewable energy generation presents many difficult challenges, in both technical and policy dimensions. For example, dynamic models of variable energy resources, storage, and demand must be integrated into large-scale system-level models to measure and control the operation of the grid. Improved methods of forecasting variable generation are needed. Strategies for dealing with variability in generation through a combination of demand response, storage, and diversity must be developed. Case studies and empirical data are essential to ensure that models and strategies for managing the grid are valid and are practical to deploy at scale. Appropriate policies and regulation to facilitate economic deployment of new technologies while maintaining system reliability must be understood. The experience of the telecom industry in undergoing technology, policy, and regulatory change also suggests that the reengineering of the grid may lead to new game-changing business models for the industry. In noting the technological achievements of the last hundred years, the National Academy of Engineering cited the electric grid as the greatest engineering achievement of the twentieth century. If we are successful in meeting the challenges posed by renewable energy integration, future generations may well think of the future grid powered by renewable energy as one of the great engineering achievements of the twenty-first century.

Lawrence Jones has produced an impressive collection of works with the help of distinguished experts around the world who address the many aspects of renewable energy integration. This book

2014 Published by Elsevier Inc.

makes a significant contribution by presenting a rich set of case studies that present empirical results from real-world application and advance our understanding of state-of-the-art approaches to practical management of electric grids powered by renewable energy.

Dr. George Arnold
Washington DC

Introduction

Our changing global climate, access to clean energy and water, food security, and population growth are all among the web of complex interdependent challenges facing the world. The first, global climate change is perhaps the most pressing threat to humanity and the future of the planet. Recent reports such as the *Climate Change 2014: Impacts, Adaptation and Vulnerability* from the Intergovernmental Panel on Climate Change (IPCC), and the *2014 US National Climate Assessment* document that climate change is real, and that the related impacts are already beginning to affect different parts of the world. Everyone can relate to the apparent physical and financial impacts of this anthropogenic phenomenon. These impacts will be exacerbated in the coming the decades, unless aggressive measures are taken to mitigate negative consequences and adapt our responses to the new environment.

One response to climate change is for the world to transition to a low-carbon energy future through increased use of renewable energy sources. Across the world, governments are scaling up their efforts to spur more investments in wind, solar, and other forms of low-carbon energy. As a result, electricity generated from wind and solar energy is growing rapidly worldwide. There is strong correlation between successful integration of renewable energy and the employment of policies and regulatory mechanisms that catalyze private sector investment. Businesses, therefore, also need to adapt radically practical strategies, as outlined in the book, *The Big Pivot*, by Andrew Winston. However, as the penetration of renewables increases, one quandary that continues to receive much attention focuses on how to gracefully integrate these nonconventional, variable forms of energy into existing grids and eventually emerging grids, as well as into electricity markets.

My interest in renewable energy began as a child growing up in the West African nation of Liberia, located about 6° north of equator. I was fascinated by the sun and the power of its radiation. In 1998, this interest shifted more toward wind energy while I was working on my doctoral thesis a PhD at the Royal Institute of Technology (KTH) in Stockholm Sweden. In 2000, I cofounded the International Workshop on Large-Scale Integration of Wind Power and Transmission Networks for Offshore Wind Farms at KTH. During the past decade, I have collaborated with many leading experts around the world on the subject of renewable integration. In 2010/11 I was fortunate to serve as the principal investigator of the 2010/11 Global Survey on Renewable Energy Integration in Power Grids, funded by the U.S. Department of Energy (DOE). Surveys were conducted on 33 grid operators of electric power systems in 18 countries about their operating policies, best practices, examples of excellence, lessons learned, and decision support tools. The power grids encompassed by the study are located in varying geographies with diverse climate weather patterns and many of the utilities surveyed operate under dissimilar regulatory frameworks, whether within regulated or deregulated electricity markets. The survey findings are published in the report, Strategies and Decision Support Systems for Integrating Variable Energy Resources in Control Centers for Reliable Grid Operations.

The idea behind this book began in 2011 during a brief conversation I had with Tiffany Gasbarrini, who at the time was a Senior Acquisitions Editor at Elsevier. Tiffany had listened to me present findings from the aforementioned report at the annual conference of the American Wind Energy Association (AWEA). When she expressed her interest in publishing the report as a book, I replied that was not possible due to the terms of the contract with the DOE. Given the huge interest in renewable integration worldwide, she asked if I would consider writing a completely new book on the subject. Initially I declined. However, she was persistent and made multiple requests for over a year. Finally, after getting the green light from my family (although they knew this would consume time during the weekends), I agreed to serve as editor. Thus began the fulfilling journey of preparing what has now emerged as a groundbreaking book. One of the most valuable experiences shaping this endeavor has been collaborating and cocreating with 59 renowned experts from around the world.

During the past decade, results from numerous studies have shown that integrating renewable energy is a multidisciplinary undertaking. Therefore, it was clear from the outset that for this volume to help advance the goal of supplying more of the world's electric energy from renewable sources, it would have to cover a broad spectrum of topics, featuring carefully selected material based on the myriad experiences of contributing authors working around the world with diverse power systems.

There is no one-size-fits-all solution for integrating renewable energy that applies to all power systems. This book constitutes a singular attempt to provide a distilled examination of the intricacies of integrating renewables into power grids and electricity markets. It offers informed perspectives from an internationally renowned cadre of researchers and practitioners on the challenges and solutions derived from demonstrated best practices as developed by operators of power systems and electricity markets across the globe. The book's focus on practical implementation of strategies provides real-world context for theoretical underpinnings and enables the development of supporting policy frameworks. It considers a multitude of wind, solar, wave, and tidal integration issues, thus ensuring that grid operators with low or high penetration of renewable generation can leverage the victories achieved by their peers.

As suggested by the book's subtitle, the three underlying attributes that characterize the integration of renewables in power grids and markets are variability, uncertainty, and flexibility. Similarly, this book's journey involved coping with all three. There was *variability* in the style of writing, the systems presented in the case studies, and how authors perceived different, but often interrelated topics. *Uncertainty* was inherent in the process, given that until the authors submitted the first draft of the chapters, I could not entirely know what to expect. Analogously to integrating renewable generation, as the editor, initially all I could do was to try to *forecast or predict* the content I was going to receive, based on my historical knowledge of the authors' works and expertise. In the end, my best efforts to integrate and interweave all 35 chapters into a synchronous and seamless flow of ideas that conveys key coherent themes and approaches bore fruit only by allowing some *flexibility* to the contributors. Elsevier also gave me a great deal of flexibility.

OVERVIEW OF CHAPTERS

The book you are holding (and will hopefully read) is the result of successfully integrating what initially seemed to be a disjointed collection of chapters into a pioneering volume consisting of 10 interrelated parts, comprising 35 cohesive and carefully written chapters which circumscribe the practical management of variability, uncertainty, and flexibility.

Part 1, Policy and Regulation consists of three chapters that examine the policy and regulatory issues, and the future outlook of renewables in the Europe Union, the United States of America, and Africa. Chapters 1, 2, and 3: The Journey of Reinventing the European Electricity Landscape—Challenges and Pioneers by Helena Lindquist; Policies for Accommodating Higher Penetration of Variable Energy Resources—U.S. Outlook and Perspectives by Steven Fine and Kiran Kumaraswamy; and Harnessing and Integrating Africa's Renewable Energy Resources by Ijeoma Onyeji, review the evolution of renewables in the different regions. The authors provide suggestions about what it will take to accommodate even higher penetration of renewables.

Deriving from parts of the European Union and the United States where the percentage of renewable generation capacity is now in double digits, the first two chapters offer good insights about the specific policies and regulatory incentives which catalyzed private sector investments, but also alleviated operational electricity barriers to integrating new variable generation in those jurisdictions.

Africa, with its vast and largely untapped endowment of renewables, is beginning to see greater interest in using solar, geothermal, wind and other forms of renewables to address the vexing problem of lack of access to electricity. Onyegi provides a thorough discussion of the continent's renewable landscape, the potential challenges, and the opportunities of tapping clean energy resources. Whether grid-connected or off-grid systems, appropriate long-term policies along with adequate investment in infrastructure and human capacity are crucial to harnessing the renewable energy in African countries.

Chapters in **Part 2, Modeling of Variable Energy Resources**, address the important first step to integrating renewable generation in power systems. Modeling renewable energy must address emerging behavior at new spatial and temporal scales. Modeling must also consider factors associated with resource availability, economic viability, and other grid-related issues.

In Chapter 4, Challenges and Opportunities of Multidimensional, Multiscale Modeling and Algorithms for Integrating Variable Energy Resources in Power Networks, Santiago Grijalva reviews the relevant spatial, temporal, and scenario dimensions. He discusses the state of the art in three key areas of multiscale, multidimensional analysis: (1) modeling and analysis, (2) optimization and control, and (3) data management and visualization. An integrated multidimensional, multiscale framework for design, deployment, operation, and management of networks with high levels of renewable energy is proposed.

The case study in Chapter 5, Scandinavian Experience of Integrating Wind Generation in Electricity Markets, by Anders Plejdrup Houmøller describes the experiences of integrating renewables in the Nordic region. The four countries—Denmark, Finland, Norway, and Sweden—have four separate but fully integrated power systems and a single wholesale electricity market. The chapter focuses mainly on Denmark, which gets a substantial amount of its energy from wind power. Wind energy growth in Denmark is expected to grow unabated over the next decade as the country seeks to achieve the goal of 100% carbon-free energy by the year 2050.

In Chapter 6, Case Study—Renewable Integration: Flexibility Requirement, Potential Overgeneration, and Frequency Response Challenges, Mark Rothleder and Clyde Loutan provide an overview of analyses the California Independent System Operator (ISO) conducted to better understand the challenges related to renewable integration. The foundation for performing such a study relies on successfully modeling the renewable resources. The study includes recommendations that the generation fleet needs to have a flexible component. Load shifting and storage technologies are necessary to mitigate potential overgeneration conditions, and inertial response is a characteristic that asynchronous resources should provide to maintain grid reliability.

The chapters in **Part 3, Variable Energy Resources in Power System and Market Operations**, examine system and market impacts of variable generation. Operation of power systems and electricity markets requires that the aggregate generation and load must be balanced instantaneously and continuously for the electric power system to perform reliably. Integration of more and more variable energy resources (VER) introduces new challenges to power system and market operations.

In Chapter 7, Analyzing the Impact of Variable Energy Resources on Power System Reserves, Brendan Kirby, Erik Ela, and Michael Milligan discuss the different reserve types and requirements in general, and address how to determine impacts. Furthermore, the task of determining the additional reserves required is explained in-depth. The authors make clear that the task is complicated by the nonlinear nature of reserves, as well as by the lack of data and experience.

In Chapter 8, Advances in Market Management Solutions for Variable Energy Resources Integration, Xing Wang provides an overview of wholesale electricity markets and market management systems. Enhancements are being made in both market design and market analytical tools in terms of managing operational uncertainties introduced by VER integration. The chapter focuses on two such areas of market enhancement. First, it explores the feasibility of a ramp market in real time, balancing operation to create the right market incentives for resources to provide enough ramping energy to compensate for VER volatility. Second, it examines how to manage short-term VER uncertainty by applying robust optimization to look ahead unit commitment.

In Chapter 9, Reserve Management for Integrating Variable Generation in Electricity Markets, John Dumas and David Maggio present a case study from the Electric Reliability Council of Texas (ERCOT). As an independent system operator, ERCOT is tasked with ensuring the balance of electrical supply and demand. Over the last several years, this function has been complicated by the additional uncertainty and volatility introduced by increased installment of renewable energy resources, particularly wind generation. The chapter discusses many of the steps that ERCOT and its market participants have made to address and minimize these issues. Particular focus is put on the evolution of wind power at ERCOT, including the use of a probabilistic wind power output ramp event forecast tool, and enhancements made to their methodology for determining the minimum level of ancillary services required to maintain system reliability.

In Chapter 10, Grid and Market Integration of Wind Generation in Tamil Nadu, India, Anish De and Puneet Chikara discuss results from a case study of wind farms in Tamil Nadu conducted to identify the micro-practices in variable renewable energy (VRE) management in the state. Renewable energy in India has grown at a fast pace over the past few years. Compared to other Indian states, Tamil Nadu has considerable experience in operating a power system with VRE resources. The study analyses the intrastate impacts of wind integration, and examines commercial mechanisms, such as Feed-in-Tariff, as well as other regulatory and policy measures meant to encourage renewable integration.

Part 4, **Forecasting Renewables**, deals with the fundamental task of managing uncertainty via accurate prediction of output from intermittent sources of power. Forecasting of renewable energy has increasingly taken on the sine qua non role in the operations of power systems with more wind, solar, and other forms of variable generation. The chapters in this section discuss the state of the art and state of the science of forecasting, and how the forecast information can be integrated in utility control centers.

In Chapter 11, Forecasting Renewables Energy for Grid Operations, Audun Botterud discusses the potential applications of renewable energy forecasts for system operators, renewable power producers, and other participants in the electricity market, followed by a brief overview of forecasting methods for wind and solar energy. The author argues that good estimates of forecast uncertainty are also of major importance to better handle renewable energy in grid operations. The conclusion highlights that improved forecasting systems and corresponding decision support tools are key solutions to enable a clean, reliable, and cost efficient future electricity grid driven by renewable resources.

In Chapter 12, Probabilistic Wind and Solar Power Predictions, Luca Delle Monache and Stefano Alesandrini review several state-of-the-science techniques supporting probabilistic power predictions for wind and solar generation, and demonstrate how to verify and evaluate such predictions. Deterministic prediction has provided useful information for decision making in the conventional operational paradigm. However, under the new grid regime of increased operational uncertainty, deterministic approaches have fundamentally limited value. For one, the prediction represents only a single plausible future state from the continuum of possible states that result from imperfect initial conditions and model deficiencies. Accurate knowledge of that continuum is considerably more useful to decision making.

In Chapter 13, Incorporating Forecast Uncertainty in Utility Control Centers, Yuri Makarov, P. V. Etingov, and J. Ma examine different sources of uncertainty and variability, demystify the overall system uncertainty model, and propose a possible plan to transition from deterministic to probabilistic methods in planning and operations. Uncertainties in forecasting the output of intermittent generation and system loads are not adequately reflected in existing industry-grade tools used for transmission system management, generation commitment, dispatch, and market operation. An example of uncertainty-based tools for grid operations is presented.

Part 5, **Connecting Renewable Energy to Power Grids**, takes on the salient issue of physically connecting renewable generation plants to power grids.

In Chapter 14, Global Power Grids for Harnessing World Renewable Energy, Spyros Chatzivasileiadis, Damien Ernst and Goran Andersson introduce the concept of a Global Grid that all regional power systems feed into one electricity transmission system spanning the entire globe. The Global Grid will facilitate the transmission of "green" electricity to load centers, serving as a backbone. This chapter elaborates on four stages that could gradually lead to the development of a globally interconnected power network. Quantitative analyses are carried out for all stages, demonstrating that a Global Grid is both technically feasible and economically competitive. Real price data from Europe and the USA are used to identify the potential of intercontinental electricity trade, showing that substantial profits can be generated through such interconnections.

Chapter 15, Practical Management of Variable and Distributed Resources in Power Grids, by Carl Barker discusses the use of High Voltage Direct Current (HVDC) technology to transmit electricity. Barker reviews the latest HVDC technology, the known as Voltage Source Converter (VSC). Introduced in 2000, VSC technology further increases the flexibility of DC transmission; hence the realistic prospect of DC transmission networks on- and offshore. These DC transmission networks may replace AC transmission applications in the future or may act as a backbone reinforcing the existing AC grid. This new infrastructure may be critical for the power systems of tomorrow to adapt to the increased use of renewable sources, e.g., the Global Grid discussed in Chapter 14.

In Chapter 16, the case study Integration of Renewable Energy—Indian Experience is presented by Sushil Kumar Soonee and Vinod Kumar Agrawal. Integration of renewable energy sources is a priority area in India at all levels, from policy to regulation, implementation, and grid integration. The chapter explores Indian policies, as well as regulatory initiatives, transmission planning and power system operator's perspectives on integration of renewable energy in the country. The experience of the Renewable Energy Certificate (REC) mechanism introduced in India is discussed.

Part 6, **System Flexibility**, addresses several pertinent issues related to the ability of a power grid to accommodate the variability and uncertainty in the load and generation, while simultaneously maintaining specified acceptable levels of performance across different timescales. Increasing penetration of variable generation leads to more variability in the net load and thus the need for solutions that provide additional flexibility in the grid grows. Grid operators need the proper mix of flexible resources, ranging from the supply-side, delivery-side, and the demand-side. While the need for greater flexibility is well established, how to model and determine the amount of system flexibility that is needed for certain levels of variable generation capacity is the subject of extensive research and debate.

In Chapter 17, Long-Term Energy Systems Planning: Accounting for Short-Term Variability and Flexibility, Manuel Welsch, Dimitris Mentis, and Mark Howells describe the provision of flexibility in power systems through supply- and demand-side operating reserves and storage options. They review several modeling approaches and present the limitations of long-term models regarding their temporal resolution and the metrics applied to ensure the power system's reliability. The authors also consider selected modeling approaches which address the gap between the short timescales required to assess operational issues and the long-time planning horizons of the investment decisions calculated by these models.

In Chapter 18, Role of Power System Flexibility, Andreas Ulbig and Goran Andersson analyze the role of operational flexibility in power systems and its value for integration of high-penetration shares of renewable energy source in power grids. The chapter introduces a new method for assessing the technically available operational flexibility along with illustrative examples. An important research topic is how to quantify flexibility in a given power system. The authors discuss necessary metrics for defining power system operational flexibility, namely the power ramp rate, power and energy capability of generators, loads, and storage devices.

In Chapter 19, The Danish Case: Taking Advantage of Flexible Power in an Energy System with High Wind Penetration, Anders N. Andersen and Sune Strøm describe how wind generators and decentralized combined heat and power (CHP) plants can contribute to efficient balancing of power systems. While other grids around the world have different mixes of load, generators, and interconnection to neighboring systems, this Danish case study challenges the conventional thinking about the how variable generation impacts grid operations. Under the right conditions, wind generation can contribute to operational flexibility of power systems.

The chapters in **Part 7, Demand Response and Distributed Energy Resources**, explore the application of demand response (DR) and distributed energy resources (DER) to help accommodate higher penetration of renewable energy in power grids.

In Chapter 20, DR for Integrating Variable Renewable Energy: A Northwest Perspective, Diane Broad and Ken Dragoon provide a study on the use of demand response in the US Pacific Northwest region. This region, which has large amounts of hydropower, is experiencing the same rapid growth in variable renewable generation that is occurring in many other parts of the world. However, the limits to flexibility are now being reached and the region has embarked upon a number of demand response pilot projects demonstrating the ability of loads to both rise and fall to help accommodate new renewable resources. Despite the pilot projects' success in showing how demand can help integrate renewable energy at low cost, the development of markets and policies is likely needed to achieve widespread commercial application.

In Chapter 21, Case Study: Demand Response and Alternative Technologies in Electricity Markets, Andrew Ott discusses how the PJM Interconnection's (PJM) wholesale electricity market has evolved to promote open competition between existing generation resources, new generation resources, demand response, and alternative technologies to supply services to support reliable power grid operations. PJM has adapted market rules and procedures to accommodate smaller alternative resources, while maintaining and enhancing stringent reliability standards for grid operation. While the supply resource mix has tended to be less operationally flexible, the development of smart grid technologies, breakthroughs in storage technologies, microgrid applications, distributed supply resources, and smart metering infrastructure have the potential to make power transmission, distribution, and consumption more flexible than it has been in the past. Competitive market signals in forward capacity markets and grid services markets have resulted in substantial investment in demand response and alternative technologies to provide reliability services to the grid operator. This chapter discusses these trends and the market mechanisms by which both system and market operators can manage and leverage these changes to maintain the reliability of the bulk electric power system.

In Chapter 22, The Implications of Distributed Energy Resources on Traditional Utility Business Model, Fereidoon P. Sioshansi explores the impact of the recent rise in distributed self-generation through solar rooftop photovoltaics, CHP, fuel cells, and a combination thereof. He also examines the impacts of increased investment in energy efficiency and requirements such as Zero Net Energy (ZNE) on the business models of utilities.

In Chapter 23, Energy Storage and the Need for Flexibility on the Grid, David Mohler and Daniel Sowder present a case study on how energy storage is used to facilitate integration of renewable energy in the distribution grid. Energy storage is considered by many to the distributed resource that could transform the entire power industry as related technologies become more cost competitive. Stated differently, energy storage enables supply and demand to be balanced even when the generation and consumption of energy do not occur at the same time. The authors examine the ability to flexibly move energy across time as a tool that can be applied in many different applications on the electric grid. Several illustrative examples are also provided.

The two chapters in **Part 8**, **Variable Energy Resources in Island Power Systems**, explore some of the problems and solutions around integrating variable renewable energy on island power systems. Because of their isolated geographic locations, most island power systems have highly developed renewable energy resources. However, as these systems are typically not interconnected to other systems, they present a unique set of challenges when dealing with the intermittent nature of renewable resources.

In Chapter 24, Renewables Integration on Islands, Toshiki Bruce Tsuchida provides a summary of the paths and issues encountered by forward-leaning islands that have been planning and implementing ways to integrate renewables at penetration levels in excess of 30%, a level higher than any interconnected system in the world. Challenges ranging from long-term planning to short-term operations require island system operators to meld existing technologies and further explore innovative technology options. The challenges these island systems face when addressing renewable integration amalgamate a variety of issues considered as separate topics in interconnected grids, ranging from smart grid and distributed generation, to climate policy, system resilience, and storage technologies. For the first time in history, island systems are the test bed of innovative technologies that can potentially lead the future of large interconnected systems.

In Chapter 25, Glauco Nery Taranto and Tatiana M. L. Assis present a case study from Brazil on Intentional Islanding of Distribution Network Operation with Mini Hydrogeneration. The abundant presence of large river basins made Brazil a worldwide powerhouse of hydroelectricity. According to official numbers, the installed capacity of hydrogeneration in the country surpasses 70%, which, in a favorable precipitation year, can account for over 90% of the country's annual electric energy needs.

The highly industrialized southern part of the country, although largely hydro-based, currently faces an exhaustion of its hydro capacity, and is exploring alternative renewable sources by exploiting small hydro plants. Due to hilly terrain in large parts of the country, and to government incentives, more mini hydro plants generating a few megawatts each are being used to supply small towns in rural areas, which are essentially island power systems. This chapter focuses on a case study of a real mini hydropower plant connected to a 25 kV rural distribution feeder located within Brazil's south eastern state of Rio de Janeiro.

Part 9, Solar, Tidal, and Wave Energy Integration, explores the integration of emerging low-carbon technologies. While wind and solar PV account for most of the variable renewable energy produced today, technology advances are increasing the future outlook for electricity generation from large-scale grid-connected solar plants, wave and tidal energy, as well as mass deployment of solar PV in large power systems.

In Chapter 26, Economic and Reliability Benefits of Large-Scale Solar Plants, Udi Helman examines the economic valuation methods used by utilities and regional energy planners to determine the net costs of adding renewable energy sources. The net cost equation is the renewable plant's contracted cost or estimated levelized cost of energy plus its transmission and integration costs, minus its energy, ancillary service, and capacity benefits. Helman reviews key findings in large and growing research literature analyzing components of this net cost equation for solar resources, both on an individual project level and as components of expanding renewable portfolios. He also discusses the operational limits and the use of storage to cope with the growing presence of solar plants.

Chapter 27, State of the Art and Future Outlook for Integrating Wave and Tidal Energy by Timothy R. Mundon and Jarett Goldsmith discuss key aspects of wave energy and tidal energy as separate areas and describe the key features of each resource, including the basic principles involved in generating electric power. It highlights the variability of the resources and explains how we may use modern tools to predict power and energy output in order to successfully integrate these resources in power grids.

In Chapter 28, German Renewable Energy Sources Pathway in the New Century, Matthias Müller-Mienack presents a case study about Germany which focuses on challenges with integrating renewable and the solutions being explored by Transmission System Operators (TSOs) in the country. The particular problems with the occurrence of 50.2 Hz frequency and potential risks this poses to grid operation and the design standards for new PV devices are discussed.

Finally, the chapters in **Part 10, Enabling and Disruptive Technologies for Renewable Integration**, present new approaches and illustrative examples for applying advances in control systems, measurement and sensing devices, visualization tools, methodologies for decision support systems, data mining and analytical tools, and high-performance computing to integrate variable renewable energy resources.

In Chapter 29, Control of Power Systems with High Penetration Variable Generation, Christopher L. DeMarco and Chaitanya A. Baone consider new challenges that wind and photovoltaic pose to the existing philosophies of primary and secondary control practices in power systems operations. The authors consider the impact of potentially very different control characteristics of the new power-delivery technologies of renewables. If increased penetration of renewable generation brings a new class of control actuators to the power grid, in a complementary fashion, increasing penetration of phasor measurement units (PMUs) brings a new class of sensor technologies. The chapter approaches renewables' control design differently, seeking to answer two salient questions: (1) what are the objectives associated with primary control that are necessary to maintain stable, secure operation of the power grid with more renewables and (2) what control actions can renewables and storage provide to help meet these objectives, while remaining within their operating limits?

In Chapter 30, Enhancing Situation Awareness in Power Systems under Uncertainty, Mica R. Endsley and Erik S. Connors address the increasingly important need for high levels of situation awareness (SA) to ensure the reliable and sustained operation of today's highly interconnected electric power systems. As the industry advances toward a more renewables and energy-friendly smart grid, compensating for the complications and challenges that variable generation imposes on operator's SA within transmission and distribution control centers becomes increasingly important. This chapter discusses how to overcome uncertainty and variability with renewable resources through the use of SA-Oriented Design to ensure that system operators are presented with the right information in an effective and integrated manner, so that they can develop the proper mental models to fully understand the state of a complex and dynamically changing system, project future changes, and respond in a timely manner.

In Chapter 31, Managing Operational Uncertainty through Improved Visualization Tools in Control Centers with Reference to Renewable Energy Providers, Richard Candy presents a case study from South Africa. The installation of renewable generation in power systems is tantamount to throwing large stones into a quiet pond. The consequences are similar in that both cause waves and surges that disrupt the harmony of the system. In order to account for the disruption and unpredictability of renewable energy providers, additional tools are needed to both visualize and manage the uncertainty and to soften the impact on the control staff. Classical SCADA and alarm systems simply do not have the ability to deal with the situation. The chapter describes a completely different approach to designing and implementing visualization tools in control centers, one in which the environmental factors that drive the uncertainty in renewable energy and the SCADA monitoring systems are combined, on a common platform, to provide the control staff with predictability and full situational awareness of these disruptors.

In Chapter 32, Dynamic Line Rating (DLR): A Safe, Quick, and Economic Way to Transition Power Networks toward Renewable Energy, Peter Schell describes dynamic line rating (DLR) and the role it can play in making renewable energy integration quicker, cheaper, and safer. He explains how DLR fits within the bigger picture of future grid planning and operation with high penetration of renewable energy resources. The current-carrying capacity of an overhead line depends on the ambient conditions. The more the line is cooled, the more current can be carried without exceeding safe operational limits. In today's power systems, with their increasing variability and need for flexibility, this weather-dependent variable, i.e. dynamic capacity, can be put to good use. But DLR is not as simple as it seems. The chapter concludes with an overview of the key challenges and possible solutions for DLR implementation.

In Chapter 33, Monitoring and Control of Renewable Energy Sources Using Synchronized Phasor Measurements, Luigi Vanfretti, Maxime Baudette, and Austin White provide an overview of how synchrophasor technology can be applied for developing real-time PMU applications, which help in monitoring and control of unwanted dynamics that are a product of renewable energy sources interacting with the power grid. Different paradigms for PMU-based monitoring and control systems are described and software environments for developing PMU applications are discussed. An example is presented to illustrate the development of real-time monitoring tools for the detection of subsynchronous wind farm oscillations, and the testing and validation of experiments performed using historical data and laboratory experiments. The chapter concludes with an outline for the development of new PMU applications that aid in the monitoring and control of renewable energy sources and their interaction with the grid.

In Chapter 34, Every Moment Counts: Synchrophasors for Distribution Networks with Variable Resources, Alexandra von Meier and Reza Arghandeh address how the direct measurement of voltage phase angle might enable new strategies for managing distribution networks with diverse, active components such as wind and solar plants. Historically, power distribution systems did not require elaborate monitoring schemes. With radial topology and one-way power flow, it was only necessary to evaluate the envelope of design conditions. But the growth of distributed energy resources such as renewable generation, electric vehicles, and demand response programs introduces more short-term and unpredicted fluctuations and disturbances, and the increased likelihood of two-way flow of power. This suggests a need for more refined measurement, given both the challenge of managing increased variability and uncertainty, and the opportunity of recruiting diverse resources for services in a more flexible grid. Specifically, von Meir and Arghandeh discuss high-precision synchrophasors, or micro-phasor measurement units (μPMUs), that are tailored to the particular requirements of power distribution in order to support a range of diagnostic and control applications, from solving known problems to unveiling as yet unexplored possibilities.

Finally in Chapter 35, Big Data, Data Mining, and Predictive Analytics and High-Performance Computing, Philippe Mack explores how we can extract value from, and make use of, the huge volume of data generated in power systems operations in order to better integrate large amounts of intermittent renewable energy resources. After a brief history of the evolution of what is now known as "Big Data," Mack provides an overview of sources of data in grid operations. He then discusses various data mining techniques, as well as tools for predictive analytics and high-performance computing. Examples of applications of predictive analytics to power systems and renewable integration are presented.

The book concludes with the *Epilogue* in which J. Charles Smith provides a future outlook on integration of variable resources in power grids and electricity markets. Although no one likes to make predictions, Smith makes bold, yet reasonable, projections of the near future based on society's increasing demand for a decarbonized energy future. In addition, technology advances and continued reduction in costs make wind, solar, and emerging low-carbon energy technologies well positioned to supply a larger portion of the insatiable energy demand in the world. Smith stresses the fact that we need not wait until all challenges to integrating renewables are addressed. Instead, we must continue to make progress down the road to meet all of the challenges of the future by using what we already know. Smith concludes with the words of the singer/songwriter Bob Dylan, "The answer, my friend, is blowin' in the wind," but then he adds that as we look over our shoulder, we should be aware that here comes the sun!

WHO SHOULD READ THIS BOOK

The intended audience for this book includes transmission and distribution grid operators and planners; electrical, mechanical, power, control, sustainability, and systems engineers; energy economists; government regulators and utility business leaders; researchers; students; and investors in, and developers of, renewable energy technologies and projects.

HOW TO READ THIS BOOK

This book covers a broad set of both theoretical and practical aspects of renewable integration. It is intended to be more of a reference volume than a standard textbook. Therefore, depending on the specific area of interest, the reader is encouraged to skip to the relevant chapter. Each chapter contains references to additional related literature that will provide more detailed information for interested readers.

Questions and feedback about the book can be sent to editor@renewenintegrate.com.

Policy and Regulation

The Journey of Reinventing the European Electricity Landscape—Challenges and Pioneers

Helena Lindquist

CEO and Founder of LightSwitch AB

1. Background

The oil crisis of 1973 became an important wake-up call for Europe. The dependency on energy imports from OPEC countries exposed the vulnerability of the economy to external events beyond the political control of European nation-states. Increasingly, concern was also growing for the effect of the fossil-based energy system on the environment and its link to global climate change. This prompted some pioneering countries in Europe to invest in research to develop wind and solar power technologies, most notably Denmark and Germany.

In the middle of the 1980s, the European Community (now European Union (EU)) began developing policies aimed at reducing oil consumption and greenhouse gas emissions while promoting the use of renewable energy (henceforth referred to as RE) for electricity, heating/cooling, and transport. In 1997 the European Commission published a white paper calling for the community to source 12% of its total energy, including 22% of electricity from renewables, by 2010 [1]. However, these targets were nonbinding and did not give the market enough confidence to make the necessary investments. The far-reaching Renewable Energy Roadmap published by the European Commission in 2007 [2], which was later enacted in the EU climate and energy package of 2009, kick-started real growth in the RE sector, most notably in wind and solar power. The often cited 20-20-20 targets of the EU Renewable Energy Directive form the core of the legislative framework and consist of three main pillars: (1) a binding target to reduce greenhouse gas emission by 20% by 2020 based on 1990 levels; (2) a binding target to increase the amount of energy consumption originating from renewable sources to 20% by 2020; and (3) a nonbinding target to improve energy efficiency by 20% in relation to projections for 2020 [3]. The EU directive meant that the EU committed to reaching these targets as a collective. These were then divided into separate targets in National Renewable Energy Action Plans for individual member states, taking into account their respective starting points and potential. The drivers behind this significant policy initiative were not only the concern for the growing threat of climate change, but—equally important—geopolitical considerations about security of supply and the dependency of foreign imports of oil and gas, not least affecting the EU's vulnerable relationship with Russia. It should be emphasized that developments in Europe are intrinsically linked to, and preconditioned by, externalities of the globalized economy and energy system.

Renewable Energy Integration. http://dx.doi.org/10.1016/B978-0-12-407910-6.00001-6

2. The post-2020 Europe

In its 2013 progress report [4] the EU Commission states that the Renewable Energy Directive has initially produced the desired effects in terms of stimulating RE deployment. However, the Commission expressed great concern that many key barriers to RE growth had not been removed as projected. The directive has not been implemented in full in all member states, and several states have deviated from their National Renewable Energy Action Plans, prompting the Commission to launch several infringement cases in regard to nontransposition of the directive (e.g. Austria, Finland, and Slovakia).

Approaching the target year 2020, however, the polemic discussion about the post-2020 policy framework is intensifying at all levels in European politics, among stakeholders in the power sector as well as in society at large. The European Commission has recognized that market stakeholders need clarity regarding the long-term policy framework to be able to invest in new infrastructure. However, due to the stark differences between member states in terms of their energy situation, challenges, and ambitions, it is not easy to find a solution acceptable to all parties. Arguably, the current policy design is not sufficient to significantly progress the European energy transition and target serious market failures [5].

The list of question marks and interrelated challenges shaping the developments of the energy transition is overwhelming. Around the middle of the 2010s, the following issues are dominating the debate at the macro level.

- The long-term EU RE policy is ultimately dependent on a global agreement on curbing greenhouse gas emissions.
- The role of shale gas in the global energy system (particularly in the United States and China) divides European countries. Those with serious security-of-supply issues, such as Poland, are tempted to exploit this previously inaccessible resource, thereby weakening the incentive to invest in RE and associated grid upgrades.
- The future role of nuclear power in the energy mix is another uncertain factor. Finland is constructing new nuclear reactors, while Germany has decided to phase out all nuclear production by 2022.
- The European economy is struggling to maintain growth and several countries in southern Europe have received EU emergency bailouts to avoid bankruptcy. Making the necessary investments in grid upgrades and sustained financial support to RE deployment during economic recession and growing unemployment is difficult.

3. Renewable integration in Europe: Challenges and policy responses

During the last decade the European electrical power system has undergone significant changes. Traditionally, the system was dominated by a few, often publicly owned, actors with monopoly control over both power grids and large power generation plants. Increasingly, this picture has been replaced by deregulated and unbundled markets with a large number of smaller producers and market stakeholders involved in power generation, transmission, distribution, and network development. At the same time, the share of intermittent RE, primarily wind and solar power, in the energy mix is growing quickly (Figure 1). As other RE technologies are becoming more mature, the exploitation of the vast

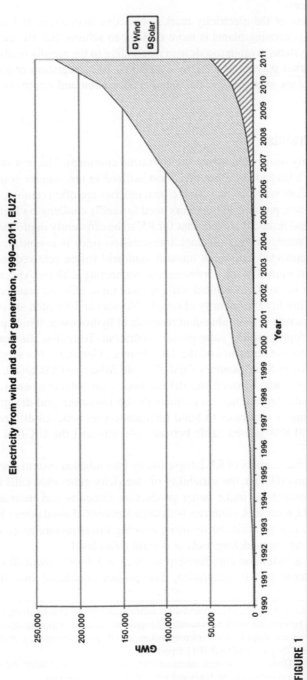

Eurostat.

FIGURE 1

Gross electricity generation (GWh) from intermittent renewable energy sources in the European Union, 1990–2011.

energy potential of the oceans will also become possible through the deployment of wave, tidal, and ocean current power devices.

Given the fragmentation of the electricity market, effective interaction and communication between grid operators and generating plants is more difficult to achieve [6]. Hence, the adaptation of technical and regulatory systems is often too slow in responding to the rapidly evolving energy sector, causing problems for market players at all levels. Efficient RE integration requires not only grid expansion and reinforcements on a grand scale, but also adaptation and reform of regulatory frameworks and market design.

3.1 Locational constraints

RE resources are not evenly distributed across the European continent. The best wind conditions are found around the British Isles and solar power is best utilized in the sunnier countries of southern Europe. In addition, locations with optimal wind characteristics are often remote, far from adequate transmission capacity. Hence, power grid operators need to handle challenging locational constraints at international and national levels. It is clear that for RE to be efficiently deployed across the whole of Europe, massive investment in long-distance transmission lines is essential. There is a strong lobby favoring an interconnected European transmission grid (often referred to as the European supergrid), stretching from Ireland in the northwest and connecting with the African continent in the south. Better cross-border networks in general will promote more efficient transportation of surplus production, improving reliability and security of supply. Norway and Sweden are often referred to as the "battery of Europe" because of their abundant resource of hydropower, which provides balancing power for weather-dependent wind and solar power generation. To realize the potential of offshore wind in the North Sea, the governments of the UK, France, Germany, Norway, and Sweden are cooperating through the *North Sea Countries Offshore Grid Initiative (NSCOGI)*, the goal of which is to connect offshore wind farms around the British Isles with balancing capacity in the Nordic countries. NSCOGI is conducting studies[1] to evaluate the best offshore grid design scenarios. There are several large-scale projects underway to build interconnectors between different countries—for instance, the planned 1400 MW subsea cable between Norway and the UK that is expected to be operational by 2020.

In regard to the operational aspects of RE integration by transmission system operators (TSOs) and distribution system operators (DSOs), the variability of electricity generation calls for the use of new technologies and methodologies to make better production forecasts and increase grid flexibility. Although these are complex processes, countries with large amounts of wind power like Denmark have shown that great progress can be achieved by using existing infrastructure more efficiently and by relatively modest investments in modeling tools at control room level.

The EU has co-funded a 56 million euro flagship research and development (R&D) project, called "TWENTIES,"[2] on the topic of RE integration. The project concluded that the European grid

---

[1] Initial results indicate that a "meshed approach (i.e. coordinated offshore and interconnector design, which implies multilateral cross-border cooperation between the North Seas countries) is preferred in front of a radial solution (i.e. point-to-point connection of offshore wind farms and shore-to-shore interconnectors, which implies continuing with mainly uni- or bilateral solutions between countries)." Source: NSCOGI 2012 report [7].

[2] "TWENTIES" is an acronym for Transmission system operation with large penetration of Wind and other renewable Electricity sources in Networks by means of innovative Tools and Integrated Energy Solutions.

infrastructure could be used much more efficiently than is the case today and that there are a number of tools available to allow for higher penetration of wind power in the system.[3]

In a large study for the European Commission analyzing the level of progress in different member states in regard to RE integration, the following issues were highlighted. At transmission level, the principal challenge relates to the need for infrastructure developments to keep up with new generation capacity coming online and the need to increase the level of interconnections in the system, as well as the need for smart grid concepts for increased flexibility. Although a lot of focus at the policy level has been on the TSOs' role in integrating RE, there are many challenges at the distribution grid level that arguably deserve more attention, such as grid planning, DSO control, and harmonization of technical regulations [6].

In addition to technical adaptations and grid reinforcements to allow RE integration, it is critical that stable and transparent administrative arrangements and cost-sharing rules for using the grid be reformed as the power system evolves. In its 2013 Renewable Energy Progress Report, the EU Commission states that "the increase of transparency and equitable grid connection and the development of cost sharing rules will provide the incentives on all producers to improve system-wide efficiency and not to make production or location decisions in isolation [4]."

The European Network of Transmission System Operators for Electricity (ENTSO-E) is a key organization working at a practical level with developing the European transmission network to accommodate ever-increasing levels of RE. Through Regulation (EC) 714/2009, the EU requested ENTSO-E to "adopt a nonbinding Community-wide ten-year network development plan" (TYNDP) [9]. This would ensure transparency and support decision-making processes at the regional and European levels, as well as establishing a base for the selection of "Projects of Common Interest."[4] The TYNDP is updated bi-annually and involves wide-ranging stakeholder consultations to ensure coherency and commitment from market stakeholders. One of the key responsibilities of ENTSO-E is the development of grid codes to ensure that the transmission grid is able to cost-efficiently accommodate ever more RE, in accordance with the priorities defined by the European Commission and the framework directives adopted by the Agency for Cooperation of Energy Regulators (ACER). The grid codes are designed to handle cross-border issues, e.g. connection of new generation plants, market integration, interoperability between transmission grids, and grid congestion problems. According to the regulation, these grid codes will, in regard to cross-border operations, be supranational and binding legislative instruments, taking precedence over national rules.

Further, ENTSO-E has a special Working Group for RE helping TSOs analyze the long-term consequences of a high penetration of RE on wholesale markets, which affects the functioning of the European power system as a whole. Important issues include assessing EU legislative proposals, suggesting policy recommendations for RE deployment, market integration, and design [10].

3.2 Market design

Even though there is a lot of potential for RE in the current European grid infrastructure, the electricity markets of individual member states are still designed for a power system based on more predictable

---

[3]E.g. that "wind farms can provide wide area voltage control, and secondary frequency control services to the system; and that virtual power plants enables reliable delivery of ancillary services, like voltage control and reserves, by intelligent control of distributed generation including wind farms and industrial consumption," p. 9, TWENTIES project final report [8].
[4]Regulation (EU) 347/2013 on guidelines for trans-European energy infrastructure highlights the important role of TYNDP.

production from centralized, conventional sources like coal and nuclear power. The work to establish a single European market for electricity started already, back in 1986, but there are still a number of outstanding issues to resolve. Although far from perfect and probably not up to the job of cost-efficiently integrating large shares of RE in the power system, the European electricity market is scheduled for "completion" in 2015, with the final implementation of the third energy package [11].[5] Although the EU single market for electricity represents an important step toward harmonization, member states still have a large degree of freedom to structure their own domestic markets, including subsidy systems for renewables. For the vision of a European common retail market for electricity to become a reality, the EU still needs to overcome political disagreements and fragmented legislation and to solve some technical issues. The harmonization model applied by Norway, Sweden, and Finland (cooperating through NordREG), resulting in the setting up of a single end-user market based on a supplier-centric model for the region by 2015 [12], could provide much-needed inspiration for the rest of the EU.

It is clear that existing market designs do not cater well to the needs of an increasingly complex and interdependent power system with high penetration of RE across several countries. The EU Renewable Energy Directive set targets for member states' production of RE, and it also requires priority grid access for such production. However, the rules regulating the integration of renewables in the marketplace are neither mandatory nor quantified. Article 16 of the directive states that "Member States shall take the appropriate steps to develop transmission and distribution grid infrastructure, intelligent networks, storage facilities and the electricity system, in order to allow the secure operation of the electricity system as it accommodates the further development of electricity production from renewable energy sources, including interconnection between Member States." The wording is open to a variety of interpretations from member states. The European Commission is concerned that the "current failure to modernize the grid as the energy mix is changing is causing problems for the development of the internal market, technical problems related to loop flows, grid stability and growing power curtailment, and investment bottlenecks resulting from delayed connection of new power producers" [5]. Granting priority grid access to renewable power, without the appropriate compensatory mechanisms for other types of power producers, means that the economics of conventional electricity production is seriously undermined, and in times of large RE production there have been instances of negative prices, e.g. in Germany. As the share of RE increases, the issue of load curtailment is becoming more pressing.

Since there is no uniform EU system for subsidizing RE, investment decisions are highly influenced by competing subsidy regimes in different member states (rather than the optimal geographic location for wind production and cost-efficient grid access). In addition, subsidy schemes for RE have in several instances changed very rapidly, sometimes even retroactively, which has caused large problems for the RE industry. This has been particularly acute in regard to solar photovoltaics (PV). Rigid national support schemes were generally unable to adapt quickly enough to the rapidly falling costs of solar PV, which generated large profits for investors and accelerated the pace and scale of installation in some countries during times of economic recession (e.g. in the UK and Spain). Fast and

---

[5]A legislative package consisting of two directives and three regulations with the overall purpose of opening up the internal gas and electricity markets of the EU. The package entered into force on September 3, 2009. Core elements of the third package include "ownership unbundling, stipulating the separation of companies' generation and sale operations from transmission networks; the establishment of a National Regulatory Authority (NRA) for each Member State; and the Agency for the Cooperation of Energy Regulators (ACER) providing a forum for NRAs to work together" (Source: European Commission, DG Energy [13]).

unpredictable changes to national support schemes have seriously increased the level of risk, and hence the costs, for RE investments. Since RE projects are still highly dependent on subsidies to be commercially viable, many industry players think that regulatory risk is by far the most serious obstacle to further deployment. To make matters worse, when government funding is cut back in light of the economic decline in Europe, the cost of lending from the private market is also much higher. The European Commission recognizes that member state intervention in energy markets may be necessary to achieve climate targets and security of supply, but in order to prevent distorting the functioning of the internal electricity market the Commission provides guidance to member states on how to design and reform national support schemes for renewables.

4. The story of three pioneers: History and future

The UK, Germany, and Denmark are in several respects frontrunners when it comes to setting ambitious targets for RE and designing policy instruments to enable the energy transition. Their energy policies all have different backgrounds and outlooks but may offer important learning and inspiration for other countries that are struggling to transform their energy systems.

4.1 Denmark

With more than 90% of its energy mix based on imported oil, Denmark in the 1970s was particularly vulnerable to the instability of the international oil market and hence became one of the pioneers in creating policies to support the development and deployment of RE sources, most notably wind power. Denmark's proactive energy policy has been delivered through four main packages. Taxes on electricity were used to support R&D in RE and hence pushed Denmark into a world-leading position in wind power technology.

In 2013, wind power accounted for almost 30% of the domestic electricity supply, up from 12% in 2001 [14]. The current national target for wind power is to reach 50% of electricity consumption by 2020 and for the electricity mix to be 100% renewable by 2035. Grid integration is a challenge also in Denmark, and the TSO, Energinet, is actively working to accommodate more RE in the system. To improve grid integration the following measures are in focus for the period up to 2020. In the short term, focus will be on "expansion of interconnections, reinforcement and expansion of the existing power grid, downward regulation of generation aided by negative spot prices, market coupling, and better wind power forecasting." In the medium term, work will be done to improve the "geographic distribution of onshore wind farms in the gas system, onshore grid, demand response, flexible electricity generation and Smart Grid." In the long term, it will be important to develop and deploy different types of energy storage [15].

4.2 Germany

Germany is Europe's largest electricity market and, given its central location on the continent, the country is highly interdependent on its nine neighboring countries for both imports and exports of electricity. Similar to Denmark, Germany also committed to RE at an early stage and introduced a feed-in tariff (FiT) system for RE already in 1979. The Renewable Energy Sources Act (EEG) adopted in 2000 gave the RE market a real push and the reinforced FiT scheme made it attractive to invest in wind and solar power. In only 10 years the share of RE generation capacity rose from 6% to 25%. The investments in RE have been largely driven by small and medium-sized companies and have had strong support in local communities. The FiT scheme has also created the world's biggest market for solar PV, with over 35 GW of capacity installed.[6] In May 2012, Germany set a world record for solar power generation, which met almost 50% of the nation's midday electricity needs.

Following the nuclear disaster in Fukushima, Japan, in 2011, the German government took the bold decision to phase out all nuclear power production by 2022. The sudden change in energy policy is referred to as the *Energiewende* (Energy Transition) and sets high-level targets, such as an 80% reduction of emissions by 2050 (compared to 1990 levels) and 80% electricity from RE by mid-century. Not surprisingly, these ambitions are having dramatic effects on many levels of the electrical power system, in terms of changes to the energy mix, market consequences, and operational challenges for grid operators. The Energiewende has increased the urgency of expanding and adapting the power grid to cope with more RE, especially since a considerable amount of RE will be produced by offshore wind farms in the north of the country and transmitted to the more densely populated areas in the south and west. The German government has recognized that the grid is not developing sufficiently fast and therefore adopted the Act on Accelerating Grid Expansion (NABEG) in 2011. The goal of the act is to enable upgrading, optimization, and expansion of existing grids [16].

4.3 United Kingdom

In comparison to Denmark and Germany, the UK could be considered a late starter in terms of reforming its energy system. The EU Renewable Energy Directive became the real starting point for a radical overhaul of the UK energy landscape. Proportionally, the UK has been set the highest target for RE in the Directive—from 13% in 2005 to 15% RE by 2020 [3]. To meet the RE targets, the bulk will come from offshore wind power. Due to the ambitious plans for offshore wind, the UK has developed into a world leader in the field. The government estimates that the potential contribution from offshore wind could be 40 GW by 2030. The UK have given generous subsidies to renewables, and with the banded Renewable Obligation providing a higher payment for immature technologies, this has given a real push to the development of offshore wind, wave, and tidal energy. The investment in RE is costly, and in light of the deep economic recession starting in 2008 the UK has been forced to radically reform its electricity market to ensure that the necessary investments will be made. Facing large-scale closures of old generation plants, the government estimates that a further 110 billion pounds of capital investment will need to be invested in low-carbon (RE and nuclear) technologies and grid upgrades until 2020. To meet this challenge, the UK, which was one of the pioneers of deregulating its electricity market in the 1990s, is now fundamentally reforming the system, introducing more state control to make sure the country will meet its RE obligations. The main elements of the electricity market reform (EMR) include a mechanism to support investment in low-carbon generation through FiTs with contracts for difference and a mechanism to support security of supply, if needed, in the form of a capacity market [17]. These mechanisms are to be supported by a carbon price floor, an emissions performance standard, electricity demand reduction, and measures to support market liquidity and

---

[6]In effect, German consumers have in large part been paying for the development of solar technology, while production of solar panels is now increasingly taking place in China. Due to the dramatic drop in prices of PV and the spiraling cost of FiTs, the German government has in recent years been forced to curb subsidies.

access to market for independent generators. One of the major challenges for the UK has been to maintain investor confidence regarding government energy policy. During the EMR process it has become clear that regulatory risk is deemed to be a key barrier to investment in offshore wind.

In regard to RE integration, issues around planning consent and long lead times for projects have been important. The "Connect and Manage" grid access regime introduced by the regulator Ofgem has delivered good results and advanced the connection dates of many RE projects [18]. Given the challenging location of the bulk of the planned RE generation far out in the North Sea, building an offshore grid and cross-border interconnectors, e.g. to Norway, is naturally a key priority for the TSO, National Grid.

5. Trends & future outlook

The penetration of RE in European power grids is rapidly increasing, and if not tackled appropriately, grid integration risks becoming a serious bottleneck preventing the EU from meeting its climate targets. European grid operators must learn how to effectively manage the uncertainties associated with the inherently variable nature of these energy sources. There are a large number of tools and technologies already available (as will be described in later chapters in this book) to increase flexibility in transmission and distribution grids.

Arguably, "business as usual" is no longer a realistic option for realizing the energy transformation. Therefore, the question about the level of reliability in European electrical power grids deserves more attention, i.e. the idea of decreasing reliability for certain customer groups in order to increase flexibility in the grid, thereby allowing for the integration of more RE. Such changes would have important consequences for the design of grid codes and market incentive systems.

Given the urgency of the challenge, investment in research and demonstration of workable solutions to increase grid flexibility (e.g. energy storage and smart grid solutions) should be given priority in national and EU budgets. The EU Horizon 2020 framework program for the period 2014–2020 is designating a significant share of the budget to energy-related challenges; however, this needs to be complemented by national initiatives to achieve the critical mass of activity needed to accelerate the pace of development. To ascertain that existing infrastructure is employed at its full potential it is equally important to promote learning from best practice examples [19]. European countries have a lot to gain from a mutual exchange of knowledge about innovative solutions with other parts of the world, e.g. the United States.

Committing the funds for large-scale infrastructure investment in times of economic decline is not easy. Carrying out grid upgrades and energy efficiency improvement measures across industry and the built environment requires enormous levels of capital investment, but such projects are also highly labor intensive. Hence, if policy makers realize the potential in linking infrastructure investment needs with surging unemployment and struggling economies, the energy transformation could contribute to a positive economic development in Europe. Alternative financing and cost-sharing models, e.g. Public-Private Partnerships (PPPs), could further enable and accelerate investments in grid infrastructure.

The interrelationships between technology, policy, and global market developments are extremely complex and it is therefore imperative that decision makers and sector stakeholders take a long-term and holistic view when evaluating policy options and making grid investment decisions. The sudden rise of shale gas in the global energy system is an example of a potentially game-changing development influencing the evolution of RE markets and, by extension, also related grid integration issues. Given the scale of the energy system transformation and the number of stakeholders involved, the EU plays a crucial role in promoting the simultaneous and coordinated development of grid infrastructure, regulatory reform, and a functioning European single market for electricity.

References

[1] European Commission. White paper on energy for the future: renewable sources of energy. COM(97)599; November 26, 1997.

[2] European Commission Communication. Renewable energy road map: renewable energies in the 21st century: building a more sustainable future. COM(2006) 848 final; January 10, 2007.

[3] Parliament European. European Council. Directive 2009/28/EC of the European Parliament and of the Council of 23 April 2009, on the promotion of the use of energy from renewable sources and amending and subsequently repealing Directives 2001/77/EC and 2003/30/EC; April 23, 2009.

[4] European Commission. Report from the Commission to the European Parliament, the Council, the European Economic and Social Committee and the Committee of the regions – renewable energy progress report. COM(2013) 175; March 27, 2013.

[5] Koelemeijer R, Ros J, Notenboom J, Boot P. (PBL Netherlands Environmental Assessment Agency), Groenenberg H, Winkel T (Ecofys), EU policy options for climate and energy beyond 2020 www.pbl.nl/en/publications/eu-policy-options-for-climate-andenergy-beyond-2020; 2013 [accessed 22.12.13].

[6] Zane EB (Eclareon), Brückmann R (Öko-Institut), Bauknecht D (Öko-Institut), Jirous F, Piria R, Trennepohl N (eclareon) et al. Integration of electricity from renewables to the electricity grid and to the electricity market – RES-Integration, www.oeko.de/oekodoc/1378/2012-012-en.pdf; 2013 [accessed 03.01.14].

[7] Website of the North Seas Countries Offshore Grids Initiative (NSCOGI). Key initial findings from the NSCOGI 2012 report. www.benelux.int/NSCOGI/ [accessed 30.12.13].

[8] TWENTIES project final report.www.twenties-project.eu; October 2013 [accessed 03.01.14].

[9] Regulation (EC) No 714/2009 of 13 July 2009 on conditions for access to the network for cross-border exchanges in electricity

[10] Website of ENTSO-E. www.entsoe.eu [accessed 22.12.13].

[11] Website of the European Commission, DG Energy, Single market for gas & electricity. http://ec.europa.eu/energy/gas_electricity/legislation/third_legislative_package_en.htm [accessed 30.12.13].

[12] Website of Nordic Energy Regulators (NordREG). www.nordicenergyregulators.org [accessed 30.12.13].

[13] Website of European Commission, Directorate-General for Energy. www.ec.europa.eu/dgs/energy/index_en.htm.

[14] Website of Danish Energy Agency, (Energistyrelsen). www.ens.dk/en [accessed 22.12.13].

[15] International Renewable Energy Agency (IRENA). 30 years of policies for wind energy: lessons from 12 wind energy markets www.irena.org; 2013 [accessed 15.12.13]. p. 64.

[16] Website of German Energy Transition (Heinrich Böll Stiftung). www.energytransition.de [accessed 22.12.13].

[17] UK Department for Energy and Climate Change. Electricity market reform: policy overview; November 2012.

[18] Website of UK Department for Energy and Climate Change. www.gov.uk/electricity-network-delivery-and-access [accessed 15.12.13].

[19] Jones L. Strategies and decision support systems for integrating variable energy sources in control centres for reliable grid operations, global best practices examples of excellence, and lessons learned www1.eere.energy.gov/wind/pdfs/doe_wind_integration_report.pdf; 2010 [accessed 30.12.13].

Policies for Accommodating Higher Penetration of Variable Energy Resources: US Outlook and Perspectives

2

Steven Fine, Kiran Kumaraswamy

Renewable energy resources have witnessed explosive growth in the United States over the last several years, prompted by falling capital costs and greater efficiencies as well as by federal incentives and state renewable portfolio standard (RPS) policies. Many renewable sources, most notably wind and to some extent solar, are variable in their nature and present a new set of challenges in integrating them on the power grid. The policies to integrate these resources are often implemented at the balancing authority level and include, but are not limited to, the implementation of sub-hourly markets, increased ancillary or midterm capacity markets for "fast ramp" capacity, and energy imbalance markets to allow cross-region transmission to accommodate the variable power.

1. Recent renewable deployment trends

Over the last few years, renewable energy resources have witnessed an explosive growth in the United States. As shown in Figure 1, renewable energy sources, particularly wind, solar, and biomass, have been growing at a much higher rate than the conventional energy sources. In 2012 alone, more than 12 GW of new wind capacity was added, increasing the total installed wind capacity in the United States to more than 57 GW.

While a number of factors, such as steep reduction in cost of wind turbines and abundance of good-quality wind, have contributed to the impressive growth of wind energy in the United States, favorable federal and state policies deserve the most credit for this spurt in wind energy development. These include the federal production tax credit (PTC) and state RPS policies.

1.1 Federal policy impacting growth: renewable electricity PTC

The federal PTC is a per-kilowatt-hour tax credit for electricity generated by utility-scale wind turbines, geothermal, solar, hydropower, biomass, and marine and hydrokinetic renewable energy plants, which was originally enacted as part of the Energy Policy Act of 1992. The US Congress then extended the provision many times, most recently by the American Taxpayer Relief Act of 2012 in January 2013.[1] The PTC has been a major driver of wind energy development over the past decade, and its

---

[1]Database for State Incentives for Renewables and Efficiency, http://dsireusa.org/incentives/incentive.cfm?Incentive_Code=US13F [accessed 04.02.13].

Renewable Energy Integration. http://dx.doi.org/10.1016/B978-0-12-407910-6.00002-8

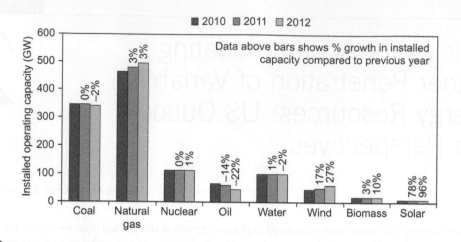

FIGURE 1

Total installed operating capacity in the United States by fuel type.

Federal Energy Regulatory Commission—Energy Infrastructure Update and ICF.

recent extension to include projects "under construction" by the end of 2013 was expected to encourage new wind projects in 2013 and 2014 as well.[2]

Wind energy development in the United States has historically relied heavily on the PTC. As shown in Figure 2, the wind industry experienced several boom–bust cycles with strong growth during the years leading up to the expiration of the PTC and a dramatic decrease in installed wind capacity in years when the PTC lapsed.[3]

1.2 State policy impacting growth: RPSs

RPS policies place an obligation on electric supply companies to supply a specified fraction of their electricity from renewable energy sources and enumerate mechanisms that are permitted to achieve compliance, such as renewable energy credits (RECs).[4] Currently, 29 states have renewable electricity standards and seven states have renewable energy goals. The idea of a federal RPS or clean energy standard has been discussed, but has never gained the necessary traction to become law.

Along with the federal PTC, the state RPS mechanisms have been primarily responsible for inducing new renewable energy capacity in the United States over the last several years. In an

---

[2]"The Prospects for US Wind Development Following the Federal PTC Extension," White Paper by Patrick Costello, ICF International.

[3]Union of Concerned Scientists, January 4, 2013, http://www.ucsusa.org/clean_energy/smart-energy-solutions/increase-renewables/production-tax-credit-for.html [accessed 25.02.13].

[4]The Environment Protection Agency (EPA) of the United States defines REC as: "'REC' represents the property rights to the environmental, social, and other nonpower qualities of renewable electricity generation. A REC, and its associated attributes and benefits, can be sold separately from the underlying physical electricity associated with a renewable-based generation source." More details can be found here: http://www.epa.gov/greenpower/gpmarket/rec.htm.

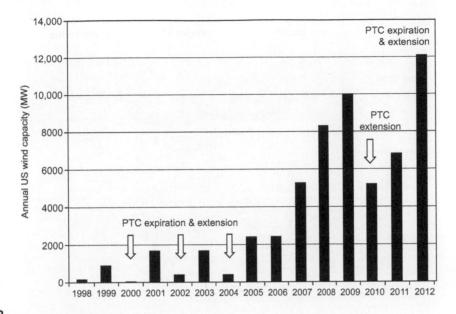

FIGURE 2

Impact of production tax credit on annual wind installation in the United States.

Union of concerned scientists, January 4, 2013.

environment of low gas prices and the potential expiration of the PTC, the continuation of state policies, in particular, will continue to be important in promoting the ongoing development of renewable energy resources.

2. Technical challenges posed by wind generation for power system operation & planning

The grid-based power system is continuously changing and is impacted by various factors that change in increments in time from seconds to minutes, minutes to hours, seasonally, and year to year. In the different time scales of operation, balance between load and available generation must be maintained. In the very short time scale—seconds to minutes—stability and reliability of the system is primarily maintained by automatic control systems such as automatic generation control (AGC).[5] In the intermediate to longer time scales, system operators and planners are responsible for maintaining system reliability through the use of operating and capacity reserves.

The inherent variability of wind generation and the uncertainty in wind forecasts have impacts on the various time scales of the power system that range from a few seconds to years. Figure 3 shows the impact of wind on various aspects of a power system in different time scales.

---

[5]AGC is an automatic control system for constantly adjusting the output of generating units at different power plants, in response to moment-by-moment variations in the load.

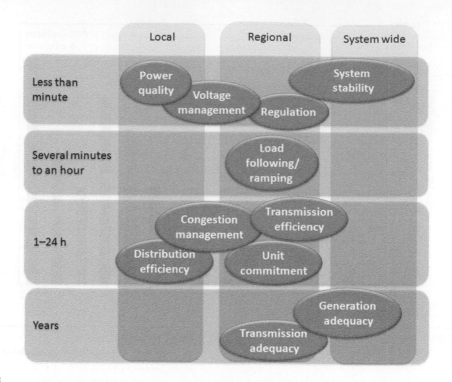

FIGURE 3

Impact of wind energy on the various aspects of power systems.

ICF.

Some of the key inferences from Figure 4 are summarized below and help identify the specific operation and planning challenges of wind generation that we explore in this chapter.

- The stability of the power system is governed primarily by synchronous generators supplied from conventional base load and dispatchable resources. Displacement of conventional generators by wind generators can adversely impact system stability by reducing the system inertia,[6] posing challenges to maintaining power system stability.
- Variations in wind energy and demand within an hour are much more significant for the system. Adequate fast ramp reserve capacity is required to manage variability of wind generation over this time frame. System operators deploy appropriate backup resources with quick-start capability to follow the variable nature of wind generation.
- Integration of wind generation is likely to increase the commitment of quick-start units to ensure available generation to handle deviations from forecast levels. On the other hand, the increased use of renewable generation typically displaces generation from thermal resources such as gas- and coal-fired power plants.

---

[6]Inertia is the characteristic of power systems by virtue of which they arrest the rate of change of frequency during sudden load and generation changes.

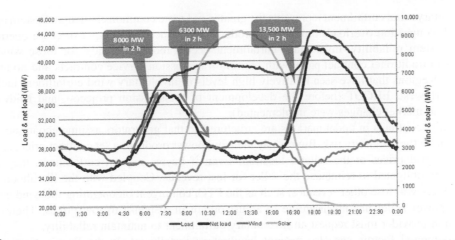

FIGURE 4

California Independent System Operator (CAISO): ramping needs.

R.11-10-023: RA Flexibility Workshop Flexible Capacity Procurement Proposal, CAISO.

For example, in the California ISO system grid operators expect that in the year 2020, with high levels of renewable penetration, conventional units may face a ramping need of over 13,000 MW in 2 h. In large part, this is driven by an increase in solar photovoltaics (distributed generation (DG)), which have a significant drop-off in capacity in the late afternoon hours (3–5 PM). Adequate—and significant—levels of quick-ramp capacity would be needed to ensure reliable system operations under such circumstances. Typically, wind resources are located far from the load, which requires additional investments in transmission infrastructure. New transmission lines remain underutilized if built to accommodate the full capacity of a wind farm, due to relatively low wind capacity factors. If lines are undersized, curtailment of wind generation occurs.

3. Economic challenges associated with high wind energy: the potential for curtailment

Wind curtailment, where the system operator cuts the amount of wind generation that can be sold onto the power grid for a specified amount of time, is usually caused by one of two factors: a transmission system that is incapable of accommodating the full dispatch of wind facilities (involuntary curtailment) or a mismatch between supply and demand (voluntary curtailment).

Voluntary curtailment manifests itself in nodal pricing markets, such as PJM, New York, New England, and Electric Reliability Council of Texas (ERCOT), in the form of low or even negative power prices (meaning generators must actually pay to produce wind power). Wind power producers respond by reducing their output. Operators that also own coal or nuclear facilities may voluntarily curtail wind power production because wind power plants require less time and money to shut down and restart in comparison to conventional generation facilities.

Involuntary wind curtailment represents the greater and more common challenge. System operators faced with congestion overloads order wind producers to reduce their output or cease operations to preserve system reliability. Involuntary curtailment occurs most frequently for those wind power producers in traditional contract paths with nonfirm transmission service contracts and no minimum price bids. Regions that have seen relatively high levels of involuntary wind curtailment include the Pacific Northwest, ERCOT (before the development of transmission projects), and Midwest ISO (before some market changes were implemented).

Typically, contracts in nodal markets are structured so that generators assume economic risk and buyers assume the reliability risk. Buyers pay the wind power producer a set contract price, regardless of the spot-grid prices at the seller's location. Even if buyers request curtailment, they may be contractually obligated to reimburse the seller for lost revenue. Coupled with the likelihood that the wind producer will lose other revenues, such as the PTC, contingent on operating the wind generator, the wind power seller has no incentive to adjust output in response to low prices. Therefore, the transmission provider must request an involuntary curtailment to maintain reliability.

Environmental factors can also prompt involuntary curtailment. In the Pacific Northwest, for example, transmission providers facing overloads caused by rising water levels occasionally curtail wind, thermal, and other energy sources in favor of hydropower to protect and preserve wildlife habitat.

3.1 The repercussions of discarded energy

When usable energy is curtailed, all parties involved suffer financial consequences. These include:

- *Lost revenues*—Wind curtailment prevents lenders and other investors that invest in wind power production from receiving the returns they expected.
- *Lost incentives*—Wind power producers qualify for valuable PTC that equal 2.2 cents per kilowatt-hour on an after-tax basis for the first 10 years of electricity production. They can also earn renewable energy certificates Renewable Energy Credits (RECs) or a green power premium effectively bundled in the energy sales for their generated power. But wind producers only earn these rewards if they operate their facilities and distribute their wind power on the grid.
- *Higher energy prices*—Wind power producers may need to charge higher prices to help offset the risks of revenue lost to curtailment or negative pricing. Negative pricing occurs when wind power producers opt to pay grid operators to take their power so they can still earn PTC and REC benefits.
- *Challenges to business operations*—Wind power producers face competition from generators of coal, gas, hydropower, and other energy sources that face little to no exposure to curtailment risks. Furthermore, the risks and uncertainties of curtailment can make it difficult for new wind power producers to secure funds from financing agencies.

Certain markets like the Midwest ISO have instituted programs called "Dispatchable Intermittent Resources" to ensure that wind units follow price signals and change their dispatch accordingly. As of 2013, almost all the wind generation in Midwest ISO is dispatchable, which has reduced the instances of involuntary curtailment and made almost all of the curtailment to be on economic basis. Such market changes are encouraging for wind generators, primarily because they provide greater transparency in control actions taken by the grid operator. Transparency is particularly helpful while making decisions on the size, location, and market for power plant developers and utilities that subscribe to renewable capacity. It also helps plan, mitigate, and hedge against the risk of curtailment more effectively.

However, this development still does not take away the risk of curtailment of wind energy; it just makes it more manageable. Additional transmission development and broadening the market's footprint are some of the fundamental ways of reducing overall levels of curtailment. By leveraging resources across a broader area, the volatility associated with wind can be effectively minimized.

4. Transmission development for wind integration: challenges and success stories

Central station renewable resources are typically located far from networked transmission lines, requiring developers to design and construct transmission systems specifically designed to transfer the power generated by the renewable energy systems. Construction of transmission lines is a challenging process for the following reasons:

1. Long lead time associated with obtaining necessary permits and approvals (typical high-voltage lines face a 5–7 year lead time)
2. Cost allocation within utilities in the region
3. Environmental issues and local citizen activism (Not in My Back Yard, or NIMBY, issues)
4. Siting and permitting challenges with local state and municipal agencies

Most recent developments at the federal level, in the form of Federal Energy Regulatory Commission (FERC) Order 1000, have created an incentive for transmission development toward achieving public policy objectives like renewable energy targets. Consequently, it is possible that in the near future additional transmission may be developed in certain regions to facilitate renewable integration, with the costs allocated among multiple parties that benefit from the project.

However, there have also been several challenges for transmission—most notably the explosive growth of distributed energy resources (DERs). DERs can provide reliability benefits to the system in the form of deferred Transmission and Distribution upgrades if targeted in specific localities. The class of DERs that include energy efficiency, demand response (DR), and combined heat and power applications have the effect of reducing peak demand by providing appropriate incentives to customers to lower energy consumption during stressed grid conditions. In markets like PJM, demand resources already contribute about 7.5% of the anticipated peak demand in a year. Large transmission projects planned in this region, like the Mid-Atlantic Power Pathway, have been deferred because of weak growth in peak demand, partly due to an increase in DERs and specifically DR (see Chapter 40).

4.1 Policy developments could incentivize transmission system development to accommodate renewable generation

Several states across the country have successfully developed transmission facilities to accommodate large amounts of traditional and renewable generation in a short span of time. Understanding the success factors and the challenges faced through the transmission development process can be a critical step in the development of transmission infrastructure to accommodate renewable generation.

4.1.1 Texas planning example

ERCOT's Competitive Renewable Energy Zones (CREZ) are a major success story of planning transmission for integrating large amounts of utility-scale wind power into the electric grid. By identifying regions with good wind energy potential and sufficient land area to generate electricity, ERCOT was able to develop a transmission expansion plan that would enable delivery of about 18,000 MW of wind energy to load centers. The plan was approved by the Public Utilities Commission of Texas in 2008 and all the transmission facilities identified in the plan were expected to be complete by 2013. Figure 5 shows a timeline of key milestones in CREZ.

As previously stated, development of new transmission is a fairly long process due to several contentious issues related to cost allocation and transmission siting. In early 2014, the final segment of the CREZ projects were placed into service implying that CREZ was completed in 8 years. Eight years is relatively a very short duration for the scale of transmission development ($\sim$2300 miles of 345 kV lines) that will take place in CREZ. Some of the key factors that have played a role in the success of Texas-CREZ are listed in Figure 6.

4.1.2 California renewable and transmission development example

Another major transmission development project for integrating large amount of utility-scale wind energy resources is the Tehachapi Renewable Transmission Project. Southern California Edison (SCE), one of the three investor-owned utilities in California, is currently constructing transmission facilities that will enable about 4500 MW of wind energy from the Tehachapi region to be delivered to the distant load centers. The project is expected to be completed by 2015 at a cost of around $1.8 billion.

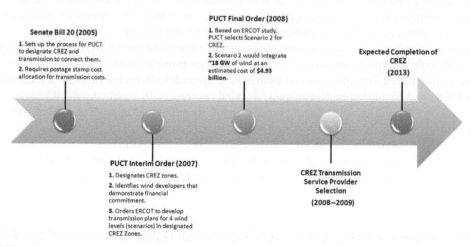

Senate Bill 20 (2005)
1. Sets up the process for PUCT to designate CREZ and transmission to connect them.
2. Requires postage stamp cost allocation for transmission costs.

PUCT Interim Order (2007)
1. Designates CREZ zones.
2. Identifies wind developers that demonstrate financial commitment.
3. Orders ERCOT to develop transmission plans for 4 wind levels (scenarios) in designated CREZ Zones.

PUCT Final Order (2008)
1. Based on ERCOT study, PUCT selects Scenario 2 for CREZ.
2. Scenaro 2 would integrate "18 GW of wind at an estimated cost of $4.93 billion.

CREZ Transmission Service Provider Selection (2008–2009)

Expected Completion of CREZ (2013)

FIGURE 5

Evolution of Competitive Renewable Energy Zones.

Hunt Transmission LLC, Sharyland Utilities, and ICF.

The major driver behind this project is the aggressive RPS goal set by California. Wind energy resources in Tehachapi could play a significant role in achieving the RPS goals, but transmission constraints from the California high desert to load centered in the Los Angeles Basin needed to be overcome. This transmission project helps alleviate such constraints.

Cost recovery was the major impediment for this project due to the "chicken and the egg" problem—major transmission additions and their cost allocation processes cannot be approved by the California Public Utilities Commission (CPUC) until sufficient generator commitments have been obtained; and without sufficient transmission, generator commitments are hard to obtain. Due to FERC

Index	Factors	Responsible players	Impact
1	Government mandate for CREZ and clearly stated cost allocation method (postage stamp) through Senate Bill 20 (SB 20).	Legislature and State Government	Reduced investment risk for wind power and transmission investors and developers.
2	SB 20 requirement for PUCT to designate best areas for renewable energy development that satisfy the following criteria: 1. Availability of best quality renewable resources. 2. Availability of sufficient land area for generating power from renewable resources. 3. High level of financial commitment from renewable energy developers.	Legislature and State Government	Credible selection of CREZ zones that would increase benefit to cost ratio.
3	Transmission Optimization Study by ERCOT to identify optimal transmission plans for 4 levels of wind generation development in PUCT identified CREZ zones.	PUCT and ERCOT	The study enabled the selection of a wind generation scenario in CREZ zones that maximized wind power generation at optimal transmission development cost.
4	Competitive selection of transmission developers.	PUCT	Those transmission plans and developers were selected for implementing CREZ which would maximize consumer benefits at lowest possible cost.
5	Allowing utilities selected for developing CREZ transmission to include associated costs in their rate base.	Legislature and State Government	Further reduction in financial risk associated with transmission development.

FIGURE 6

Texas success story—development of Competitive Renewable Energy Zones.

jurisdiction over wholesale transactions on SCE's transmission facilities, the CPUC only had a "backstop" authority to approve cost recovery for the project through an increase in electricity rates of SCE's retail customers. In order to ensure that the burden of the project is also shared by all the generators that benefit from the transmission development, the CPUC encouraged the California Independent System Operator (CAISO) to prepare a policy, and urge FERC to approve it, that will extend the cost allocation backstop authority to include location-constrained generators. As a result, CAISO prepared and FERC approved the Location Constrained Resource Interconnection (LCRI) policy in 2007. Through the LCRI, retail customers of SCE have initially underwritten the complete project, and as new wind generation projects in Tehachapi use the transmission facilities, they will bear the burden of the project cost.

4.1.3 Midwest ISO multivalue project cost allocation mechanism

Midwest ISO uses a multivalue construct for cost allocation of high-voltage transmission projects within its footprint. FERC approved this mechanism in December 2010 to implement a regionwide cost allocation for special high-voltage transmission projects that provide multiple benefits, known as multivalue projects (MVPs). Midwest ISO primarily planned to build MVPs to evacuate clean energy generated largely in the Northwest region by wind farms to load centers in the Midwest and Mid-Atlantic region; in addition, it provides reliability and various economic benefits distributed over multiple price zones.

By December 2013, Midwest ISO will extend its operation to the integrated Entergy region (a new South region), which will establish similar benefits to customers in that region. Midwest ISO proposed a 5-year transition plan to extend its cost allocation practices from the Midwest region to the new South region in a reasonable manner. The MVPs provide benefit to the whole region and therefore the costs are spread uniformly across the entire region. The existing MVPs were planned without the South region, and so customers in the South are not required to pay for the MVPs' cost.

4.1.4 Merchant transmission examples for variable energy resource integration: atlantic wind connection and tres amigas

4.1.4.1 Atlantic wind connection

The Atlantic Wind Connection (AWC) is a subsea high-voltage direct current (HVDC) transmission line running from northern New Jersey to southern Virginia. It will cover the Mid-Atlantic region and primarily consists of two main circuits; New Jersey Energy Link and Delmarva Energy Link will connect through Bay link, as shown in Figure 7.

The AWC project will be built in three phases over a 10 year period that will connect up to 6 GW offshore wind farms built at least 10 miles off the coast. Each circuit, New Jersey and Delmarva, will have three offshore hubs and each hub will have capability to accept 1000 MW of wind power and deliver it to the PJM grid. Development of a transmission corridor to move economical wind power in the region where it is needed most is the elemental barrier for development of offshore wind farms. The AWC project will enable smooth development of offshore wind farms because wind developers would not have to plan and build individual interconnecting radial lines. Additionally, the AWC transmission line will help reduce congestion in the eastern PJM grid that essentially occurs due to power import from the West region. The offshore buried cables of the AWC project will enhance resiliency of the interconnected transmission grid during severe weather and destructive events. It will provide PJM an

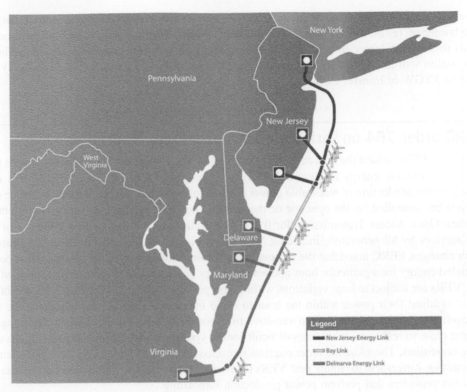

FIGURE 7

Location of Atlantic Wind Connection high-voltage direct current transmission project.

Atlantic Wind Connection.

extra margin of controllable transmission capacity over the north–south corridor to improve grid reliability under varying system conditions.

4.1.4.2 Tres amigas

The Tres Amigas is a superstation that will employ the latest in power grid technologies, including superconductor cables, voltage source converters, and energy storage systems, to interconnect the nation's three individual power grids, known as the Eastern Interconnection, the Western Interconnection, and the Texas Interconnection.

The regions in the Western Interconnection are solar-rich regions, and Texas has the best wind in the nation, but a weak transmission network is a major impediment for the development of renewable capacity in this region. Also, the variable and uncertain nature of renewable generation makes it difficult for system operators to maintain system reliability under high renewable penetration. The larger integrated grid reduces variability and uncertainty in renewable generation, and allows for lower reserve margin requirements since the greater geographic distribution of renewable resources effectively reduces the aggregate variability. The Tres Amigas, located in New Mexico's eastern region near the border of Texas, will unite America's isolated interconnection grids; it will allow renewable resources to utilize a larger integrated grid and move clean power over a long distance. The Tres Amigas station will have 5 GW of power transfer capacity in the initial phase but eventually could move up to 30 GW of transfer capacity.

5. FERC order 764 on variable energy resource integration

On June 2012, FERC issued the final rule on Order No. 764 which is intended to remove barriers for the integration of variable energy resources (VERs) into the grid. As per the order, VER is a device employed for the production of renewable power that cannot be stored and has variability in the output that cannot be controlled by the operator or the owner. The order requires transmission providers to amend their Open Access Transmission Tariff (OATT) to allow 15-min intrahour scheduling of transmission services by all generators, including VERs, that want to make such intrahour scheduling additions or changes. FERC noted that the resources pay for imbalance charges for any deviation between its scheduled energy for a particular hour and actual energy generated in that hour. FERC pointed to the fact that VERs are subject to huge variations within short periods and that the existing rules do not allow resources to adjust their power within the hour to avoid imbalance charges. Consequently, the FERC ruling requires transmission providers to transition from hourly to 15-min transmission scheduling. With VER input close to real time, it becomes difficult for the system operator to manage operations during real-time conditions. Therefore the order requires transmission providers to amend the Large Generator Interconnection Agreement to require new VERs to provide meteorological and forced outage data to the transmission providers that perform power production forecasting.

On December 20, 2012, FERC issued Order No. 764-A to address rehearing and clarification requests received from multiple parties. In the order, FERC confirmed that the intrahour scheduling reforms apply to all the transmission customers scheduling transmission service under the OATT, which includes entities using point-to-point service or opting for network integration service. In addition to that, FERC also cleared that transmission providers must average the imbalances of each 15-min scheduling interval over the entire hour for customers that schedule transmission service at 15-min intervals so as to reduce the imbalance charges for these customers. FERC also clarified that firm transmission schedules hold curtailment priority over nonfirm schedules. In summary, FERC's actions through Order 764 should facilitate the integration of higher levels of VERs on the system.

6. The future of renewable development in the united states

The future of central station renewable resources faces several challenges going forward. These include the facts that many RPS policies are currently scheduled to level off in terms of their percentage requirements, absent any further action, by 2020; low natural gas prices will continue to put pressure on the ability of renewables to achieve grid parity; and the extension of the PTC remains uncertain.

These challenges are countered, to varying degrees in different markets, by the falling levelized cost of renewables as a result of lower capital costs and the greater efficiencies, and by the fact that many public and private entities seek to incorporate more renewables onto their systems, both due to the desire to incorporate nonemitting resources onto their system and as a hedge against future potential increases in the cost of fossil-fired generation. The continued development of both DER and central station renewables is dependent upon their successful integration onto the grid. Effective federal and state policies related to development of transmission facilities needed for renewable integration and market design to accommodate the variability concerns would be critical items to watch for in years to come.

Harnessing and Integrating Africa's Renewable Energy Resources

3

Ijeoma Onyeji
New Energy Insights, United Kingdom

1. Introduction

Africa's long-term economic growth and competitiveness fundamentally depends on reliable access to energy services, yet the population of sub-Saharan Africa lags far behind the rest of the world in terms of electricity access rates. While the average 2011 rates in Latin America, the Middle East, and developing Asia are 5%, 9%, and 17%, respectively, sub-Saharan Africa ranks lowest, with 68% of its population lacking access to electricity [1,2]. So far investments on the continent have been far below requirements [3,4], explaining why Africa's energy sector has been fraught with deficiencies such as low access and insufficient capacity, poor reliability, and extremely high costs. These and other shortcomings in the power sector have significantly retarded Africa's economic growth and put at risk its long-term economic prosperity.

The grim reality of energy poverty in most African countries, however, is in stark contrast to the continent's rich endowment with renewable and exhaustible energy resources. Africa's reserves of renewable energy resources are the highest in the world and are estimated to be sufficient for meeting the continent's current as well as incremental future needs [5]. The significant decrease in costs of renewables over the recent years offers huge potential for Africa's power sector; the question remains how best to make use of it.

In this chapter we briefly revisit some of the key issues to consider for successful integration of renewable energy in Africa: Section 2 gives an overview of the contrast between energy poverty and resource abundance in Africa; Section 3 puts Africa in the context of the global energy shift toward renewables; Section 4 discusses some of the key aspects going forward; and Section 5 concludes.

2. Background and context

2.1 Africa's energy challenge

Today, about 600 million people in Africa, roughly 57% of the African population, lack access to electricity. Average annual per capita consumption of power in sub-Saharan Africa is 536 kWh, compared to the global average of 3044 kWh [6]. Electrification rates are correspondingly low (Figure 1): the average rate in sub-Saharan African countries is 32%, with an average of 18% in rural areas; overall only about 32% of the sub-Saharan African population has access to electricity [1,2].

Renewable Energy Integration. http://dx.doi.org/10.1016/B978-0-12-407910-6.00003-X

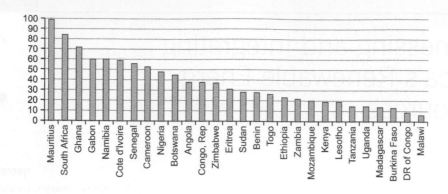

FIGURE 1

Electrification rates of sub-Saharan African countries, 2011 (%).

Refs [1,2].

Africa's current consumption is only about 25% of the global average of per capita energy consumption, its mix consisting of hydropower, fossil fuels, and biomass (predominantly traditional uses) [7].

The African energy sector faces two key challenges: (1) low access to electricity; and (2) insufficient and unreliable power supply. The limited and unreliable access to energy significantly inhibits socioeconomic advancement, which fundamentally depends on reliable access to modern energy services. The lack of it is detrimental to economies not only in terms of lost productive output, but also in terms of the cost it represents to households and the constraints it imposes on quality of life by limiting economic, educational, and social activities, preventing the development of human capital to its full potential.

Moreover, Africa is undergoing unprecedented growth and is currently the fastest-growing continent in the world. African economies are expanding by an average annual growth rate of 4%. Of the 20 fastest-growing countries in the world in 2012, 13 were in Africa [8]. At the same time, its population is projected to triple over the next 40 years, and its urbanization rates are expected to rise by 20% points by 2050 [7,9,10]. Sustaining such fast growth puts enormous pressure on the continent's energy system and requires a significant upgrade and expansion of generation capacities and power grids to increase energy supply. Demand projections predict average per capita power consumption growth of 120% to 304 kWh/day between 2008 and 2050 [9,10]. The task at hand is clearly a huge and complex one. It involves not only undertaking the necessary investments but also tackling other inhibiting factors such as capital constraints, skill shortages, and weak institutional capacity, as well as making important trade-offs in order to decide on the appropriate energy sources to exploit for power generation. Still, given the costs imposed on African economies by energy poverty, the price to pay for sustaining business as usual is likely much higher than the cost of improving access to and reliability of the energy supply [9,10].

2.2 Availability of natural resources

Africa's low energy consumption levels are in stark contrast to its abundance of renewable energy resource wealth. The continent's reserves of renewable energy resources are in fact the highest in the world, with a renewable energy potential sufficient even to meet its future energy needs [5,9–11] (Table 1, Figure 2).

Table 1 Technical Potentials for Power Generation from Renewables in TWh (Values Subject to ±50% Uncertainty)

	CSP	PV	Wind[a] Total	Wind[a] CF 30–40%	Wind[a] CF > 40%	Hydro	Biomass	Geothermal
Central Africa	299	616	120	16	6	1057	1572	
Eastern Africa	1758	2195	1443	309	166	578	642	88
Northern Africa	935	1090	1014	225	69	78	257	
Southern Africa	1500	1628	852	100	17	26	96	
Western Africa	227	1038	395	17	1	105	64	
Total Africa	**4719**	**6567**	**4823**	**667**	**259**	**1844**	**2631**	**88**

Abbreviation: PV, photovoltaic.
[a]Africa's large wind potential, both on- and off-shore, is located mainly in coastal areas. Eighty-seven percent of high-quality resources with world-class wind resources are located in the northern, eastern, and southern coastal zones [11,12]. The rest of mainland Africa's wind intensity is too low to be exploited for power generation.
Refs [9,10].

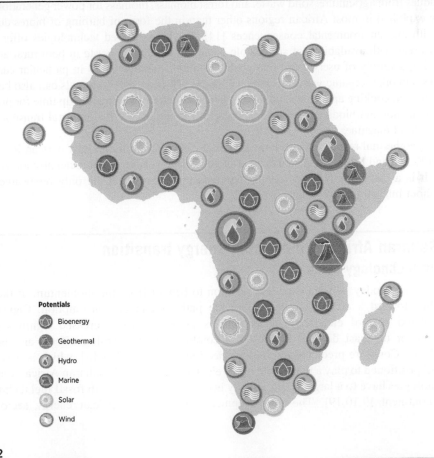

Potentials
- Bioenergy
- Geothermal
- Hydro
- Marine
- Solar
- Wind

FIGURE 2

Renewable Energy Potential in Africa.

Source: (IRENA 2013).

Hydropower is currently the most important renewable source in Africa. It makes up about 16% of Africa's power generation mix and 94% of the continent's renewable power production [9,10]. The untapped potential for large projects is mainly located in the lower Congo River and the upper Nile. Africa's economic potential of hydropower is only about half of its technical potential, yet it would be sufficient to provide a substantial share of total African power demand [9,10].

Solar energy has significant potential for sub-Saharan Africa, as the continent's solar power radiation intensity is 3000–7000 W/h/m^2, which is higher than required to support average domestic loads. Various solar photovoltaic (PV) applications offer compelling solutions for heating in households as well as commercial sectors in both rural and urban settings. The majority of the PV market (90%) consists of lighting, operation of simple appliances, and residential rooftop systems, with high potential for growth as PV panels are a particularly suitable solution for the off-grid market [7,12]. Other applications include power provision for off-grid schools, health centers, and other social institutions, as well as solar energy in telecommunications and broadcasting [13].

Bioenergy is an important source of renewable power in sub-Saharan Africa. Currently, about 79% of the sub-Saharan African population relies on traditional use of biomass [1,2]. Despite the abundance of biomass residues from agriculture, solid waste, and forest biomass, biomass for power generation has not been widely exploited in most African regions other than in the form of burning of bioresidues, with negative health and environmental consequences [14]. Yet biogas-based technologies offer suitable solutions for areas with availability of sustainable quantities of organic waste in both rural and urban regions, serving a variety of users, from households to municipalities. Cities in particular can benefit from sewages and other organic wastes to produce biogas [7]. Biogas and biofuels can also be used to fuel modern (clean) cooking appliances, improving health conditions and freeing up time for productive activities. Bioethanol and biodiesel can be cost-competitive options for fueling rural transportation as well as agricultural machines.

Excellent geothermal resources with large potential can be found in the East African Rift system, especially in Kenya and Ethiopia. The advantage of geothermal power is that it can provide electricity 24 h a day. It is a cost-effective and reliable baseload technology and the only renewable energy resource without intermittency challenges [11,12].

3. Sub-Saharan Africa in the global energy transition

3.1 Falling technology costs

Sub-Saharan Africa today is in an excellent position to benefit from the momentum of the global paradigm shift toward a cleaner, more sustainable path of energy consumption. Capital costs and the levelized cost of electricity of renewables globally have decreased significantly and continuously over the past decade, driven by a combination of learning effects and incentive schemes [15,16]. Costs are predicted to continue declining in future [17,18], which is why renewables are well positioned to play a major role in the electrification of sub-Saharan Africa. Renewable energy technologies have to a large extent already become competitive with fossil fuel technologies across the continent [9,10,19]. These developments present the African energy sectors with unprecedented opportunities to leapfrog to systems dominated by renewables, avoiding the fossil fuel-heavy path of industrialization.

3.2 A good fit

In light of recent technological and cost developments, renewable energy has the potential to go a long way in solving Africa's two key challenges. For one, it plays an important role in providing access to electricity in remote settings, far from the centralized grid. An estimated 70% of the sub-Sahara African population without access to electricity lives in far-flung, rural areas [15] with a dispersed character and low levels of commercial energy consumption of rural populations [36]. These areas are typically associated with poor-capacity utilization of transmission and distribution utilities and other energy infrastructure involved Onyeji. Despite rapid urbanization, around 40% of Africa's population is projected to still live in rural areas by 2050 [7]; Africa's rural economy and large agricultural sector will continue to play a key role in the continent's economic growth.

Decentralized renewable power solutions are particularly well suited to overcome many of these challenges, including high grid extension costs, transmission losses, and low-income segments of the rural population. Off-grid solutions using renewable technologies play a significant role because they allow for greater flexibility in supply expansion and are even applicable at the village or household level. They have the added advantage of lower project lead times. Where grid extension is not (financially) justifiable, the development of mini-grids or, in cases of limited electricity demand of isolated households, individual energy systems fed by renewable power present optimal solutions [7]. By increasing the population with access to energy, particularly in remote areas, and as more people have access to modern energy, renewable energy opens avenues to kick-start businesses and industries. Productive uses of renewable energy include water treatment and supply (desalination, pumping), heating (e.g. food, water), and cooling (e.g. refrigeration of agricultural and medicinal goods and products) [7].

At the same time, renewables have the ability to address the second key challenge of expanding power supply and increasing its reliability, with added benefits that cannot be provided by simply augmenting traditional fossil fuel-based power. Renewable energy resources are indigenous and, hence, enhance energy independence by shielding economies from the volatility of international fossil fuel prices and its negative repercussions on economies. Moreover, energy services based on renewable energy sources stimulate positive environmental impacts and help mitigate the effects of climate change [7,20].

The possibility for renewable energy to contribute considerably toward mitigating some of Africa's major energy challenges is tremendous; yet it begs the question of how best to promote the uptake of renewables on the continent. In the following section we consider some key aspects.

4. The way forward

4.1 Financing

Despite falling technology costs, it remains challenging to bring renewable energy to rural areas, the main reason being large up-front costs associated with renewable energy projects. Local banks could act as crucial facilitators by offering financial products tailored to rural communities' cash profiles, but lack of understanding of technologies as well as issues related to installation and maintenance remain a major barrier [1,2]. Although financial flows related to the energy sector in developing countries are significant, they are still not adequate to the task of delivering energy access to those deprived. The sub-Saharan African region in particular has not been part of the global rise of gross fixed capital formation for electricity and gas distribution in general in the past decade [3]. Comparing financial flows with estimated requirements for extending energy access to those lacking it suggests that significant funding for energy access in Africa will have to be sourced internationally [35]. According to the International Energy Agency, an estimated $34 billion in investments is required annually to achieve global universal energy access by 2030; 60% of these investments ($19 billion a year) is required in sub-Saharan Africa [21]. The cost of the transition to a low-carbon economy and additional associated barriers, such as lack of capacity, make the transition to more renewable energy in Africa unlikely without international support [22].

The main sources available for financing the expansion of renewable energy in Africa are multilateral (World Bank Group and regional development banks, as well as multilateral funds such as Organization of Petroleum Exporting Countries Fund for International Development and Scaling-up Renewable Energy Program for Low-Income Countries) and bilateral sources (mainly from members of the OECD Development Assistance Committee), sources from governments in developing countries, and private sector sources [21]. Bloomberg New Energy Finance estimates the cumulative clean energy investment in Africa by development banks over the period 2007–2012 to be $14.6 billion, of which the African Development Bank (AfDB) was the largest source ($4.3 billion), followed by the World Bank Group with $2.9 billion [17,18]. Power Africa, a 5-year multistakeholder initiative recently announced by US President Barack Obama, aimed at doubling the number of people with access to electricity in sub-Saharan Africa by unlocking substantial renewable energy resources in the region, has already garnered more than $21 billion in financial support, including direct loans, guarantee facilities, and equity investments by its financial partners (including the US government, AfDB, World Bank, African governments, private sector participants, and a number of US government agencies) [23]. International financial institutions such as the Global Environmental Facility, the European Investment Bank, and, more recently, the Clean Technology Fund and the Sustainable Energy Fund for Africa have played a critical role in most of the African power sector's recent activity [22,24].

While the public sector is responsible for building more enabling business environments, it falls on the private sector to ensure that the continent's abundant renewable energy resources are developed in a sustainable way. Attractive returns are one of the most important drivers for private investors in the African renewable energy space [19]. Asian investors are increasingly targeting African renewable energy projects, and Chinese power generation companies in particular are widely expected to play a larger role in the African renewable energy sector in the future [19]. Calls for augmented African private sector investments in the renewable energy sector are equally becoming louder. Nigerian entrepreneur and philanthropist Tony Elumelu, for example, has made an Africa-wide call to action, promoting the idea of *Africapitalism*—the philosophy that the African private sector has the power to transform the continent economically and socially through long-term investments in key sectors such as infrastructure and power, which not only generate high returns but also help alleviate some of Africa's most pressing challenges [25]. Nonetheless, inadequate financial resources and low skill levels, as well as insufficient technical capacity in African countries, largely still make development finance institutions critical players in matching governments' development goals with the private sector's profit motive [24].

4.2 Regional cooperation: electricity markets, power pools, and grid infrastructure

When considering an increase in the share of renewable power fed into a centralized grid, it becomes imperative to consider implications for grid stability and reliability, as challenges can arise at as little as 10% power supply from renewables. Diversification of renewable resources, electricity storage, and backup capacity all become important options for ensuring grid stability [9,10]. In addition, grid interconnections with neighboring countries are required to profit from a more diversified power generation mix, as well as to balance operations of regional power markets. The Ethiopia–Djibouti interconnector, the first cross-border power connection in East Africa, for instance, connects the grids of the two countries; it enables Djibouti to replace 65% of its fossil fuel-produced power with electricity from Ethiopia's renewables, thereby reducing the costs and polluting emissions related to power generation [7].

Building regional long-distance grids that link abundant resources with centers of high demand enables the expansion of power trading even beyond immediate neighbors. The East African region, for example, has significant renewable energy potential, which in addition to fueling the region's growth can be used to help meet incremental power demand in southern Africa that is currently met largely by unsustainable coal-fired power production [7]. To this effect, the energy project portfolio of the Program for Infrastructure Development for Africa (PIDA)[1] includes the development of four transmission corridors, among other cross-border energy market-enhancing projects. The four corridors include a West African power transmission corridor (from Guinea to Ghana), a north–south transmission link (from Egypt to South Africa), a North African transmission link (from Morocco to Egypt), and a central corridor (from Angola to South Africa) [7,26].

The level of integration for such an undertaking requires high legal, institutional, and technical capacities within regional power pools. There are five power pools operating in Africa: the Southern African Power Pool, West African Power Pool, Central African Power Pool, East African Power Pool, and Arab Maghreb Union. While these pools enable significant cross-border flows of electricity, the increased integration of renewables, especially on a large scale, will require an increase in inter-regional power pooling. Large hydropower projects, such as the Grand Inga Dam in the Democratic Republic of Congo (one of a number of priority energy projects to be implemented by 2020 as part of PIDA), may not be viable considering the demand of one country alone, but may be highly valuable when addressing aggregate demand from several countries. It thus requires bulk inter-regional transmission networks to reap the full benefits of regional power sharing with large-scale projects. Potential benefits to countries dependent on fossil imports, those with less availability of economical renewable resources, or those with small loads hardly able to obtain economies of scale are significant [7]. Moreover, the integration of smart electricity systems could greatly amplify benefits by increasing the accuracy of balancing mechanisms, thereby improving quality and reliability of power and

---

[1]PIDA is a joint effort by the African Union Commission, the NEPAD Planning and Coordinating Agency, the AfDB, the United Nations Economic Commission for Africa, and Regional Economic Communities to promote regional economic integration by bridging Africa's infrastructure gap [26].

reducing losses and costs, while also offering African countries' energy sectors opportunities to leapfrog traditional power systems [27,28].

In addition, the recently approved new methodology for interconnection between electricity systems for energy exchange by the Clean Development Mechanism board enables cross-border electricity transmission projects to benefit from the sale of Certified Emission Reductions—a significant new source of revenue for countries whose renewable energy resources exceed their national needs and could therefore be exported. These additional revenues would contribute to the financial viability of concerned projects [24,29,30].

4.3 Renewable energy support policies

Aside from the key role governments play in the development of renewables by fostering regional integration and cooperation, they play an essential part at the national level. There are plenty of opportunities for governments to encourage advancement toward the widespread adoption of clean energy supply on a national level, notably through implementation of power sector regulations and playing field-leveling policies (e.g. fossil fuel subsidy removal), as well as the creation of attractive investment and financing conditions in the sector. The Renewable Energy Policy Network for the 21st Century (REN21) makes annual assessments of the global renewable energy policy landscape [15]; Figure 3 shows the number of countries in sub-Saharan Africa having implemented individual policy instruments.

In spite of the overall absence of coherent, consistent, and favorable policies targeting renewable energy development in most African countries, a growing number of these policy tools are being

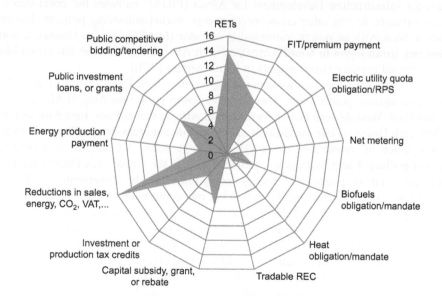

FIGURE 3

Number of Sub-Saharan African Countries to have implemented Renewable Energy Support Policies by early 2013.

Source: (REN21 2013)

adopted on the continent [15]. In early 2012, for example, Nigeria, Rwanda, and Uganda adopted new Feed-in-tariff (FIT) policies, adding to the list of African countries already operating with FITs, including Ghana, Senegal, Kenya, and Tanzania [15]. In the same year the 15 countries of the Economic Community of West African States (ECOWAS), one of the most active regions in Africa with respect to the promotion of renewables, adopted new targets: 10% of the power mix from renewables by 2020 and 19 by 2030 [15]. Among the most implemented policies in Africa are renewable energy targets and reductions in sales, energy, CO_2, value-added tax, or other taxes.

There are a number of examples of policies in respective renewable energy sectors. The South African government set a target of 10,000 GWh of renewable energy by the end of 2013 and contributed to this goal by supporting technologies such as solar water heating or residential heat pumps through schemes that significantly reduce their cost. In Rwanda the government has been promoting biogas as an alternative fuel for cooking and lighting in households and institutions since 2009, encouraging every Rwandan that owns at least two cows to build a biogas plant. The Mauritanian government played an instrumental role in the development of bagasse co-generation in the country by implementing legislation specifically for the sugar industry that improved the business environment and provided tax incentives for investments in power production [7]. Still, overall weak institutional capacity, inadequate technical skills, and financial barriers remain major inhibiting factors, leaving the private sector and lending facilities reluctant to capitalize on the sector.

Going forward, it is important for African governments to remove financial barriers, one of the key obstacles being the high cost of equipment for renewables. Facilitating access to renewable energy technology can play an important role in the continent's development process. Since most African countries rely on imports of equipment, it is vital to reduce the tax burden on renewable energy technology imports, which could significantly boost respective sectors. Efforts should be made simultaneously to further develop local equipment manufacturing capacities, which would help reduce costs further. Emphasis should also be placed on the removal of bottlenecks in the approval process for setting up micro-renewable energy ventures, as these are complementary solutions to grid extension not only in the short term but also in the long term and, moreover, are vital for the achievement of universal energy access.

Finally, it is essential for the successful integration of renewables that governments focus efforts on human capacity building. In general, African countries are short of technical skills. Some of the consequences when it comes to renewable technology are the poor maintenance of imported systems and lack of adequate after-sales services [31]. Governments have a key role to play in creating an institutional environment that fosters the development and strengthening of relevant skills, knowledge, and experience, which in turn can support the components of innovation systems, including tertiary education or international links between local businesses, universities, and international technology experts [29,30]. The public sector itself equally lacks adequately skilled personnel to undertake effective decision-making, monitoring, and evaluation exercises. Emphasis on human capacity building is essential as the integration and (commercial) sustainability of renewable energy in African countries crucially depends on a skilled work force and a high level of institutional quality [31].

4.4 Innovative business models

Even with often unfavorable local conditions, there is increasing recognition of the huge potential offered by the opportunity to provide millions of energy-deprived households in the low-income segment, the so-called bottom of the pyramid (BoP), as well as deprived businesses with modern energy services. While the public sector is largely responsible for creating an enabling environment for the investment in and widespread adoption of renewables, it remains the private sector's task to figure out viable and sustainable ways of financing and actively delivering sustainable energy solutions across Africa.

Recently emerging business models have come up with innovative strategies of catering to local contexts, specifically targeting low-income segments. These include: (1) creating unique products, services, and technologies to meet BoP needs; (2) localizing value creation, e.g. through franchising or agent strategies involving local networks of vendors and suppliers; and (3) sachet marketing, i.e. offering products and services in single-use or other small units to make them affordable to the BoP [32]. An example of a business model including these strategies is pay-as-you-go solutions made possible by mobile money for purchase of small solar home systems. These leasing models target the aspect of affordability with consumer financing that allows customers to pay in small increments, e.g. on a daily or weekly basis, until the products are paid off and ownership is handed over to them [33]. The model deals with the aspect of remoteness of the rural BoP by using technology that enables consumers to pay remotely and operators to control (switch on and off) systems remotely. Another common strategy is for businesses to engage in partnerships, such as public-private partnerships, partnerships with nongovernmental organizations (NGOs), and partnerships with microfinance entities [32]. Examples include businesses collaborating with governments and NGOs to develop distribution networks for clean cook stoves. Other businesses partner with microfinance institutions (MFIs) to expand the energy product lines they offer, focusing on microloans for small solar and improved cook stoves, which are financed by the aggregation and sale of resulting carbon credits produced by products bought through MFI loans [34].

Local and international entrepreneurs will be essential for African countries to provide all of their populations with access to modern energy services. Much more private sector investment will be necessary to scale viable models. With demonstration of market success come falling risk profiles, new opportunities for higher profits, and more commercial investment in the low-income sustainable energy sector. In the meantime, government and aid agencies can play an important role in supporting market sectors that the private sector is currently unwilling or unable to invest in Ref. [34].

5. Conclusion

Africa has a momentous opportunity to leapfrog to modern renewable energy. Despite the general recognition by African governments that renewable power provides huge opportunities given the continent's enormous resources, the lack of coherent, consistent policies; technical skills; institutional capacity; infrastructure; and financial incentives remains a major barrier for widespread adoption. Coordinated efforts, including government leadership at the national and international level, support from the international community, and increasing entrepreneurship in the sector, will go a long way in guiding Africa towards a cleaner, more sustainable energy future.

References

[1] IEA. World Energy Outlook 2013. Paris (France): OECD/IEA; 2013.
[2] IEA. World Energy Outlook Resources, http://www.worldenergyoutlook.org/resources/; 2013 [accessed December 2013].

[3] Bazilian M, Nussbaumer P, Gualberti G, Haites E, Levi M, Siegel J, et al. Informing the Financing of Universal Energy Access: An Assessment of Current Financial Flows. Electr J 2011;24:57–82.

[4] Practical Action. Poor people's energy outlook 2012: energy for earning a living. Rugby (UK): Practical Action Publishing Ltd; 2012.

[5] WEC. Survey of energy resources. London (UK): World Energy Council; 2010.

[6] World Bank. World Bank data. The World Bank Group; http://data.worldbank.org/; 2013 [accessed 19.12.13].

[7] IRENA. Africa's renewable future – the path to sustainable growth. Abu Dhabi: International Renewable Energy Agency; 2013.

[8] AfDB Group. Annual development effectiveness review 2013-towards sustainable growth for Africa. Tunis: African Development Bank Group; 2013.

[9] IRENA. Power sector costing study update, www.irena.org; 2012 [accessed December 2013].

[10] IRENA. Prospects for the African power sector. Abu Dhabi: International Renewable Energy Agency; 2012.

[11] Musaka AD, Mutambatsere E, Arvanitis Y, Triki T. Development of wind energy in Africa. Tunis: African Development Bank Group; 2013.

[12] Suberu MY, Mustafa MW, Bashir N, Muhamad NA, Mokhtar AS. Power sector renewable energy integration for expanding access to electricity in sub-Saharan Africa. Renew Sust Energy Rev 2013;25:630–42.

[13] Ondraczek J. The sun rises in the east (of Africa): a comparison of the development and status of solar energy markets in Kenya and Tanzania. Energy Policy 2013;56:407–17.

[14] Amigun B, Musago JK, Stafford W. Biofuels and sustainability in Africa. Renew Sust Energy Rev 2011;15.

[15] REN21. Renewables 2013 global status report. Paris: Renewable Energy Policy Network for the 21st Century; 2013.

[16] WEC. World energy perspective – cost of energy technologies. London: World Economic Council in Partnership with Bloomerg New Energy Finance; 2013.

[17] BNEF. Development banks – braking the $100 bn-a-year barrier. BNEF white paper series; September10, 2013.

[18] BNEF. Global renewable energy market outlook 2013. London: Bloomberg New Energy Finance; 2013.

[19] Backer, McKenzie. The future for clean energy in Africa. Chicago: Backer & McKenzie; 2013.

[20] UNECA. Integrating renewable energy and climate change policies: exploring policy options for Africa. Addis-Ababa: United Nations Economic Commission for Africa; 2011.

[21] IEA. World Energy Outlook 2011. Paris: OECD/IEA; 2011.

[22] Weischer L, Wood D, Ballesteros A, Fu-Bertaux X. Grounding green power – bottom-up perspectives on smart renewable energy policy in developing countries. Climate & energy paper series 2011. The German Marshall Fund of the United States; 2011.

[23] USAID. About power Africa, http://www.usaid.gov/powerafrica/about-power-africa; December 13, 2013 [accessed 30.12.13].

[24] Matthee R, Gombault T. Accenture. Johannesburg: Accenture; 2013.

[25] Heirs Holdings. Africapitalism http://www.heirsholdings.com/africapitalism/; 2014 [accessed 07.01.14].

[26] PIDA. PIDA energy projects. Programme for infrastructure development for Africa, http://www.au-pida.org/energy; 2013 [accessed 19.12.13].

[27] Bazilian M, et al. Smart and just grids: opportunities for sub-Saharan Africa. London: Imperial College London; 2012.

[28] Fadaeenejad M, Saberian AM, Fadaee M, Radzi MAM, Hizam H, AbKadir MZA. The present and future of smart power grid in developing countries. Renew Sust Energy Rev 2014;29:828–34.

[29] AfDB. New methodology on cross-border electricity transmission approved by UN climate change board. African Development Bank, http://www.afdb.org/en/news-and-events/article/new-methodology-on-cross-border-electricity-transmission-approved-by-un-climate-change-board-9732/; September 18, 2012 [accessed December 2013].

[30] AfDB. Technology transfer for green growth in Africa. In: African development report 2012-towards green growth in Africa. African Development Bank; 2012.

[31] UNIDO. Scaling up renewable energy in Africa. Vienna: United Nations Industrial Development Organization; 2009.

[32] WRI, IFC. The next 4 billion – market size and business strategy at the base of the pyramid. Washington: World Resources Institute & International Finance Corporation; 2007.

[33] Wogan D. Pay-as-you-go solar energy finds success in Africa. Sci Am 2013;22:11.

[34] KPMG. Innovative business models for sustainable energy access in Africa: the REACT experience. KPMG; 2013.

[35] Brew-Hammond A. Energy access in Africa: challenges ahead. Energy Policy 2010;38:2291–301.

[36] Onyeji I, Bazilian M, Nussbaumer P. Contextualizing electricity access in sub-Saharan Africa. Energy Sust Dev 2012;16:512–27.

Modeling of Variable Energy Resources

CHAPTER

4

Multi-Dimensional, Multi-Scale Modeling and Algorithms for Integrating Variable Energy Resources in Power Networks: Challenges and Opportunities

Santiago Grijalva[1,2]

[1] *National Renewable Energy Laboratory (NREL), Golden, CO, USA,* [2] *Georgia Institute of Technology, Atlanta, GA, USA*

1. Power system dimensions and scales

Power system dimensions can be broadly classified into three groups: temporal, spatial, and scenario dimensions. Spatial dimensions and past temporal dimensions are continuous, but future temporal and scenario dimensions (contingency, pattern, and profile) exhibit nuances, including sparsity, variable granularity, and combinatorial properties.

1.1 The spatial dimension

Spatial dimensions in power systems are related to the size (and power levels) of transmission, distribution, microgrid, commercial, industrial, or residential systems. Emerging problems often require geo-referencing to these spatial scales. For instance, transmission expansion, renewables electric vehicle integration, and outage management systems [1,2] require knowledge of where the sources, storage, and sinks of electric energy are located and relevant factors associated with those locations.

The electricity system is physically interconnected across all spatial scales, but traditionally, systems at various scales have been studied separately. These scales are illustrated in Figure 1. Clearly, aggregation and exchange of information is at the core of handling spatial dimensions in emerging electricity systems. For instance, the load of customers is aggregated at the feeder level; a town's total load is aggregated as a single 69-kV bus in a transmission model; generation reserve in ancillary service markets is provided at the zonal level. Current technologies are limited in their ability to allow information to be dynamically aggregated at different levels without resorting to model conversions [3]. Often, the data from more detailed models have to be aggregated and transferred "by hand" to other applications.

Example 1: Transmission and Distribution Multi-Level Information: The independent system operator (ISO) model delivery points to distribution utilities as single loads that aggregate distribution networks. When distribution systems experience operational changes, internal events, or

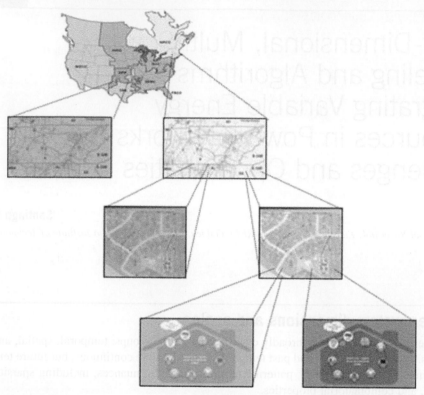

FIGURE 1 The spatial dimension

Illustration of the spatial dimensions in power systems from interconnections and ISO levels down to distribution utilities, small cooperatives and municipals, aggregators, microgrids, and homes. Although not shown, the spatial scale can go even further to represent individual devices, such as electric vehicles, solar photovoltaic, and appliances. For coordination and organization purposes, most of the electricity industry around the world has been organized in a hierarchical manner, with varying levels of discretion regarding ownership, voltage levels, aggregation, and control.

dangerous conditions, these problems are not directly observable to the ISOs. Various ISOs have expressed the need to model and gather real-time information from the sub-transmission and distribution system to enhance reliability and awareness [4,5] (i.e., to provide finer granularity modeling). Although ISOs have traditionally been able to forecast loads within a 2% error, deployment of energy resources and storage in the distribution side may increase the error substantially. Several modeling challenges limit the integration of transmission and distribution models, including single- versus three-phase modeling, non-standardization of operational and planning models, node/breaker versus bus/branch modeling, data privacy, and operational liability. Utility engineers and academics have begun to question the historical separation of transmission and distribution scales, including the fundamental assumption of a balanced system at the bulk transmission level [6].

1.2 **The temporal dimension**

Temporal dimensions are associated with natural frequencies of relevant behavior. An example is electricity markets, which operate with long-term contracts, the day-ahead (hourly) energy market, and the real-time or spot market. Another example is power system transient effects, which cover the range from lightning propagation (microsecond) to transient stability (ms/s) to long-term dynamics (min/hour).

Example 2: Energy Scheduling for High Penetration of Wind Generation: Wind energy is highly variable and difficult to predict. Wind variability becomes particularly critical in the seconds to minutes time frame because of the need of conventional reserve to respond to potentially drastic changes of wind production. Various systems in the United States have experienced emergency conditions and events because of unexpected reduction of wind production, which resulted in automatic load shedding [7,8]. The limitations of the existing wind forecasting tools and the computational intensity of scheduling software have resulted in most utilities with growing penetration of wind deciding to operate the systems conservatively, using high levels of regulating, contingency, and supplemental reserve [9,10]. In particular, unit commitment has traditionally been performed using a 1-h or 0.5-h temporal granularity, whereas real-time economic dispatch runs every 5 min based on snapshot information only. A better understanding of the temporal multi-scale behavior of wind, including how to incorporate significant uncertainty in reserve scheduling to ensure reachability, is highly needed. The economic dispatch software must be replaced with short-term stochastic *Energy Scheduling* software with granularity in the range of several seconds and a horizon from minutes to a few hours [11,12].

Example 3: Dampening Renewable Energy Variability with Flexible Load Control: Demand response today encompasses a set of consumer actions to disconnect, reduce, or shift demand on critical events or based on economic incentives, such as dynamic pricing. Demand shifting can be generalized with the use of flexible loads, such as electric vehicles. Electric vehicles are stationary 21–22 h per day, providing ideal resources to dampening renewable energy variability even with grid to vehicle technologies only [13].

1.3 **The scenario dimension**

Because of increased uncertainty about future system conditions, utilities must analyze many scenarios to ensure secure and reliable system operation. Consider, for instance, a bulk transmission operations planning process that requires determining interface limits for the following week. Those limits will depend on hour of the day, temperature, $N-2$ contingency considerations, remedial action scheme modeling, generation profile, including wind and solar scenarios, hydrolevels, and load patterns [14,15]. For a typical study, the number of contingencies may be on the order of thousands, and the total number of scenarios that need to be evaluated can be on the order of 10^6 to 10^8. The resulting security dimension data sets contain the system conditions and operating regions that should be avoided.

Example 4: Contingency Information Complexity: To illustrate how the scenario dimension arises, consider Figure 2, which shows the visual pattern of thermal overloads of transmission devices in a small system obtained by evaluating $N-1$ contingencies. Large-scale systems are hundreds of times larger, and $N-2$ security analysis is usually required.

Linear techniques are often used as part of contingency analysis, available transfer capability, and security-constrained optimal power flow (SCOPF). The later instantaneous deterministic optimization tool (SCOPF) may need to evolve into dynamic stochastic scheduling, which would create temporal interdependencies and further increase the number of scenarios [16].

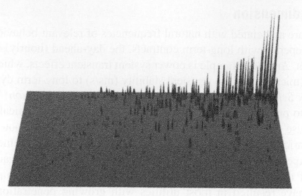

FIGURE 2 The scenario dimension

Clustered $N-1$ thermal overloads in a 112-branch electric system. The size of the horizontal plane is 112×112. The vertical axis represents the megawatt overload of a transmission component on the single outage of another transmission device. Branch/outage dimensional sorting, based on total megawatt of overload, results in the shown clustering. Realistic system size is on the order of 10^4 branches, which results in 10^6 to 10^8 data points for an $N-2$ security analysis of the system at a single point in the temporal dimension.

Example 5: Scenario Analysis in Transmission Planning: Maintaining a high level of power system reliability requires the evaluation of many plausible contingencies, planning alternatives, and operational scenarios. The number of these scenarios increases drastically in power systems with significant penetration of renewable energy. Methods for reducing the problem dimensionality by filtering nonrelevant scenarios are essential for robust transmission planning. In particular, systematic ways to prioritize searches, reduce scenario spaces [17], and dynamically update multi-dimensional information are topics of ongoing research.

1.4 Power system scales

The spatial scales of emerging sustainable power systems range from continental interconnections, to ISO/control areas, distribution utilities, microgrids, buildings, homes, and appliances. Emerging energy management systems for energy control are progressively being developed and deployed at each level, from Wide-Area Energy Management Systems to appliance and device power management [18].

Temporal scales range from long-term planning to scheduling and maintenance, day-ahead operations, real-time control, power system dynamics, and electromagnetic transients.

2. Modeling and analysis

2.1 Multiple scales and scale invariance

Multi-scale processes are physical, engineering, biological, or social processes characterized by coupled phenomena that occur in disparate spatial and temporal scales. Usually, interactions at all scales affect the ultimate behavior of the complete system. Multi-scale modeling and analysis considers the following [19,20]:

1. A library of models that provide a consistent description of phenomena at the various relevant scales (e.g., transmission and distribution).
2. Simulation methods that account for steady-state and dynamic scale interactions.
3. The representation of uncertainty and its propagation across scales.

One of the central challenges in multi-scale modeling is the determination of what information on the finer scale is needed to formulate equations or relations for the effective behavior on the coarser scale. Multi-scale modeling provides a balance so that the models become computationally feasible without losing much information. This requires making rational decisions about the appropriate scales used. An example is weather forecasting: phenomena on the micron scale have impacts on planetary scale weather through the physical processes of cloud formation, precipitation, and radiation, in conjunction with nonlinear advective processes linking disparate scales. In global computational weather models, all processes beyond the resolution of the numerical model are accounted for through the incorporation of parameterizations.

Recently, there has been interest in moving from rather empirical approaches to more proper stochastic treatment of the problem of parameterization of weather processes [21]. A practical example is multi-scale aspects in forecasting of solar power. Solar power production forecasting for utility applications is usually based on satellite images and computation that uses granularity of 3–9 km^2 per cell and is usually available in the day ahead and few hour-ahead scales. On the other hand, local, higher-resolution, spatiotemporal solar forecasting is needed for end consumer applications. Inaccurate forecasting can result in highly suboptimal energy scheduling of both conventional and flexible loads at the residential level.

2.2 Multi-scale frameworks

Several analytical frameworks are commonly used to represent multi-scale systems. For instance, the power system spatial scales can be organized in a hierarchical structure, in which each lower-level subsystem is seen as a component belonging to the upper level in the hierarchy. A network bus modeled by an ISO can represent the distribution system of a town or even an entire city. The electric system in the lower level is governed by physics, similar to the higher-level system, but at a lower scale. When the various levels of the hierarchy have similar characteristics, the system is said to exhibit *self-similarity* [22,23]. One of the advantages of multi-scale systems with self-similarity is that the modeling of the system at various levels can be done by using the same reference model, which enables expansion and collapse of lower-level information. Power systems exhibit self-similarity. In particular, any power system from large interconnection to home networks can, in theory, be modeled for steady-state analysis using a generic three-phase, four-wire model.

The need for information and coordination of systems at different scales is apparent during coordinated maintenance and restoration processes involving transmission and distribution. It can be argued that multi-scale self-restoration and self-healing objectives would require automatic interchange of such information and decision logic. It can also be argued that ambitious objectives, such as frequency regulation supported by thousands of electric vehicles (EVs), will require capabilities for seamless expansion of the lower-level models into the upper coarser model.

2.3 Inference and system identification

In many power system applications, it is desirable to identify the properties or parameters of a system without knowing the detailed model of the system. Examples of such applications are, among others: external model determination in ISO systems, load modeling in utility and industrial applications, load transient behavior, and phasor measurement unit (PMU)–based disturbance detection [24]. Although single-scale system identification is reasonably mature, inference and system identification in multi-scale power systems has not been pursued. Such development will include the specification of broader assumptions about the scales or the identification of the actual relevant scales of an unknown system. This result points to a critical need for operational inference related to events and disturbances in external systems. For instance, a utility can detect an internal failure in a microgrid network. A microgrid can potentially detect cascading problems in the utility system and take preventive actions, including disconnection from the grid on impending blackout. Similar methods should be investigated to provide highly desired information about the parameters and behavior of the load and of distributed solar PV, for a variety of applications, including control and scheduling. Smart meter and production instrumentation may be leveraged with this purpose.

3. Optimization and control

Control of bulk power systems uses a centralized supervisory control and data acquisition (SCADA) and EMS paradigm, which has successfully supported electricity system operations for several decades. With the exception of automatic generation control, the control requires a human in the control loop. An equivalent control loop takes place in real time in distribution grids through a distribution management system (DMS). The corresponding functions of DMS systems have yet to achieve the same level of penetration as EMS systems. For instance, many distribution utilities operate with SCADA-only systems.

The next logical step in automation down in the spatial dimension, which has received significant attention by the power system community in the past few years, is the microgrid [25,26]. The natural technological trend is the development of the equivalent of microgrid energy management systems (μGEMS). Microgrids pose several new control and operational requirements, such as capability of isolated operation, self-restoration, integrated control of industrial loads, and self-reconnection. In addition, the microgrid must be able to operate unmanned (e.g., the μGEMS must reach a level of sophistication sufficient to bypass the human operator altogether).

There is significant interest in deploying microgrids that can operate in an isolated manner supplied by renewable sources and storage. Renewable energy sources are also highly variable at the microgrid level, which requires more accurate, local Distributed Energy Resources (DER) forecasting and advanced real-time dispatch of inverter-based sources. Beyond microgrids, similar control systems continue to be developed and deployed at the building and residential levels through building energy management systems and home energy management systems [27,28], respectively. These systems also have the requirement of being unmanned. Deployment of the various categories of EMSs at various scales is the basis for a shift toward a distributed control paradigm.

EMSs were designed based on the notion of real-time control of a just-in-time product and its mostly deterministic instantaneous optimization embodied by the SCOPF application. There are three emerging aspects associated with the temporal scale that will alter the operational philosophy of EMSs at all scales from bulk operation to homes:

1. renewable energy variability,
2. ability to store growing amounts of energy, and
3. information-driven demand response and demand shifting.

In addition to these strong drivers, both renewable energy source output and demand response capabilities exhibit varying degrees of uncertainty. Thus, the traditional instantaneous optimization paradigm and the corresponding algorithms must be replaced by decentralized stochastic dynamic energy scheduling, fully capable of handling uncertainty, and the mentioned temporal issues [29].

The area of multi-scale optimization has been the subject of considerable research among the finite element, multigrid, and combinatorial optimization communities. Hierarchical optimization, which is a type of multi-scale optimization, has been the subject of research by the power system community in the past decade [30,31]. Computational solutions are obtained by decomposing a system into subsystems and optimizing them while considering the coordination and interactions between the subsystems. Hierarchical optimization takes the interaction into account by placing a *coordinator* above the subsystems to manage it. As a result, the entire system is optimized while considering the interaction between the scales by means of coordinated information. Most of the work related to power system hierarchical control and optimization addresses two-level problems.

Multi-scale system control refers to a formal framework that uses multi-scale modeling and theory to realize desirable control and system response at all scales. Hierarchical control is gaining impetus among the smart grid and automation community [32,33], but a comprehensive framework at all the relevant spatial levels is yet to be established. Furthermore, the system control must address control at all the relevant temporal scales. This is necessary for a variety of emergent problems, such as distributed frequency regulation, demand response, ancillary services provision, disturbance detection, and wholesale/retail markets.

We have mentioned that the properties of self-similarity and scale-free systems may have implications for power system multi-scale control methods. Normalization, traditionally used in power system analysis, suggests scale-free characteristics. This would also pose the question of whether multi-scale hierarchical control, networked control, or a fusion of the two is possible in multi-scale electricity networks with high penetration of variable generation.

4. Data handling and visualization

4.1 Multi-scale data

Modernization of the electricity industry is resulting in massive data being acquired and stored in areas such as smart metering, substation intelligent electronic devices, and PMUs. These data should not be understood as being isolated, but as part of a multi-dimensional, multi-scale data set that will ultimately enable the realization of the various grid objectives. Information architecture and data set modeling are initiatives under the umbrella of "smart grid" that have not been adopted with a system-wide perspective. For instance, although some utilities read smart meters every 15 min and are able to pass the readings to the control center every hour, others are able to read once per hour and report once a day. Other utilities have realized the advantages of faster information to enable control, and have retrofitted communications systems to support readings every 15 s and use near-real-time information for application such as feeder voltage/variable control.

There is great value in the emerging data for understanding the behavior of new subsystems, such as consumer behavior, and, more fundamentally, to better understand the type of data acquisition that is ultimately needed to design advanced controls and realize reliable electricity networks with high penetration of renewables.

4.2 Visualization

Electricity grid visualization has become increasingly important and a topic of significant research in the past decade [34–36]. Visualization has been the mechanism to drastically increase situational awareness in bulk energy control centers. Systems with two-dimensional (2D) spatial visualization of the controlled region are the norm. Typically, the visualization requirements for these systems are at

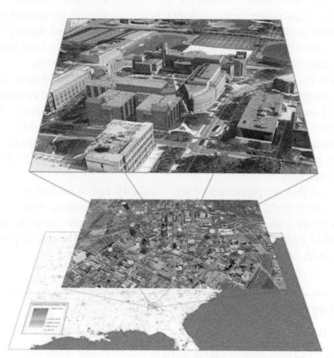

FIGURE 3 Conceptual multi-scale visualization for coordination of renewables and distributed storage

A scattered cloud system results in highly variable solar production near afternoon commuting time. The Atlanta, GA, area exhibits potential problems because of loss of stationary storage (red contouring driven by aggregated security metrics). Variability absorption capabilities at 2×2 mile^2 resolution are illustrated in the upper 3D layer using cylinders (virtual generation). Fuchsia stacks are EV V2G energy, blue stacks are frequency regulation, and green and yellow stacks are 5 and 12 min EV cleared reserve, respectively. Demand response at the building level is triggered to help dampen the variability.

the level of 10^3 to 10^4 measurements every 5–10 s. Zooming and panning are usually too slow for real-time response in this environment. In general, interactive visualization in real time has not been largely deployed. Spatiotemporal and spatial-security three-dimensional (3D) visualization has been proposed, but it has not been widely deployed for either operations or planning because of performance limitations, even with graphics processing units (GPUs).

Visualization at the distribution level is driven by geo-referencing requirements and hence is coupled with geographic information systems (GISs). The norm is a 2D spatial GIS representation. Spatial-security techniques have been proposed for distribution systems. Cluster GPU applications have been proposed but have not been implemented at control centers. The four-dimensional (spatial + temporal + security) representation has been proposed by the author for small data sets. Multi-scale 2D spatial visualization is rudimentary, available only for snapshot data sets and in off-line modes.

Because of the performance and design of the GPU, highly parallelized and efficient computation is possible. Several general purpose GPU libraries have been developed to provide a better match to the computing domain. In recent years, numerous algorithms and data structures have been ported to run on GPUs. Scientific visualization generally has a spatial 2D or 3D mapping and can thus be expressed in terms of the graphics primitives of shader languages. In general, the current GPU programming model is well suited to managing much spatial data. An example of a GPU-generated image is shown in Figure 3. GPU visualization for large grids has been reported to be approximately 30 times faster than CPU-based visualization. GPU clusters can theoretically yield visualization speeds in the order of 10^3 times the current levels, opening many research and application possibilities. Visualization of multi-scale, multi-dimensional systems is an enabling technology for the understanding and study of emerging sustainable power systems. Although, in the past, visualization has been seen as a "presentation" tool, there is growing acknowledgment that visualization tools are becoming indispensable for advanced analytics.

5. Integrated multi-dimensional analytics platforms

We now discuss an example of a multi-dimensional analytical platform for the study of offshore wind integration. Achieving cost reduction and high penetration of offshore wind requires providing stakeholders with actionable information on the implications of deploying and interconnecting large-scale offshore wind systems to the existing electricity grid. The variables that affect offshore wind investment decisions can be classified in three areas:

1. Characteristics of offshore wind turbines (capacity, turbine technology, support structures, and collector technologies),
2. Factors associated with the candidate sites for offshore wind farm installation (resource potential, leasing blocks, permits, environmental constraints, and human use), and
3. Factors associated with the interconnection of the resource to the electricity grid (voltage level, connection points, required transmission expansion, congestion, ancillary services, and induced oscillations).

To develop optimal offshore wind solutions, information systems and computational tools must handle multiple factors and the combinatorial nature of the interconnection optimization problem. Wind resource, human, economic, ecological, and policy factors are collected and consolidated into

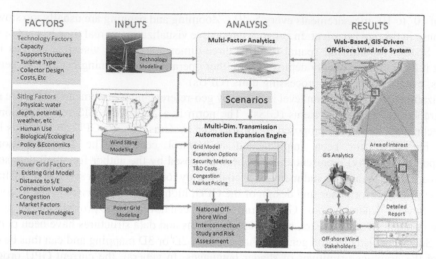

FIGURE 4 Integrated, multi-dimensional analytics tool for offshore wind development

Tool aggregates multi-dimensional data into inputs associated with technology, resource, and electricity grid infrastructure. Computational tools provide multi-dimensional results of best locations for development.

geospatial data sets. The data are integrated with detailed US electricity grid models, allowing for GIS-based multi-factor scenario development. Multi-dimensional transmission planning algorithms are coupled with multi-factor GIS modeling. The interconnection study hence uses a native DER transmission planning module, which supports hundreds of candidate electric grid connection points, candidate topologies, transmission expansion alternatives, and spatiotemporal bulk power system scenarios (wind variability, load profiles, and N-k security evaluation).

The model includes critical factors, such as explicit representation of multi-scale temporal wind variability, cost modeling, reserves impact, frequency regulation requirements, and transient and voltage stability constraints. Economic, security, policy, and environmental considerations are monetized and embedded in the optimal transmission expansion algorithm. The method integrates cost input vectors associated with offshore wind to determine locational marginal prices of both energy and reserve, enabling accurate assessment of the impact on electric grid congestion and transmission expansion requirements. This integrated analytics platform is illustrated in Figure 4.

6. Conclusions

Realizing sustainable electricity systems requires study of emerging behavior that is essentially multi-dimensional and multi-scale. This chapter provided an overview of the challenges and the opportunities in multi-dimensional and multi-scale electric power modeling and computation for high penetration of renewable energy. The chapter has studied three broad areas: (1) modeling and analysis; (2) optimization, dynamics, and control; and (3) data handling and visualization. It has provided an example of an integrated analytics platform for offshore wind development.

There are numerous unresolved challenges of multi-scale, multi-dimensional analysis for a power system with high penetration of renewable energy. Emerging data at various scales will be instrumental in increasing our understanding of emerging behavior. The ultimate problem to be addressed is how millions of spatially distributed sources can be controlled in a secure manner, possibly requiring a shift toward formal decentralized control architectures.

References

[1] Sridharan K, Schulz N. Outage management through AMR systems using an intelligent data filter. IEEE Power Eng Rev 2001;21(8). http://dx.doi.org/10.1109/MPER.2001.4311582.

[2] Dong Y, Aravinthan V, Kezunovic M, Jewell W. Integration of asset and outage management tasks for distribution systems. IEEE Power & Energy Society General Meeting; 2009. http://dx.doi.org/10.1109/PES.2009.5275527.

[3] Marzinzik CM, Grijalva S, Weber JD. Experience using planning software to solve real-time systems. In: Proceedings of the Hawaii international conference on system sciences; 2009.

[4] CEATI International Inc, Power System Planning and Operations Interest Group. Invitation for proposals, "improved T&D interfaces and information"; Dec. 2010.

[5] Castanheira L, Ault G, Cardoso M, McDonald J, Gouveia JB, Vale Z. Coordination of transmission and distribution planning and operations to maximise efficiency in future power systems. In: 2005 International conference on future power systems; 2005. pp. 1–5. http://dx.doi.org/10.1109/FPS.2005.204305.

[6] Meliopoulos APS, Cokkinides G, Galvan F, Fardanesh B, Myrda P. Advances in the super calibrator concept - practical implementations. In: HICSS 40th Hawaii international conference on system sciences, 2007; 2007. http://dx.doi.org/10.1109/HICSS.2007.49.

[7] Ela E, Kirby B. ERCOT event on February 26, 2008: lessons learned. Technical Report, NREL/TP-500–43373; July 2008.

[8] NERC. Electric industry concerns on the reliability impacts of climate change initiatives. Special Report; November 2008.

[9] Constantinescu EM, Zavala VM, Rocklin M, Lee S, Anitescu M. Unit commitment with wind power generation: integrating wind forecast uncertainty and stochastic programming. Argonne National Laboratory; September 2009.

[10] Albadi MH, El-Saadany EF. Overview of wind power intermittency impacts on power systems. Electr Power Syst Res 2010;80(6):627–32.

[11] Ruiz PA, Philbrick CR, Sauer PW. Day-ahead uncertainty management through stochastic unit commitment policies. Seattle, WA. In: proceedings of 2009 IEEE power systems; March 2009.

[12] Powell W, George A, Lamont A, Stewart J. SMART: a stochastic multiscale model for the analysis of energy resources, technology and policy; 2009.

[13] Hernandez JE, Divan D. Flexible electric vehicle (EV) charging to meet renewable portfolio standard (RPS) mandates and minimize greenhouse gas emissions. In: IEEE energy conversion congress and exposition (ECCE). pp.4270–4277, September 12–16, 2012.

[14] Grijalva S, Dahman S, Patten K, Visnesky A. Large-scale integration of wind generation including network temporal security analysis. IEEE Trans Energy Convers March 2007;22(1).

[15] Dosano RD, Hwachang S, Byongjun L. Network centrality based N-k contingency scenario generation. In: IEEE transmission & distribution conference & exposition: Asia and Pacific; 2009. http://dx.doi.org/10.1109/TD-ASIA.2009.5356963.

[16] Qiming C, McCalley JD. Identifying high risk N-k contingencies for online security assessment. IEEE Trans Power Syst 2005;20(2):823–34.

[17] Razali NM, Hashim AH. Backward reduction application for minimizing wind power scenarios in stochastic programming. In: 2010 4th international power engineering and optimization conference (PEOCO); 2010. pp. 430–4. http://dx.doi.org/10.1109/PEOCO.2010.5559252.

[18] Hubert T, Grijalva S. Realizing smart grid benefits requires energy optimization algorithms at residential level. In: IEEE conference on innovative smart grid technologies (ISGT), Anaheim, California; January 17–19, 2011.

[19] Benveniste A, Nikoukhah R, Willsky AS. Multiscale system theory. IEEE Trans Circuits and Syst I 1994;(1): 2–15.

[20] Guenard G, Legendre P, Boisclair D, Bilodeau M. Multiscale codependence analysis. Ecology October 2010; (10):2952–64.

[21] Uchida K, Senjyu T, Urasaki N, Yona A. Installation effect by solar heater system using solar radiation forecasting. In: 2009 IEEE transmission & distribution conference & exposition: Asia and Pacific; 2009. pp. 1–4. http://dx.doi.org/10.1109/TD-ASIA.2009.5356904.

[22] Maver J. Self-similarity and points of interest. IEEE Trans Pattern Anal Mach Intell 2010;32(7):1211–26.

[23] Vanderkerckhove C. Macroscopic simulation of multiscale systems within the equation-free framework [Ph.D. thesis], Katholieke Universiteit Leuven, June 2008.

[24] Borghetti A, Bosetti M, Di Silvestro M, Nucci CA, Paolone M. Continuous-wavelet transform for fault location in distribution power networks: definition of mother wavelets inferred from fault originated transients. IEEE Trans Power Syst 2008;23:380–8.

[25] Colson CM, Nehrir MH. A review of challenges to real-time power management of microgrids. In: 2009 IEEE power & energy society general meeting, Calgary, Canada; July 2009.

[26] Eto J, Lasseter R, Schenkman B, Stevens J, Klapp D, Volkommer H, et al. Overview of the CERTS microgrid laboratory test bed. In: 2009 CIGRE/IEEE PES joint symposium on integration of wide-scale renewable resources into the power delivery system; 2009.

[27] Xudong Ma, Ran Cui, Yu Sun, Changhai Peng, Zhishen Wu. Supervisory and energy management system of large public buildings. In: 2010 international conference on mechatronics and automation (ICMA); 2010. pp. 928–33. http://dx.doi.org/10.1109/ICMA.2010.5589969.

[28] Zhao P, Suryanarayanan S, Simões MG. An energy management system for building structures using a multi-agent decision-making control methodology. In: IEEE industry applications society annual meeting (IAS); 2010. http://dx.doi.org/10.1109/IAS.2010.5615412.

[29] Wunsch D, Powell W, Barto A, Si J. Toward dynamic stochastic optimal power flowIn Handbook of learning and approximate dynamic programming; 2004. http://dx.doi.org/10.1109/9780470544785.ch22. pp. 561–598.

[30] Han S, Han SH, Sezaki K. Design of an optimal aggregator for vehicle-to-grid regulation service. In: Innovative Smart Grid Technologies (ISGT); 2010. http://dx.doi.org/10.1109/ISGT.2010.5434773.

[31] Careri F, Genesi C, Marannino P, Montagna M, Rossi S, Siviero I. Definition of a zonal reactive power market based on the adoption of a hierarchical voltage control. In: 2010 7th international conference on the European energy market (EEM); 2010. http://dx.doi.org/10.1109/EEM.2010.5558672.

[32] Mori H, Saito M. Hierarchical OO-PTS for voltage and reactive power control in distribution systems. Bologna, Italy. In: Proceedings of 2003 IEEE PES power tech, vol. 2; June 2003. pp. 353–8.

[33] Ming Chen, Nolan C, Xiaorui Wang, Adhikari S, Fangxing Li, Hairong Qi. Hierarchical utilization control for real-time and resilient power grid. In: 21st Euromicro conference on real-time systems; 2009. http://dx.doi.org/10.1109/ECRTS.2009.19. ECRTS '09.

[34] Overbye TJ, Klump R, Weber JD. Interactive 3D visualization of power system information. Electric Power Comp Syst December 2003;31(12):1205–15.

[35] Venkatesh A, Cokkinides G, Meliopoulos AP. 3D-visualization of power system data using triangulation and subdivision techniques. In: 42nd Hawaii international conference on system sciences; 2009. pp. 1–8.

[36] Milano F. Three-dimensional visualization and animation for power systems analysis. Electric Power Syst Res Dec. 2009;79(12):1638–47.

Scandinavian Experience of Integrating Wind Generation in Electricity Markets

5

Anders Plejdrup Houmøller

CEO, Houmoller Consulting ApS, Middelfart, Denmark

1. Introduction

Scandinavia is part of the Baltic-Nordic electricity market. The Baltic States are Estonia, Latvia and Lithuania. In this chapter, the Nordic countries considered are Denmark, Finland, Norway and Sweden (Iceland is excluded, as there is no grid connection between Iceland and other countries). Figure 1 shows the Nordic-Baltic countries.

The Baltic-Nordic countries have a common electricity exchange: Nord Pool Spot. The common exchange is used to create a common Baltic-Nordic electricity market (to the extent the grid bottlenecks allow). The Baltic-Nordic electricity market is widely acknowledged for being one of the world's best functioning liberalized electricity markets.

Table 1 below shows the wind power data for the three Scandinavian countries. Denmark has the world's highest wind penetration. In 2012, the Danish consumption of electricity was 34.3 TWh, whereas the wind turbines produced 10.3 TWh. Hence, the wind turbine production corresponded to 30% of the consumption of electricity. High wind penetration is a unique feature for the Danish electricity market. Hence, Denmark has the world's highest wind penetration while simultaneously

FIGURE 1 The Baltic-Nordic countries

Renewable Energy Integration. http://dx.doi.org/10.1016/B978-0-12-407910-6.00005-3

Table 1 Data for the Three Scandinavian Countries

Countries	Installed Wind Capacity in MW (at the End of 2012)	Wind Production 2012 in TWh	Electricity Consumption 2012 in TWh	Population in Millions
Denmark	4162	10.3	34	5.6
Norway	703	1.6	128	5.0
Sweden	3745	7.1	142	9.5

www.ewea.org, World Bank, IEA Wind, Danish Energy Agency, Swedish Energy Agency and www.nordpoolspot.com.

having a well functioning liberalized electricity market. Norway and Sweden have well functioning liberalized electricity markets while simultaneously having a market-based subsidy system for wind energy and other renewables (i.e, green certificates). Norway and Sweden have introduced green certificates as the subsidy system for electricity from renewable sources. Green certificates are described in the sections 8.2 and 8.3.

This chapter will first describe how the common Baltic-Nordic electricity market works. In the following sections, it is explained how the three Scandinavian countries have adopted different strategies for integrating wind energy in the liberalized market and round out the discussion with remarks on the task of creating well functioning, multinational electricity markets with a high degree of renewables.

2. The transmission system operators

Figure 2 illustrates the electricity market. The water illustrates the electrical energy. The container walls illustrate the grid. The producers feed energy to the grid, and the consumers tap energy from the grid.

In the European Union (EU), each country has a transmission system operator (TSO). The TSO has two tasks:

1. Own and operate the transmission grid (i.e., the high-voltage grid).
2. Maintain a balance between supply and demand. Hence, the TSO is responsible for the security of supply.

Three of the EU countries have more than one TSO (Austria, Germany, and Great Britain). In these countries, each TSO has a control area inside the country, for which the TSO performs the two tasks previously mentioned.

FIGURE 2 The electricity market

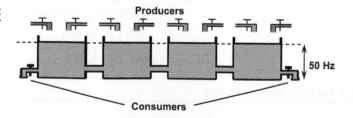

Producers

50 Hz

Consumers

3. The Baltic-Nordic spot market

Elspot is the Baltic-Nordic electricity exchange's day-ahead auction market, where electrical energy is traded. Players, who want to buy energy from the spot market, must send their purchase bids to Nord Pool Spot at the latest at noon the day before the energy is delivered to the grid (i.e., gate closure time is 12-o'clock Central European Time the day before the day of operation). Correspondingly, participants who want to sell energy to the spot market must send their sale offers to Nord Pool Spot at the latest at noon the day before the day of operation. The bids and offers are sent electronically to Nord Pool Spot: the participants send the bids to Nord Pool Spot via the Internet.

When the exchange's prices are calculated, the purchase bids are aggregated to a demand curve. The sale offers are aggregated to a supply curve (Figure 3). The exchange price can be read at the point where the two curves intersect, if there's neither market coupling nor market splitting.

Note: precisely at the two curves' intersection point, the exchange's purchase volume is equal to the exchange's sales volume. Without market coupling or market splitting, the exchange needs to arrange the price calculation, so the two volumes are equal (please also refer to Chapters 6 and 7). An exchange price is calculated for each hour of the next day. The spot market is a day-ahead market, because this is trading of energy produced and consumed the following day. This way of calculating the price is called a double auction, because both the buyers and the sellers submit bids (for many other auction types, only the buyers submit bids).

Hence, the spot market is an example of a day-ahead auction market (because the word "double" is cut out from the type description). At noon, the calculating the day-ahead prices starts. When the calculation is finished, the exchange prices are published. These day-ahead auction prices are called "spot prices." Figure 4 shows an example of spot prices. Also, when the calculation is completed, Nord Pool Spot reports to the participants how much electricity they have bought or sold for each hour of the following day. As of December 2013, the standard Elspot trading fee was 0.035 EUR/MWh. The fee is paid by both buyers and sellers.

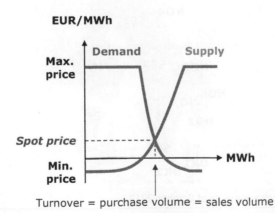

FIGURE 3 Spot price calculation when there's neither market coupling nor market splitting

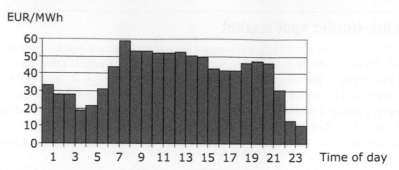

FIGURE 4 Spot prices in Eastern Denmark on Monday, August 6, 2012

Other European countries also have electricity exchanges operating day-ahead auction markets. In general, such exchange markets are called "spot markets," and the exchanges are called "spot exchanges." Hence, Elspot is an example of a spot market, and Nord Pool Spot is an example of a spot exchange.

4. Price zones

Due the grid bottlenecks, the Baltic-Nordic area is divided into several price zones. A price zone is a geographical area, within which the players can trade electrical energy a day ahead without considering grid bottlenecks. In Figure 2, you see four price zones. Figure 5 shows the Baltic-Nordic price zones as of October 2013. For example, when a producer in Eastern Denmark (DK2) sends bids to

FIGURE 5 The Baltic-Nordic price zones as of October 2013

Nord Pool Spot, the producer must specify that these bids belong to the price zone DK2. A spot price is calculated for each price zone for each hour of the following day. For example, Figure 4 showed the 24 spot prices for Eastern Denmark for August 6, 2012. Naturally, there are often hours when neighboring price zones have the same spot price.

5. Day-ahead grid congestion management: market splitting

Apart from calculating day-ahead prices, the spot market is also used to perform day-ahead congestion management in the Baltic-Nordic area. This day-ahead congestion management is called market splitting.

To explain market splitting, let us consider a grid bottleneck with a capacity of 600 Mega-Watt. We will consider one given hour of the following day. Assume the price-calculating computer during the calculation of the spot prices discovers there will be different prices on the two sides of the bottleneck during this hour: One side of the bottleneck will be a low-price zone, whereas the other side will be a high-price zone. In this case, the computer will automatically insert an extra purchase of 600 MWh in the low-price zone and an extra sale of 600 MWh in the high-price zone and then carry on with the price calculation. The extra sale and the extra purchase are made the day before the day of operation. The next day, when the given hour is reached, the extra purchase in the low-price zone will cause a production surplus of 600 MWh in this zone: in the low-price zone, there are producers who will produce the extra 600 MWh, because they have sold this to the exchange. However, in the low-price zone, there is no corresponding local consumption. Because of the production surplus, electricity must flow out of the low-price zone.

Likewise, the extra sale in the high-price zone will lead to a production deficit of 600 MWh in this zone: in the high-price zone, there are consumers who will consume the 600 MWh. However, in the high-price zone, there is no corresponding local production. Because of the production deficit, electricity must flow toward the high-price zone. Hence, once the computer has inserted the extra purchase in the low-price zone and the extra sale in the high-price zone, the laws of nature will do the rest: the next day, during the hour in question, electricity will flow from the low-price zone into the high-price zone. Please refer to Figure 6.

As a result, the extra purchase will increase the price in the low-price zone. Likewise, the extra sale will decrease the price in the high-price zone. Thus, market splitting also implies that the bottleneck capacity is used to level out price differences as much as possible. By means of market splitting, the spot market is used to perform the day-ahead congestion management on interconnectors linking the price zones in the Baltic-Nordic area. Implicit auction is the common term for market coupling and market splitting.

5.1 Day-ahead congestion management: market coupling

In case of market splitting, you have the same electricity exchange on the two sides of an interconnector linking two price zones.

Consider a border where two electricity exchanges meet. The two electricity exchanges can perform the day-ahead congestion management on the border using the principle previously

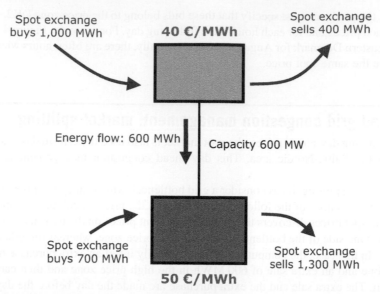

Spot exchange
buys 1,000 MWh

40 €/MWh

Spot exchange
sells 400 MWh

Energy flow: 600 MWh   Capacity 600 MW

Spot exchange
buys 700 MWh

50 €/MWh

Spot exchange
sells 1,300 MWh

FIGURE 6 Implicit auction (market splitting or market coupling)
An example for 1 h of operation.

described: when the electricity exchanges during the calculation of the spot prices realize there is a price difference on the border, extra electricity is bought from the exchange in the low-price zone, and extra electricity is sold to the exchange in the high-price zone. When we arrive at the hour in question the next day, the extra purchase will create a production surplus on the low-price side of the border. Correspondingly, the extra sale will create a production deficit on the high-price side of the border. Hence, the energy will flow out from the low-price zone toward the high-price zone. Again, please refer to Figure 6.

5.2 Calculation of the spot prices with market coupling or market splitting

Figure 7 illustrates the spot calculation for a price zone, when there's neither market coupling nor market splitting. For such a price zone, the spot exchange must arrange the spot price calculation, so the exchange's purchase volume equals the exchange's sales volume. However, with market coupling/splitting, the spot exchange must buy extra energy in the low-price zones and sell this extra energy in the high-price zones.

Figure 7 illustrates the spot price calculation for a simple example with only two prices zones. In the low-price zone, the exchange sets the spot price higher than the price given by the intersection of the local demand-and-supply curves. By increasing the price in the low-price zone, the exchange gets a purchase surplus in the low-price zone. Correspondingly, in the high-price zone, the exchange sets the spot price lower than the price given by the intersection of the local demand-and-supply

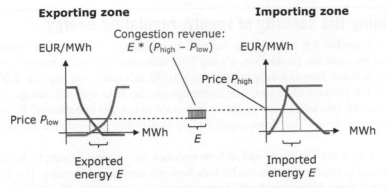

FIGURE 7 Spot price calculation when there's implicit auction (market splitting or market coupling)

curve. By lowering the price in the high-price zone, the exchange gets a sales surplus in the high-price zone.

In this way, the spot exchange reduces the price differences between the zones. If the grid capacity between the two zones is high, the exchange can level out the prices and create a common spot price for the two zones. However, Figure 7 illustrates a case, where the grid capacity is too low for this: there's a spot price difference ($P_{high} - P_{low}$) between the two zones. This creates an arbitrage revenue, because the exchange is buying the transferred energy E at price P_{low} and selling E at price P_{high}. This arbitrage revenue is called the congestion revenue (or congestion rent). The spot exchange hands over the congestion revenue to the owners of the cross-border capacity (normally, transmission system operators).

5.3 Integration of wind energy in the Danish electricity market

The implicit auction system ensures the Danish electricity market is integrated with the German, Norwegian, and Swedish markets as much as the grid bottlenecks allow.[1] Also, as previously mentioned: the high wind penetration is a unique feature of the Danish electricity supply system. Because of the high wind penetration, when there are high winds in Denmark, the country can export electricity. The implicit auction system ensures the surplus energy is offered to markets with the highest willingness to pay.

The Danish experiences clearly illustrate how an interstate electricity market furthers the integration of the intermittent wind energy. Without exchange of energy with the neighboring states, the costs of integration of lots of wind energy in the Danish supply system would have been much higher than what has actually been the case.

---

[1] However, the current version of the implicit auction system has severe shortcomings. Unfortunately, the planned future version of the implicit auction system also will have serious inefficiencies. These shortcomings and inefficiencies are discussed in Chapter 18. For now, the text will ignore the faults in the current and planned implicit auction system: for simplicity, apart from the Chapters 18 and 19, the text will assume we have an efficient implicit auction system.

6. Maintaining the security of supply: regulating energy

Each TSO is responsible for maintaining the security of supply in his or her country.[2] Hence, if the consumption exceeds the production, it's the TSO who must buy extra energy from suppliers to re-establish the balance between supply and demand. In this case, we say the TSO is procuring up-regulation. If the production exceeds the consumption, the TSO will sell energy. If the TSO sells energy to a producer, this causes the producer to reduce his or her production. We say the TSO is procuring down-regulation. The down-regulation re-establishes the balance between supply and demand.

This energy, which the TSO buys and sells to maintain the security of supply, is called regulating energy, because this is used to regulate the balance between supply and demand. The TSO's trading of regulating energy is done during (or shortly before) the hour of operation. The hour of operation is the hour at which the electrical energy is produced and consumed.

6.1 Maintaining the security of supply: regulating capacity

Naturally, the TSO needs to ensure that there are always players who can perform up- and down-regulation. To ensure this, the TSO buys regulating capacity.

6.2 Example: 20 MW up regulating capacity bought by the TSO

In this example, the TSO has bought 20 MW up-regulating capacity from a producer. In the example, it's manually activated capacity: the producer must be capable of supplying 20 MWh of up-regulating energy, if the TSO calls the producer asking for this energy.

In Denmark, this capacity is bought the day ahead: the capacity is bought during the morning the day before the day of operation. The capacity is bought per hour: the producer has sold the 20 MW for one given hour of the next day.

Assume the producer owns a 100-MW gas-fired power plant. Because of the sale of capacity to the TSO, for the given hour, the producer can, at most, sell an energy of 80 MWh to the market. The producer needs to keep 20 MW in reserve, which can be activated in case the TSO calls.

6.3 Regulating capacity: the Danish experiences

Wind energy is inherently an intermittent energy source. Therefore, when the installed wind capacity increases, the TSO needs to buy more regulating capacity. Naturally, this is a cost. It's financed by all grid users via the grid fees collected by the TSO.

As the Danish experiences show, to keep these costs in check, it's important the TSO creates markets with efficient competition. Both at the TSO's market for regulating energy and the TSO's market for regulating capacity, there must be severe competition. Otherwise, the costs of integrating

---

[2]For simplicity, in Chapter 10, we ignore the fact that each of the countries Austria, Germany and Great Britain has more than one TSO. However, for these three countries, each TSO must for his *control area* perform the tasks described in the text.

wind energy will be unnecessarily high. To have efficient competition at the market for regulating capacity, the TSO must design the market, so it has the following features:

1. Normally, the TSO has a threshold for the minimum amount of capacity, which the players can offer at the market for regulating capacity. The threshold must be small. A small threshold makes it possible for small producers and consumers to participate at the market, thereby improving the competition.
2. The TSO must buy the capacity a day ahead. This maximizes the amount of capacity offered at the market, because industrial consumers, who can adjust their consumption, get a chance at participating at the market.

   For a consumer, it will be difficult to offer capacity, if the TSO buys either a month or year ahead. The same is often the case for an electricity producer, who owns a combined heat and power plant.

   By buying a day ahead, the TSO establishes a short-term price signal for capacity. In turn, this short-term price signal can be used by potential investors in capacity. This can be compared with the spot market, which provides a short-term price signal for electrical energy, which, in turn, is used as the basis for long-term contracts and for investment decisions.

7. Regulating energy and the security of supply: making the wind turbines part of the solution

At the outset, for the security of supply, the intermittent energy supplied from the wind turbines is perceived as a problem. However, *provided the TSO has an intelligent design of the market for regulating energy, the wind turbines can be part of the solution.*

7.1 Example: a wind turbine providing down regulation

For a given hour of operation: assume the wind turbine owner has sold 0.6 MWh at a price of 40 EUR/MWh (i.e., the expected production from the wind turbine is 0.6 MWh). The wind turbine can be remote controlled. Assume the grid is being severely oversupplied with energy. In this case, the price for regulating energy will turn negative.

Assume the price for down-regulation is −100 EUR/MWh. Now the wind turbine owner can buy 0.6 MWh from the TSO at the price of −100 EUR/MWh (and stop the wind turbine). The purchase from the TSO makes it possible for the owner to fulfil his or her obligation of supplying 0.6 MWh. At the same time, by stopping the wind turbine, the owner helps the TSO by reducing the production.

In the market for regulating energy, the renewable-friendly design features are similar to the ones for the capacity market. These features include:

1. Normally, the TSO has a threshold for the minimum amount of energy, which the players must be willing to trade, if they want to participate at the market for regulating energy.

   The threshold must be small. A small threshold makes it possible for small producers and consumers to participate at the market, thereby improving the competition.

For example, a small threshold makes it possible for wind turbines to participate at the market.

2. The market for regulating energy must be separated from the market for regulating capacity. Hence, the TSO needs not necessarily trade the regulating energy with the players, from which the TSO has bought regulating capacity. This makes it possible for wind turbines and other intermittent energy producers to provide regulating energy. Because of the uncertainty of the future energy production, these producers cannot offer the TSO regulating capacity. However, they can offer regulating energy.

Hence, *by separating the markets for regulating energy and regulating capacity, for the security of supply, the wind turbines become part of the solution.*

8. Other mechanims and policies for integrating wind in electricty markets

8.1 Multinational markets recommendable

Both for regulating energy and regulating capacity, establishment of cross-border competition is strongly recommended. To some extent, this has been done in the Nordic countries. However, an EU-wide establishment of cross-border competition will improve the liquidity and the competition at the markets. Thereby, multinational markets can reduce the costs of integrating wind energy in the electricity supply system.

8.2 Danish subsidies for wind turbines

Denmark has different schemes for subsidizing wind turbines. For the individual wind turbine, the subsidy scheme depends on when the turbine was built. However, at the outset, the subsidies are feed-in tariffs. New wind turbines, onshore and offshore, receive a price premium of 33.5 EUR/MWh for the first 22,000 full-load hours. Additional 3 EUR/MWh is provided during the entire lifetime of the turbine to compensate for the cost of balancing.[3] These subsidies are given on top of the market price. Hence, the wind turbine owner has two sources of revenue (please refer to Figure 8):

1. The revenue from the subsidy. After the first 22,000 full-load hours, this revenue is reduced substantially.
2. The revenue from the sale to the market.

(Question: which market? Answer: virtually all the wind energy is sold to the spot market.)

For special offshore wind farms, the subsidy is settled by a tender procedure. For a new offshore wind farm built near the island Anholt, the tender procedure specified that no subsidy will be given during hours when the spot price is negative. Thereby, the wind farm is encouraged to stop producing during hours when oversupply threatens the security of supply.

---

[3]The wind turbine owner has an imbalance between the sale of energy and the production of energy, when the wind turbine's production deviates from the sale to the market. This imbalance causes a settlement with the TSO.

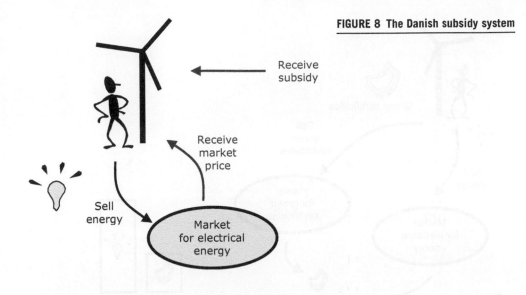

FIGURE 8 The Danish subsidy system

8.3 Green certificates: a market-oriented subsidy scheme for renewables

In Norway and Sweden, green certificates have been introduced as the subsidy scheme for renewables. Figure 9 illustrates how green certificates work: The owner of a wind turbine sells the energy produced by the turbine at the market for electrical energy. The wind turbine owner is competing against all other producers, both green producers and standard producers. However, for each MWh produced by the wind turbine, the owner also receives a green certificate. The certificate is a security, which can be sold at the market for green certificates. At this market, the turbine owner is only competing against other green producers.

The buyers at the markets are retailers (and big consumers who themselves buy at the whole-sale market). In Figure 9, a retailer is buying electricity for his or her customers. However, the retailer is obliged to buy several certificates corresponding to a certain percentage of his or her customers' consumption. The percentage is set by the politicians. Because of this obligation, there are always buyers at the market for green certificates. Hence, the turbine owner has two streams of revenues: one from the market for electrical energy and one from the market for green certificates. The size of the subsidy is determined by the competition at the market for green certificates. Hence, with green certificates, competition between the green producers determines the size of the subsidies. Also, the subsidy is technology neutral: producers using all sorts of renewables compete against each other at the market for green certificates.

For the introduction of renewables at the electricity market, the green certificate scheme allows politicians to exercise framework control: politicians can set a future target for the percentage of green certificates, which must be purchased by buyers at the whole-sale market. The market and the technological development will decide how the target is met. Thereby, we avoid political micromanagement of the electricity market.

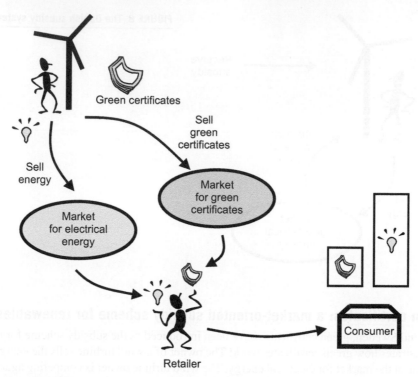

FIGURE 9 Green certificates

8.4 The Swedish-Norwegian implementation of green certificates

The Swedish system emerged on May 1, 2003, and is intended to increase the production of renewable electricity and also make the production more cost-efficient. The objective of the Swedish electricity certificate system is to increase the production of renewable electricity with 25 TWh by 2020 compared with 2002. The system replaces earlier public grants and subsidy systems. From January 1, 2012, Sweden and Norway have a common electricity certificate market. During the period until 2020, the two countries aim to increase their production of electricity from renewable energy sources by 26.4 TWh. The joint market permits trading in both Swedish and Norwegian certificates, and market players can receive certificates for renewable electricity production in either country.

Purchasers are Swedish or Norwegian parties having quota obligations. They are required to purchase certificates corresponding to a certain proportion of their electricity sales or use, known as their quota obligation. The sizes of quota obligations are set by the Swedish and the Norwegian Act Concerning Electricity Certificates, and create the demand for certificates. Each year, those having quota obligations must hold (i.e., have purchased) the number of electricity certificates needed to meet their quota obligations to the Swedish or Norwegian states. Once a year, these certificates are cancelled, which means that the holders must start to buy new certificates to meet next year's quota obligation, thus creating a constant demand for certificates. A multinational market for green certificates is an advantage. It increases the liquidity at the market for green certificates and hence secures a more efficient price discovery.

9. Efficient and nonefficient multistate markets (EU as a case)

Well-functioning, cross-border exchange of energy and ancillary services is crucial to have cost-efficient integration of wind energy, as previously described. Unfortunately, we do not yet have this in Europe. For ancillary services, the European countries have mainly national markets. For example, we do not yet have a single market for capacity in EU. For the cross-border exchange of energy, there is also much room for improvement. The EU does not yet have a single market for electrical energy. As an example of how there is much room for improvement, we can have a look at the spot markets in Northern Europe.[4] Northern Europe has a central calculation, where the spot prices and the day-ahead plans for the cross-border energy flows are calculated. The calculation is performed daily by the European Market Coupling Company (EMCC). In the calculation, the EMCC uses the method outlined in Figure 6 and Figure 7. When the EMCC performs the daily calculation, all spot bids from Northern Europe and all information on cross-border capacities are included in the calculation. Also, the EMCC software has proved to be reliable: the EMCC calculation has never crashed. Because of this combination of full information and reliable software, EMCC has always calculated reliable spot prices.

Unfortunately, the spot exchanges in Northern Europe insist on re-calculating the spot prices. When they do their re-calculation, the spot exchanges do not have full information on the spot bids in Northern Europe (they mainly have information on the spot bids from their own area). Furthermore, the software used by the spot exchanges is not reliable: the spot exchanges' re-calculations have repeatedly crashed. However, even when their re-calculations crash, the spot exchanges refuse to use EMCC's spot prices in the settlement of the spot trading. This has inflicted huge losses on market players (among others, owners of wind turbines).

For example, Nord Pool Spot's re-calculation of the spot prices for August 13, 2012, crashed. With Southern Sweden as a case, Figure 10 shows the spot prices calculated by the EMCC and by Nord Pool Spot. For this single day, the loss for the Baltic-Nordic sellers of electricity was EUR 900,000, because Nord Pool Spot used its own re-calculated prices in the settlement of the spot trading (instead of using the reliable spot prices calculated by the EMCC). Polish buyers of electricity experienced a similar loss. As can be seen, the spot prices in Northern Europe are calculated twice. The plan is to change this system in 2014. The planned, new system for spot price calculation is called price coupling regions (PCRs). With PCR, there will be no re-calculations of the spot prices.

However, PCR still has lots of redundant computers, software installations, and staff. The plan is that all the spot exchanges participating in market coupling will perform the same calculation. When a given spot exchange performs the calculation, the exchange has only aggregated and anonymized bids from the other exchanges. This means there is a lot of redundant pre-processing, because the bids submitted to a given spot exchange are aggregated and anonymized, before the

---

[4]Here "Northern Europe" means Austria, Belgium, Denmark, Estonia, Finland, France, Germany, Latvia, Lithuania, Luxembourg, Norway, Sweden and the Netherlands.

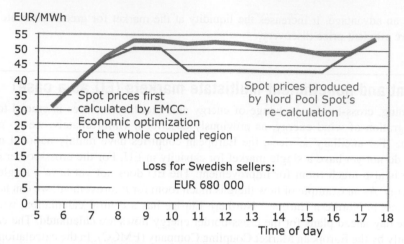

FIGURE 10 Spot prices on August 13, 2012, for Southern Sweden (SE3 and SE4 – please refer to fig. 5)

bids are sent to the other spot exchanges (and before the bids are sent to the exchange's own version of calculation). When the calculation is done, there is a redundant post-processing, because the individual spot exchanges must sort out what the calculation result means for their customers' spot trading.

In addition the spot exchanges have spent millions of euro developing the PRC software. This is an attempt at re-inventing the wheel, because the EMCC already has such software. Because the TSOs are financing the development of PCR, this re-invention of the wheel is financed by the grid users (i.e., by captive customers). Furthermore, the EMCC software has proven reliability, which contrasts strongly with the spot exchanges' track record.

10. Conclusion: The moral of the spot case from Northern Europe

As always in business, the root of the problem outlined in Chapter 18 is bad governance. For the future common European spot market, EU has not established multinational governance with clear accountability, rights, and obligations. However, as the current and planned European spot trading system illustrates, this is necessary to have well-functioning cross-border exchange of energy, and well-functioning cross-border exchange of energy is a prerequisite for a cost-efficient integration of wind energy in the electricity supply system.

Case Study–Renewable Integration: Flexibility Requirement, Potential Overgeneration, and Frequency Response Challenges

Mark Rothleder, Clyde Loutan

This case study provides an overview of analyses the California Independent System Operator (ISO) conducted to better understand the challenges related to renewable integration. This study includes recommendations that the fleet needs to have a flexible component, load shifting, and storage technology to mitigate potential overgeneration conditions, and inertial response is a characteristic that asynchronous resources should provide to maintain grid reliability. Figure 1

The ISO operates the wholesale transmission grid that serves approximately 80% of the load in California, and provides open and nondiscriminatory access that is supported by a competitive energy market and comprehensive planning efforts. The ISO is a summer peaking system, with a peak load of 50,270 Megawatts (MW) occurring on July 24, 2006. Its bulk power market allocates space on transmission lines, maintains operating reserves, and matches supply with demand. It is a nonprofit public benefit corporation that partners with more than 100 client organizations and is dedicated to the ongoing development and reliable operation of a modern grid.[1]

California has several energy and environmental policies that influence renewable integration and transmission needs. First, there is a reduction of greenhouse gas emissions to 1990 levels by 2020 that affects existing steam plants and limits available air emission credits for new power plants. Second, a state mandate requires utilities must serve 33% of their load from renewable resources by 2020. With this mandate comes an expectation that as much as 12,000 MW of distributed rooftop solar photovoltaic (PV) resources may connect to the grid by 2020. Although this large amount of rooftop solar PV has the inherent benefit of minimizing variability because of geographic diversity, it potentially lacks control and responsiveness to system contingency events. Third, there is a schedule to retrofit or retire approximately 12,000 MW of once-through-cooling[2] power plants located along the coast. Last, the state Water Resource Control Board is evaluating different levels of unimpaired water flows on some river systems that would affect hydroresource availability and dispatch patterns. Although these policies are important, the changes they bring to performing reliable grid operations need to be

---

[1]www.caiso.com.

[2]Once-through-cooling policy establishes technology-based standards to implement Federal Clean Water Act section 316(b) and reduce the harmful effects associated with cooling water intake structures from coastal and estuarine waters using a single-pass system on marine and estuarine life.

Renewable Energy Integration. http://dx.doi.org/10.1016/B978-0-12-407910-6.00006-5

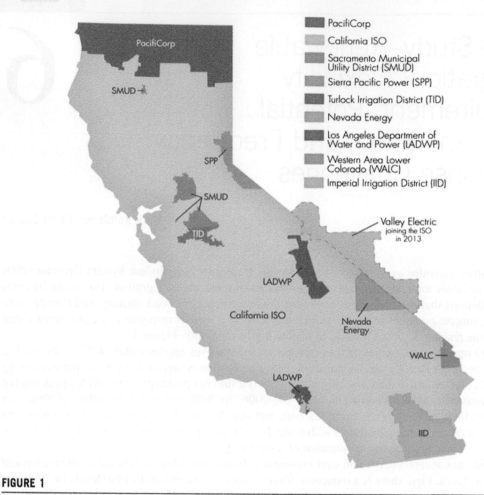

FIGURE 1

The California ISO Balancing Authority Area is shown as shaded.

analyzed and addressed to ensure a sufficient amount of capacity, with the right characteristics being in the right location and available to system operators at the right time.

Currently, there is approximately 5600 MW of wind and 3900 MW of solar resources connected to the ISO's controlled grid, and an additional 11,000 MW of variable energy resources (VERs) is expected to be connected within the next 7 years. The system is already facing intrahour VER ramps in excess of 2000 MW in 30 min. When these significant production changes occur in real time, the ISO faces challenges maintaining generation and load balance (as measured by the North American Electric Reliability Corporation (NERC) Control Performance Standard 1).[3] This indicates that the

[3]CPS1 is a statistical measure of a balancing authority's area control error variability in combination of the interconnection frequency error.

existing resource mix is not operated in a manner that ensures sufficient flexible capability is committed and available to system operators for real-time dispatch. As a result, operational practices and market timelines need to be adjusted to ensure the fleet is managed in a manner to ensure reliability can be maintained as variability of supply increases.

The availability of flexible capacity is affected as California's environmental policy on once-through-cooling resources phases in over the next 7 years because these resources play a significant role in balancing the daily load changes. This policy creates real opportunities to design the fleet makeup and incentivize new capability from a broad range of resource that includes demand response, variable resources themselves, storage technology, and regional coordination with other balancing authorities. In addition, resource commitment and dispatch would also have to play a key role to ensure adequate flexibility is available for real-time operation and meeting policy objectives while grid reliability is maintained.

It is expected during light load periods when renewable production is high that committing adequate flexible resources in real time will be challenging because resources with higher minimum load levels can result in potential overgeneration conditions. Also, with the lack of adequate resources on governor control, the ISO is concerned that arresting frequency decline and stabilizing the system after a disturbance may become more difficult because VER production is treated as "must-take" energy.

1. **ISO real-time market overview**

The ISO operates a day-ahead market (DAM) and a real-time market (RTM) that settle energy supply using locational marginal pricing (LMP) and settle load using load aggregation points. The DAM contains an integrated forward market, which is used to clear supply-and-demand bids and includes a residual unit commitment to ensure that sufficient capacity is committed to meet ISO forecast demand. The real-time market includes an hour-ahead scheduling process used to arrange for hourly intertie transactions, short-term unit commitment used to commit resources looking ahead 5 h, real-time unit commitment that runs every 15 min that commits resources and procures ancillary services, and 5-min real-time economic dispatch to meet imbalance energy requirements. The ISO relies on several processes and market software applications to manage supply-and-demand balance after the DAM closes at 10:00 a.m. the day before the operating day.[4]

The ISO also runs a congestion management application to better manage transmission constraints, which ensures efficient and feasible supply-and-demand decisions. Congestion management uses a full network model that accurately represents the ISO's balancing authority area and calculates LMPs to settle transactions. The security-constrained unit commitment is used in both the DAM and RTM to commit units, and security-constrained economic dispatch is used to balance the system generation, demand, and import and export schedules while respecting transmission constraints [1].

The RTM closes 75 min before the operating hour and economically maintains a generation and load balance every 5 min on a forward-looking basis. Within the 5-min dispatch intervals, generation-load balance is maintained through automatic generation control (AGC). Resources on

---

[4]ISO MRTU Level 200 "The Market 201" External Training: http://www.caiso.com/1860/1860902611580.pdf.

AGC control are dispatched every 4 s to maintain interconnection frequency and maintain schedules on the interties with neighboring balancing authorities. As resources on AGC control depart from their dispatch operating targets as established by the RTM 5-min dispatch application (in response to frequency and net interchange deviations), they temporarily supply or consume balancing energy [2].

The ISO market software simultaneously optimizes energy and ancillary service needs. The forecasted ancillary service regulation, spinning reserve, and non-spinning reserve are procured in the DAM, with some incremental procurement in real time to account for changes in operating conditions. In real time, any excess spinning reserve and non-spinning reserve capacity can be economically dispatched to energy regardless of whether there is a contingency event.

2. Renewable generation effects in the ISO real-time market

VERs typically submit their hourly forecasted output as schedules in the real-time market 75 min before the operating hour, which can be significantly different from actual energy production. Currently, the ISO's real-time market used for dispatching preferred resources does take into consideration VER energy production variability to procure additional flexibility on the system; however, the method of assessing flexibility capacity requirements is scheduled to be updated as additional renewable resources connect to the system. One planned change involves integrating a 15-min forecast that explicitly predicts future VER production levels, which would minimize forecast errors.

Resources dispatched through the ISO's real-time market are expected to move to their new operating target every 5 min, whereas resources not receiving a dispatch instruction are to remain at their operating level. Because VER production can change significantly within any 5-min dispatch interval, any short-term ramping shortages can create increased regulation needs and volatile market prices.

3. Flexibility requirement

The ISO has identified through technical studies the need for flexible resources to be committed with sufficient ramping capability to balance the system within an operating hour and between hours for scheduled interchange ramps. System operators must rely on ramping capability in both speed and quantity to balance the VERs' production change. Also, any underforecasting or overforecasting of demand requires dispatching flexible resources at higher or lower levels, respectively, to minimize inadvertent energy flows with neighboring balancing authorities.

As shown in Figure 2, the typical ISO load (blue curve) during the spring months has ramps that extend across multiple hours. As the penetration of VERs increases, the net load[5] (red curve) is the trajectory non-VERs would have to follow through RTM dispatches [3]. The net load comprises ramps of significant capacity and shorter duration, and on days with high VER production and light load, the

---

[5]Net Load = Load minus Wind minus Solar.

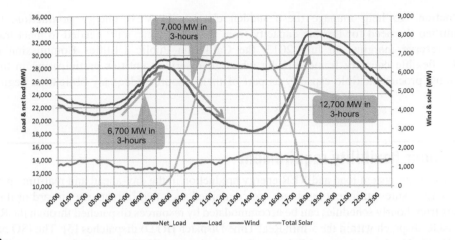

FIGURE 2

Load, net-load, wind and solar profiles: base scenario.

minimum net load occurs during the middle of the day. Figure 2 also shows that neither wind nor solar peak production coincides with the system peak load.

During the spring months, the ISO may have to cycle resources on and off more than once a day to meet the double peak shown. At times, this may not be possible because the down time between resource shutdown and start-up may be too long, which may prevent the resource from being restarted in time for the morning load ramp that begins around 4:00 a.m., or meeting system peak demand around 7:00 p.m.

To help manage the lack of fleet flexibility, the ISO is currently implementing a ramping tool[6] to predict and alert system operators of the load-following capacity and ramping requirements needed on the system in real-time. The ISO is also introducing a flexible ramp product[7] to ensure enough dispatchable capacity will be available on a 5-min dispatch basis in the real-time market.

During middays with high solar production, there is less need to commit additional supply resources, but toward sunset, an immediate need exists to replace the solar generation to continue meeting consumer demand. Many resources that could replace solar generation must be committed before this significant ramp, which begins before sunset. These resources often require several hours to a day or more to fully come on-line, which can result in more generation on-line than consumer demand, causing overgeneration conditions. By 2020, the ISO expects that increased flexibility will be needed to reliably meet two net-load peaks, which would require managing approximately 7000 MW of upward and downward ramps in 3-h time frames, and provide nearly 13,000 MW of continuous up-ramping capability to meet the evening peak, also in a 3-h time frame.

An additional condition not illustrated in Figure 2 is the variability within the hour because of clouds or storms, which requires flexible resources to smooth out the renewables' sudden changes

[6]http://www.pnnl.gov/main/publications/external/technical_reports/PNNL-21112.pdf.
[7]http://www.caiso.com/informed/Pages/StakeholderProcesses/FlexibleRampingProduct.aspx.

in production. To help manage these sudden changes in VER energy production, the ISO has quantified the need for flexible capacity from the generic capacity needed to meet traditional planning reserve margins. In July 2013, the California Public Utilities Commission (CPUC) adopted a flexible capacity framework by which load-serving entities participating in the ISO's market must procure flexible capacity as part of their resource adequacy obligation beginning in 2015 [4].

4. Intrahour flexibility requirement

The ISO forecasts system load and VER energy production in the day-ahead and real-time periods. A good forecast ensures sufficient resources are committed such that intrahourly upward or downward deviations from hourly schedules can be accommodated by resources dispatched through the RTM and through AGC dispatch within the 5-min Real Time Dispatch (RTD) dispatches [5]. The ISO partnered with the Pacific Northwest National Lab to evaluate the impact of VERs on intrahour load following or flexible capacity needs and regulation capacity needs.[8] Figure 3 illustrates the separation of load following and regulation based on real-time forecast. Load following is calculated as the difference between the forecast hourly average net load (red line) and the forecast 5-min net load (blue line), whereas regulation is calculated as the difference between the blue line and the actual load (green line) [6]. By 2020, the ISO expects that increased intrahour flexibility needed for RTD dispatch is approximately 3000 MW and the regulation capacity needed for resources on AGC control to be approximately 1000 MW.

FIGURE 3

Intrahour load-following or flexibility requirement (shown as the blue shaded area) and regulation requirement (shown as the red shaded area).

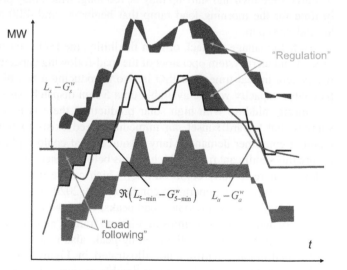

[8]http://www.caiso.com/Documents/Integration-RenewableResourcesReport.pdf.

5. Potential overgeneration problems

California's electricity supply mix comprises diverse technology types and is distinctly different than that of other states or regions; this mix is projected to significantly change as more VERs are integrated into the system in the long-term. Currently, the capacity of nondispatchable resource serving load within the ISO balancing area is greater than 10,000 MW, based on the maximum capability of each of the resources [2].

As shown in Figure 4, this high level of nondispatchable generation, coupled with high VER production during light load, can result in overgeneration conditions. Overgeneration occurs when more internal generation and imports into a balancing area exceed load and exports. Typically, before an overgeneration event occurs, the system operator will exhaust all efforts to send dispatchable resources to their minimum operating levels and will use all the decremental energy (DEC) bids available in the imbalance energy market.[9] If bids to balance the system are exhausted, the system operators can also make arrangements to sell excess energy out of market to neighboring balancing areas and can also decommit resources during its real-time unit commitment process. In addition, with a high area control error (ACE) and system frequency, which results from overgeneration, the ISO's energy management system will dispatch regulation resources to the bottom of its operating range [6].

When anticipating overgeneration, the system operator sends out a market notice and requests scheduling coordinators to provide additional DEC bids. If insufficient bids are received, the system operator may declare an overgeneration condition if they can no longer control the ACE and the associated high system frequency. During overgeneration conditions, the typical real-time imbalance energy price is negative, which reflects that the ISO will pay entities to take the excess power.

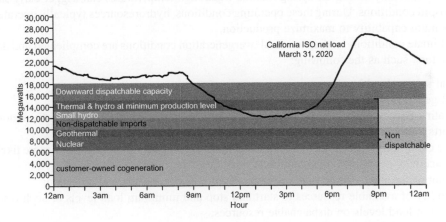

FIGURE 4

Potential overgeneration conditions.

[9]2013 Special Reliability Assessment: Maintaining Bulk Power system Reliability while integrating VERs—CAISO Approach, November 2013.

At times, the resources' minimum energy may not be needed, but the units cannot be shut down because they will be needed to meet demand later during the day. During this time, there is little that the system operator can do to accommodate this excess generation because supply and demand must match instantaneously. As the overgeneration conditions increase, so does the need to increase regional coordination. Additional help in mitigating overgeneration conditions can be met by forming an energy imbalance market with neighboring balancing authorities. Other solutions, such as advanced demand response and storage, may also be innovative solutions to help balance the system during the range of expected daily imbalance conditions. In addition, incentivizing the renewable resources to be flexible themselves may also be part of the solution.

The ISO is also exploring ways to incentivize qualifying capacity[10] to curtail production during low net load demand periods to minimize the magnitude of potential overgeneration. The ISO also plans on partnering with storage and incentiving load shifting during the hours of low net-load demand.

On days with low system demand, such as weekends, the potential can occur for the ISO's net load to decrease lower than the total production level of nondispatchable resources on the system. As shown in Figure 4, the areas labeled "Downward dispatchable capacity" can only be obtained by operating dispatchable resources above their minimum operating levels to have downward control that compensates for wind and solar production increases within an operating hour. The potential to export excess generation to neighboring balancing authorities during low system demand periods may be feasible but may not be practical because other balancing authorities may have to keep a portion of their dispatchable resources on-line to meet their load changes and comply with NERC's mandatory control performance standards.

Nondispatchable capacity can be significantly higher in years with abundant rainfall or large snowpacks, especially during the spring months when high temperatures can trigger early snow melt and hydrospill conditions. During these operating conditions, hydroresources typically operate close to their maximum capability to maximize production.

The ultimate conditions leading to actual overgeneration conditions are complicated and depend on several factors, such as the following:

1. actual system load levels;
2. VER generation production;
3. operating conditions in the rest of Western Interconnection, which determines the amount of exports that can be absorbed;
4. hydro conditions (with spring being more susceptible to spill conditions, this is more likely time to exacerbate overgeneration conditions);
5. amount of frequency-responsive resources required to be on-line and unloaded;
6. flexibility of available resources to start and stop and minimum load levels of such resources;
7. minimum load levels on dispatchable resources;
8. intrahour load-following capacity or flexible requirement;
9. other reserve requirements (regulation, spinning reserve, non-spinning reserve); and

---

[10]Qualifying facilities (QFs) are a combination of small power production facilities and cogeneration facilities. The QF categories shown in Figure 4 are comprised of gas fired QFs, and other QFs powered by renewables such as biogas, biomass, waste and oil. Cogeneration facilities are typically larger QFs that sequentially produce electricity and another form of thermal energy (such as heat or steam) in a way that is more efficient than the separate production of both energy forms.

10. amount of nondispatchable resources on-line.

Production simulation studies considering the previously described operating conditions have shown that the potential overgeneration conditions are expected to occur during the early evening hours in the spring months.

6. Inertia and frequency response

One of the reliability concerns presented by higher percentages of VERs on-line is the displacing resources that have the ability to arrest and stabilize system frequency after a grid disturbance or the sudden loss of a large generating source. Currently, photovoltaic solar generation offers no inertia and provides no frequency response, and wind generation offers virtually no recovery support unless specifically designed to do so [7].

The ISO is concerned that during periods of light load and high renewable production, the system may require subeconomic operation to meet its frequency response obligations (FROs), as proposed under NERC's Frequency Response and Frequency Bias Setting Standard (BAL-003-1).[11] An additional concern is having less inertia because VERs displace conventional resources with rotating mass and governor response. The ISO must closely monitor this displacement in real-time because the ability to arrest system frequency and meet its FRO depends on the following: the system condition just before the fault, the size of the outage, the headroom available on governor responsive resources, and the number and speed of governors providing frequency response.[12] Conventional resource displacement creates opportunities for renewable resources, demand response, storage, and smart technologies to help a Balancing Authority meet its FRO.

The ISO and General Electric International (GE) jointly conducted a frequency response study to investigate the loss of large-generation events under conditions with high wind and solar generation levels. Four extreme base cases with lighter loads and high VERs were created for the studies. They are winter low load with high ISO wind, weekend morning with high ISO wind and solar production, winter off peak with high wind production, and spring peak with high hydro and wind production. The study focused on California's frequency response for the loss of two Palo Verde[13] generating units (2690 MW) and the loss of two Diablo Canyon units (2400 MW). These operating conditions were selected because more resources would be decommitted because of the low loads and high VER production levels.

For the contingencies studied, three system equivalent frequencies were modeled (Western Electricity Coordinating Council (WECC), California, and non-California) based on the weighted speed of all the synchronous machines in the system [8]. The winter low load with high ISO wind base case is comprised of a WECC load level of 91,300 MW and 15,890 MW of wind and solar generation, of which 10,960 MW was located within California. Headroom on governor responsive resources in the base case was 13,740 MW, of which 3974 MW was located within California. As shown in Figure 5,

---

[11]BAL-003-1: Frequency Response and Frequency Bias Setting: http://www.nerc.com/pa/Stand/Project%20200712%20Frequency%20Response%20DL/BAL-003-1_clean_031213.pdf.
[12]CAISO/GE Frequency Response study can be found at http://www.caiso.com/Documents/Report-FrequencyResponseStudy.pdf.
[13]These units are located Arizona and is recognized by NERC as the largest loss of generation event in the Western Interconnection.

FIGURE 5

Frequency response to the loss of two Palo Verde units.

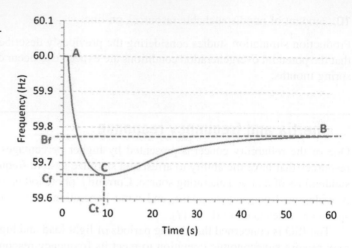

for the loss of two Palo Verde units, the frequency nadir (point C) occurred at 9.8 s at a frequency of 59.67 Hz, and the settling frequency (point B) was 59.78 Hz [4].

Figure 6 shows the electrical and mechanical power output of the governor responsive resources within California. At the time of the nadir (point Cp),[14] the resources within California were delivering approximately an extra 524 MW or 24% of the overall WECC frequency response obligation. The GE-ISO settling-based frequency response shows that California contributed 234 MW/0.1 Hz at point B. The NERC-proposed FRO for the WECC is 685 MW/0.1 Hz, which includes a 25% safety margin [9]. Based on the WECC obligation, if California was assigned 30% or 205 MW/0.1 Hz (based on load ratio), the response of 234 MW/0.1 Hz was adequate. The NERC proposed FRO for the WECC is being reviewed and may change after the review process.

FIGURE 6

California frequency response to the loss of two Palo Verde units.

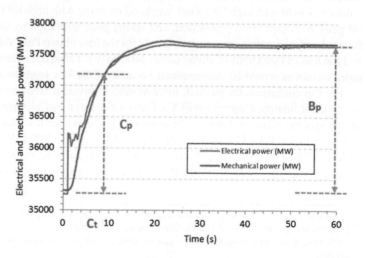

---

[14]This GE-ISO nadir metric only determine the frequency response to the frequency nadir point Cp.

7. Sensitivities

Several sensitivities were conducted to evaluate the adequacy of the committed resources to avoid unintentional load tripping. First, hourly deterministic production simulation studies were conducted for each hour of 2020. Hourly time series for wind and solar resources were developed, and resources were committed by the model to meet the hourly loads. Second, the unit commitment patterns from the production simulation studies were chosen based on low loads and high VER penetration for specific hours or potential overgeneration conditions. These generation commitment patterns and corresponding loads were then modeled in the GE stability model. The following sensitivities and results are summarized in Table 1 [8].

The system was further stressed by reducing the practical minimum headroom on governor responsive resources in the Western Interconnection to 8000 MW and further reducing the headroom to an extreme minimum level of 3000 MW. The 3000-MW was based on the WECC only carrying required spinning reserve for the interconnection as a whole. Although this is not an actual acceptable operating condition, it is a test case for determining that maintaining a minimum of the spinning reserve requirement as available headroom is not adequate to reliably operate the system.

As shown in Figure 7, the higher renewable penetration case (red line) with more wind has better frequency response than the base case (blue line) because the higher wind generation increased resource K_t headroom. For the practical headroom of 8000 MW, the frequency nadir reached 59.55 Hz, which only gives 50 mHz of margin before the first block of underfrequency load shedding relays are set to pick up at 59.5 Hz. For the extreme minimum headroom case, the frequency nadir was 59.42 Hz, which would have resulted in underfrequency load shedding picking up and unintentionally tripping load.

These sensitivities indicate that extremely depleted headroom will result in unacceptable system performance. The ratio between resources providing governor response and the other resources K_t is a good primary metric for determining the amount of governor resources that should be committed in real time [9,10].

These studies provide evidence that as high levels of wind and solar generation are integrated into the existing resource mix in the Western Interconnection, it is necessary to maintain adequate headroom on governor responsive resources to meet a required frequency response obligation. The system

Table 1 Sensitivities and Stability Results

	Impact on Frequency Nadir	Impact on Settling Frequency
Reduced inertia	Worst, sooner	No impact
Reduced headroom	Small impact	Worst
Fewer governors in operation	Small impact	Worst
More governor withdrawal	Small impact	Worst
Wind inertial control	Improve	Small impact
Wind frequency droop (governor-like control)	Improve	Improve

FIGURE 7

Frequency response for different sensitivities.

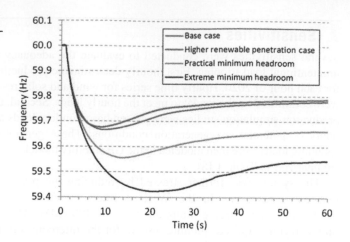

must also maintain a minimum amount of synchronized resources to arrest frequency decline after a disturbance. The speed of frequency responsive resources needs to be considered because the frequency nadir after a major loss of generation event typically occurs within 10 s after the event. For the practical minimum headroom case, the frequency nadir occurred at approximately 14 s. It is also important to sustain primary frequency response until secondary frequency response or automatic generation control dispatches resources on regulation and contingency reserve could be dispatched. The studies also concluded that withdrawal of primary response quickly contributes to degraded frequency performance of approximately 20%.

Inertial control from wind generation provides fast transient support, via controlled inertial response from wind turbines, and can significantly improve the system frequency nadir. Based on the previously described finding, the ISO collaborated with NERC to issue a joint report in which inertial response was recommended as a characteristic that asynchronous resources should provide, similar to synchronous resources, to maintain system reliability as more renewable resources are integrated into the grid [2].

References

[1] Loutan C, Young T, Chowdhury S, Chowdhury A, Rosenblum G. Impact of integrating wind power resources into the California ISO market constructnd generation. IEEE; 2010.
[2] NERC/ISO. Maintaining bulk-power system reliability while integrating variable energy resources - caiso approach; November 2013.
[3] NERC_IVGT_Task_1_4. Flexibility requirements and metrics for variable generation: implications for system planning studies; August 2010.
[4] CPUC. Decision adoptingl local procurement obligations for 2014, a flexible capacity framework, and further refining the resource adequacy program. California; July 2013.
[5] Makarov Y, Loutan C, Ma J, deMello P. Operational impacts of wind generation on California power system. IEEE Trans 2009;22(2):1039–50.

[6] Loutan C, Hawkins D, et al. Integration of renewable resources - transmission and operating issues and recommendations for integrating renewable resources on the California ISO-controlled grid; November 2007.

[7] Energy G, CAISO, PNNL, PLEXOS, Nexant. Integration of renewable resources - operational requirements and generation fleet capability at 20% RPS; August 2010.

[8] Miller NW, Shao M, Venkataraman S. California ISO (CAISO) frequency response study; November 2011.

[9] NERC. Frequency response initiative report - the reliability role of frequency response; October 2012.

[10] Undrill J. Power and frequency control as it relates to wind-powered generation; December 2010.

Variable Energy Resources in Power System and Market Operations

Analyzing the Impact of Variable Energy Resources on Power System Reserves

7

Brendan Kirby, Erik Ela, Michael Milligan

Aggregate generation and load must be balanced instantaneously and continuously for the electric power system to perform reliably. Transmission limitations impose additional restrictions on which generators must or must not operate at any specific time. Both variability and uncertainty increase the difficulty of balancing generation and load. Most individual loads are highly variable and often uncertain. Aggregate load behavior is less uncertain, but it still exhibits strong daily, weekly, and seasonal patterns that depend on weather and the business cycle. The output of conventional generators (hydro, thermal, and nuclear) is controllable under normal conditions and does not add significantly to power system uncertainty or variability, but all of these generators can and do fail suddenly and unexpectedly, which does add to system variability and uncertainty. Large amounts of wind and solar generation do not typically fail instantaneously, but the available energy does vary and is only somewhat predictable. While variability and uncertainty have always been characteristics of the power system that must be addressed to maintain reliability, variable energy resources increase both the net system variability and uncertainty.

Power system operators compensate for the variability and uncertainty of load and generation by maintaining a series of reserves: additional capacity (generation, responsive load, and storage) above that needed to meet the current net load. Multiple types of reserves are used because they differ in terms of response speed (cycles to hours), response duration (seconds to hours), and response frequency (continuous to every few days). Different amounts of each type of reserves are required to maintain reliability and those amounts vary from region to region. Terminology to describe the reserves also differs between Europe and North America, though the physics of the power system is universal.

Determining the additional reserves required to maintain reliability when significant quintiles of variable energy resources (VERs) are added to a power system is complicated by the nonlinear nature of reserves as well as by the lack of data and experience. This chapter examines reserve types and requirements in general and addresses how VER impacts are determined.[1]

1. Reserve types

Before discussing the impact of VERs on power system reserve requirements, it is necessary to briefly examine the different types of reserves, even without the presence of VERs, and the role each reserve type plays in system operation. Reserves are capacity (generation, responsive load, or storage), under system operator control, that is capable of moving up or down to maintain the net generation/load balance. Reserves are differentiated by speed and by the frequency of use (faster response vs slower,

---

[1]Much of this chapter was derived from the 2011 NREL report [1].

Renewable Energy Integration. http://dx.doi.org/10.1016/B978-0-12-407910-6.00007-7

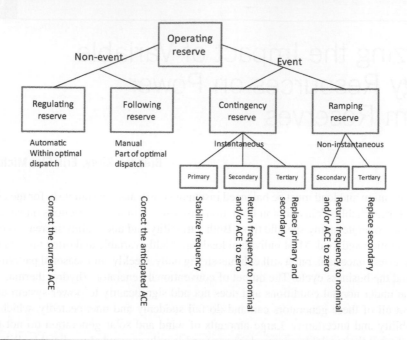

FIGURE 1

Example of operating reserve categories and how they are related[3] [1]. ACE, area control error.

continuous response vs occasional) because it is typically more expensive for resources to provide faster and continuous response.[2] Figure 1 shows how different reserve types are related and what aspect of balancing load and generation each addresses.

Regulation and following reserves respond to the non-event continuous variability of loads and generation, as shown in Figure 2. Regulation compensates for the random minute-to-minute variability, while following responds to variations ranging from 10 minutes to hours. The exact time range for each depends on the market structure for scheduling and trading energy in the specific power system with energy markets and economic dispatch accommodating the slowest variations.

Figure 3 shows how an additional set of reserves are used in a coordinated way to respond to an emergency event and represents the North American procedures. These emergencies can be caused by generation forced outages or other component outages (e.g. transmission lines or transformers), both of which cause near-instantaneous changes on the power system. During loss-of-supply events, additional supply (or demand) needs to respond to the disturbance immediately. As can be seen in Figure 3, this includes a number of different responses that vary by response time and by the length of time the response is sustained. Initially, when the loss of supply occurs, synchronous machines inherently

---

[2]More expensive regulation, for example, requires fast and continuous response, while less expensive spinning reserve requires fast response, but only occasionally.

[3]Area control error (ACE) is a measure of the instantaneous difference between a Balancing Authority's net actual and scheduled interchange, taking into account the effects of frequency bias and meter error [2].

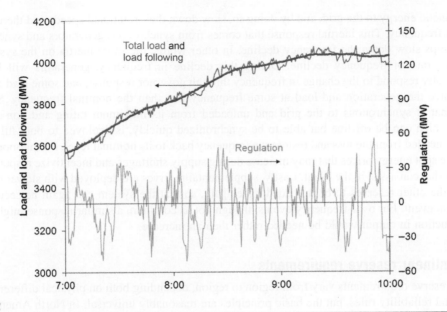

FIGURE 2

Power system operation time frames [3].

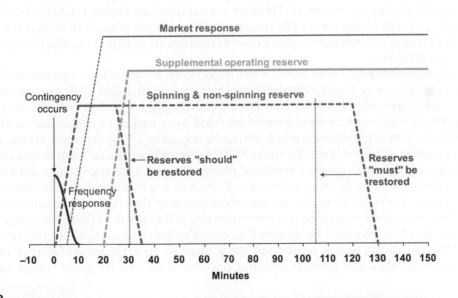

FIGURE 3

Reserve deployment [4].

supply kinetic energy to the grid, and by doing so, slow down their rotational speeds and therefore the electrical frequency. This inertial response that comes from synchronous generators and synchronous motors helps slow down the frequency decline. In other words, the more inertia on the system, the slower the rate of frequency decline. During this decline in frequency, generators will begin to automatically respond to the change in frequency through governor response, and some load response will balance the generation and load at some frequency less than the nominal frequency. Spinning reserve that is synchronous to the grid and unloaded from its maximum rating and non-spinning reserve, which can be off-line but able to be synchronized quickly, is deployed to both fill the gap in energy needed from the loss and restore the frequency back to its nominal level. Furthermore, many areas have spot energy prices that may increase during supply shortages and incentivize response from resources that can assist in the event. Lastly, supplemental reserves are deployed with slower response to allow the other reserves to be unloaded once again, so that the system can again be secure for a subsequent event. For over-frequency events, though not as common, a similar response might occur, but a reduction in output would be needed rather than an increase.

1.1 Nonlinear reserve requirements

Specific reserve requirements vary from region to region, depending both on physical differences and on regional reliability rules, but the basic principles are reasonably universal. In North America each balancing authority determines how much regulation and following reserve it requires to meet North American Electric Reliability Corporation (NERC) and regional performance standards (Control Performance Standard 1 and 2 or Balancing Authority ACE Limit). Reliability standards do not specify specific regulation or following reserve requirements, but they do set limits on the maximum number and/or quantity of imbalances. Therefore some reserves are kept to maintain balance within the reliability standards, but the specific quantity of reserve is not prescribed. However, reliability standards do specify minimum contingency reserve requirements as well as establishing contingency response requirements.

Reserve requirements tend to be nonlinear with respect to the amount of load or generation. That is, if balancing area A has twice the generation and load of balancing area B, it likely does not require twice the reserves. This is especially true for the faster, more expensive regulating reserves and for the contingency reserves. The minute-to-minute fluctuations of individual loads tend to be random and uncorrelated. Consequently, our example balancing area A will require only about 1.4 (square root of 2) times as much regulating reserves as balancing area B to meet NERC balancing requirements under normal conditions [5]. Following requirements are more correlated among individual loads since many exhibit the same daily load pattern, but reserve requirements typically increase less than linearly due to diversity.

Contingency reserve requirements are even more nonlinear than regulation requirements. Most regions base contingency reserve requirements on the size of the largest credible contingency that the region is designed to withstand.[4] The Electric Reliability Council of Texas (ERCOT), for example, has a 2300 MW contingency reserve requirement that is based on the simultaneous failure of two large nuclear units. Adding a new 800 MW coal-fired generator to the ERCOT power system would not

---

[4]The Western Electricity Coordinating Council (WECC) has an additional requirement that contingency reserves not be less than 5% of the load served by hydro and 7% of the load served by thermal generation. That requirement is likely to be changed to 3% of the load plus 3% of the generation within a balancing area.

increase the contingency reserve requirement at all, unless the new plant were uniquely situated such that it was likely to fail simultaneously with the two nuclear units.

Utilities have taken advantage of these economies of scale for reserves for over a century [6]. It was a major driver for interconnecting individual utilities early in the industry's history. It also led to the formation of power pools and reserve sharing groups. The important consequence for VERs is that reserve requirements cannot be analyzed in isolation; they must be analyzed in the context of the power system that the VER is being integrated into and in concert with all of the other VERs.

1.2 Reserve requirements vary in time

Regulation and following physical reserve requirements vary in time. Clearly, ramp-up reserves are required to follow the morning load rise and down reserves are required for the evening. Regulation requirements vary in time as well. Balancing areas that match their reserve procurements to the reserve needs saves their customers money when compared with balancing areas that establish fixed reserve requirements based on the maximum reserve needs.

2. Reserves and energy markets

Energy transaction and generation scheduling practices influence reserve requirements because they impact how well the system operator can access the flexibility that is inherent in the energy-producing resources. Two-thirds of the load in North America is served by Independent System Operators (ISOs) and Regional Transmission Organizations with 5-min energy scheduling and 5-min energy markets. System operators in these areas can rapidly adjust generation output through energy scheduling and thus reduce the amount of dedicated reserves that must be procured. A region that only allows hourly energy scheduling, that sets schedules 30 min ahead of the hour, and that uses persistence forecasts for VER production will need reserves to compensate for 90 min of VER variability. Regulation reserves are defined as the reserves that compensate for variability that is faster than the shortest energy scheduling interval. Faster energy scheduling inherently reduces the regulation requirement. A region with 5-min energy markets need only include about 15 min of VER variability in the regulation requirement. Sub-hourly energy scheduling and markets improve the economic efficiency of the power system in general and especially reduce the cost of wind and solar integration.

Figure 4 shows how 5-min energy scheduling follows the actual net load more closely than hourly scheduling. Note that the generators following the 5-min schedule are not moving with a faster ramp rate or over a larger range than the generators following the hourly schedule, they are simply moving more often. Figure 5 shows that the regulation requirements that result from hourly scheduling are much greater than what is required if 5-min energy scheduling is employed.

3. European vs North American reserve definitions

While the physical requirements to reliably balance an interconnected AC power system are universal, the terminology used to specify the reserves differs from region to region. The Union for Coordination of Transmission of Electricity (UCTE) and now the European Network of Transmission System

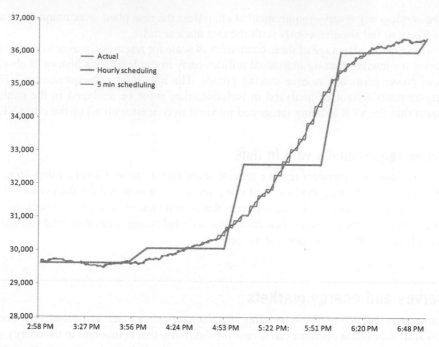

FIGURE 4

Five-minute scheduling vs hourly scheduling.

Operators for Electricity (ENTSO-E) specify reserve requirements with different terminology than that used by NERC. North American power system engineers use the technical terms of primary, secondary, and tertiary reserves, but the reserve products themselves include additional characteristics. Figure 6 shows the relationship between the two sets of reserves.

3.1 Primary reserves

In the UCTE, primary control reserve is used immediately following disturbances to stabilize the system frequency within a few seconds. Primary control reserve autonomously responds to frequency deviations through governor control. The total required primary control reserve is based on what the UCTE considers the maximum likely instantaneous power deviation in the synchronous system (largest credible contingency) and it is allocated to the individual control areas. The UCTE policy does not specify which technologies can provide primary control reserve (generation, demand response, or storage).

Primary reserve response is also understood in North America, with most generators having active governors, but specific response and specific response capability are not yet a regulatory requirement. This is an active area of discussion and it is likely that specific mandatory frequency response requirements will be developed, possibly with regional differences.

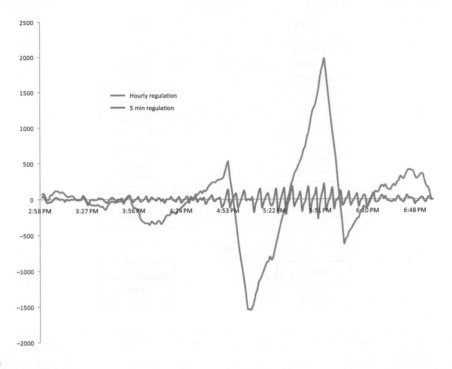

FIGURE 5

Regulating reserve needs for 5-min scheduling vs hourly scheduling.

3.2 Secondary reserves

Secondary reserve is the system operator centrally directed control of area control error (ACE) through automatic generation control (AGC) and automatic load control. This same term is used by UCTE for control under normal conditions as well as disturbance conditions. Secondary reserve requirements are based on empirical analysis or on probabilistic methods, but there is no formal compliance measure set forth by the UCTE.

NERC distinguishes between secondary reserves used under normal conditions (regulation) and secondary reserves used to respond to contingencies (contingency reserves). NERC and the regions do not have requirements to have specific amounts of regulating reserve, but they do have mandatory performance requirements for the control of ACE. NERC and the regions have requirements for specific amounts of spinning and non-spinning reserves for contingency response as well as contingency response performance requirements.

3.3 Tertiary reserves

Tertiary reserves are used to replace primary and secondary reserves following a contingency. UCTE does not restrict what technologies can provide tertiary reserves but recommends that enough tertiary control reserve be available to cover the largest loss of supply in the individual control area.

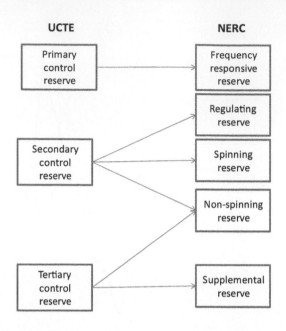

FIGURE 6

Union for Coordination of Transmission of Electricity (UCTE) and North American Electric Reliability Corporation (NERC) reserve terminology.

Reliability rules in North America do not explicitly address tertiary reserves, though they may comprise a part or all of the non-spinning reserves. Some regions also run markets for supplemental operating reserves or replacement reserves, which are available within 30 min. Sub-hourly energy markets also provide tertiary reserves.

Table 1 shows some of the policy differences on reserve requirements between Europe and North America.

4. Probabilistic methods for setting reserve requirements

While current practice for setting reserve requirements is primarily deterministic, significant academic work has been done to explore the benefits of probabilistic methods for determining reserve requirements [12,14,16]. There is interest in probabilistic methods for power systems without VERs, but the proposed methods are also expected to be helpful as VER penetrations increase. For example, most of the operating reserve requirements have been based on rules of thumb that compute the requirements based on the largest units or a percentage of the load or generation. Even if the only uncertainty is the failure of conventional units, the deterministic criteria used today do not take into account the failure rates of the generating units, the value of not shedding load, or the possibility of two or more near-simultaneous outages.

Table 1 NERC and UCTE Policy Differences

	Europe (ENTSO-E/UCTE)	N. America (NERC)
Regulating reserve	UCTE does recommend a *secondary control reserve* requirement, which is based on a statistical equation and mostly comes from load variability. However, secondary reserve is used for both contingencies and normal variations. There are no compliance measures.	NERC enforces CPS1 and CPS2 compliance measures but has no policy on what the actual *regulating reserve* requirement quantity should be. The CPS1 and CPS2 drive the requirements, which are mostly based on time of day and season.
Following reserve	No requirements.	No requirements.
Contingency reserve	Similar requirement to DCS. Return ACE to zero within 15 min. Split between *primary*, *secondary*, and *tertiary*. Enough of these reserves should be available to cover the largest contingency.	DCS requires ACE to be returned to 0 or its pre-disturbance level if negative within 15 min. Enough *contingency reserves* required to recover largest contingency. Many regions require at least 50% to be spinning.
Primary reserve	*Primary control reserve* (3000 MW) split between TSOs based on energy contribution. 3000 MW based on largest credible interconnection-wide event. Full response at 200 mHz. Response characteristics based on UFLS relay setting and 200 mHz safety margin. 20 mHz maximum insensitivity.	No requirement. Some discussions on a future requirement. Only a frequency bias requirement as part of ACE equation of 1% peak load. Governor dead bands mostly set at 36 mHz and droop at 5%, but not required.
Ramping reserve	No requirements.	No requirements.
Tertiary reserve	*Tertiary control reserve* requirement is larger than the largest contingency. There is no requirement on how soon any reserves should be replaced.	No quantifiable requirement but contingency reserve must be replaced within 105 min following contingency.

NERC, North American Electric Reliability Corporation; UCTE, Union for Coordination of Transmission of Electricity; ENTSO-E, European Network of Transmission System Operators for Electricity; ACE, area control error; TSO, transmission system operators; DCS, disturbance control standard; CPS, control performance standard; CPS1, control performance standard 1; CPS2, control performance standard 2; UFLS, under frequency load shedding.

Researchers have developed probabilistic methods where the unit commitment problem would determine the optimal amount of contingency reserves and following reserve for different load levels based on the probability of having insufficient generation due to generator outages or load forecast errors. The methods evaluate the load probability distribution and evaluate the risks involved with the current unit commitment using a capacity outage probability table by the convolution of the cumulative probability distributions. This shows the amount of risk by looking at the probability that the generation on-line is less than the load. The process would iterate until the risk level was met for the time period.

5. Determining VER impacts on reserve requirements through power system modeling

Determining the impact of large penetrations of VER on power system reserve requirements is complicated by lack of experience (penetrations are still relatively low in most locations) and the complex, nonlinear interactions of many power system characteristics on reserve requirements. This is currently an area of active research, as the power system industry gains more experience managing VERs and researchers analyze new ways to efficiently provide for these reserves in an economically efficient manner [8]. As discussed above, regulation and contingency reserve requirements do not increase linearly as new resources (VER or conventional) are added to the power system. Reserve requirements are also impacted by energy scheduling practices, power system size, ability to forecast loads and renewables, etc. Fortunately, significant progress has been made in modeling power systems with high VER penetrations.

In principle, VER reserve impacts can be determined by modeling the power system with and without VER. Time-series modeling that simulates the utility's security-constrained unit commitment and economic dispatch can duplicate actual operations with reasonable precision. Reserves can be adjusted such that power system reliability (measured by loss of load probability, for example) is the same in the with- and without-VER cases. The difference in required reserves is then the incremental reserves required with VER. Models are typically run for three or more years with time resolutions of an hour or shorter.

Accurate modeling is possible because generator characteristics are known (efficiency, fuel cost, start and stop times, start-up costs, reserve capabilities, ramp rates, etc.). Historic time series load data is also available, providing an accurate representation of system complexity. Future conditions can be modeled by escalating the load data appropriately and by changing the available generation mix to reflect the expected future generation fleet. Only the VER data is problematic since the VER likely does not already exist.

It is important to have time series VER data that matches the historic load data, since weather patterns impact both loads and renewable generation. Mesoscale weather modeling can now provide estimated historic time series wind and solar generation output over very large geographic ranges with time resolutions of 5 min and geographic spacing of 2 km or less. Vast amounts of historic time-series weather data are used to run numerical weather models that reproduce the atmospheric conditions and the resulting wind speeds at hub height and solar insolation. The historic weather data is synchronized with the historic load data. Power system modelers can then select the sites of proposed wind and solar plants such that the resulting model includes the variability and uncertainty that would have occurred had there been the high VER penetration. The large-scale renewable integration studies are able to evaluate real power systems with simulated increased variable generation penetration and practical specific changes needed for that particular system.

In recent years, a number of different entities have initiated or completed large-scale renewable integration studies specifically aimed at analyzing the impacts and costs of operating power systems with large penetrations of variable renewable generation. So far, this has mostly included wind energy, but other technologies (e.g. solar) are being included now as well.

Numerous studies have advanced the art of analyzing the impact of high VER penetration on reserve requirements. More information on these studies and summaries of the studies can be found in References [9,13,15]. Two studies are particularly noteworthy: the Eastern Wind Integration and Transmission Study and the Western Wind and Solar Integration Study.

5.1 Eastern wind integration and transmission study

The Eastern Wind Integration and Transmission Study (EWITS) evaluated the operational impacts of various wind penetrations, locations, and transmission build-out options for most of the U.S. Eastern Interconnection. The study included three scenarios of 20% wind energy, with each representing different primary locations of the wind, and one 30% wind energy scenario.

The first procedure of the study was to determine the *contingency reserves* required. As many previous U.S. studies have done, these assumed the current rule and determined that the largest contingency was not affected by the large amounts of wind generation. One and a half times the single largest hazard in each operating region determined the amount of *contingency reserves* for that region.

Many prior studies in the United States concluded a slight, but not insignificant, increase in the amount of required *regulating reserve* due to the increased variability of wind added to that of the load. In EWITS, a similar methodology to the prior studies was performed. The minute-to-minute variability separated from a 20-min rolling average of a 100 MW wind plant was used for the analysis, and the standard deviation was determined to be 1 MW. It was assumed that there is no correlation between wind plants for power output deltas in this time frame, and therefore, the total standard deviation for a balancing area was calculated by geometrically adding the 1 MW standard deviation for all 100 MW wind plants on the system. For load-only, the *regulating reserve* requirement was assumed to be 1% of the total load, and was assumed to be equal to three times the standard deviation of the load variability. Since load and all wind variability on this time frame were also considered to be independent of one another, the standard deviations of all wind and load were then geometrically added together by calculating the square root of the sum of their squares. The total standard deviation was increased by less than 1 MW when the wind was added to the load, and therefore the variability of wind was not considered as part of the *regulating reserve* for the study.[5]

Unlike other studies, it was determined that the uncertainty in the wind forecasts used for economic dispatch would impact the *regulating reserve* much more than what was shown for the variability. Economic dispatch programs that run every 5 min would use information from at least 10 min before the operating interval. Since it is too late to adjust the economic dispatch for any deviations, these deviations would all be met by units providing *regulating reserve*. Assuming a 10-min-ahead persistence forecast, the additional *regulating reserve* was determined by looking at the standard deviation of 10-min changes in wind output (load forecast for 10-min-ahead was assumed to be quite good and load forecast error was ignored). Figure 7 shows the standard deviation of the 10-min-ahead wind forecast errors as a function of the average hourly production of the total wind. The highest variability is near 50% production, where the anticipated 10-min change can be up or down, and also relates to wind turbines being at the steepest part of the wind speed–to–power conversion curve. The

---

[5]Calculations based on a balancing area with 100 GW load and 60 GW of wind power, which was about the average for the largest ISO balancing areas that were a part of the study.

FIGURE 7

Ten-minute-ahead wind generation forecast errors as function of wind production.

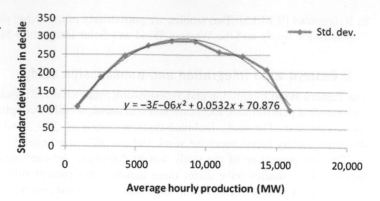

function was used for the hourly wind-related standard deviation of the *regulating reserve* requirement and was geometrically added to the load *regulating reserve* requirement discussed above. The equation is shown below, where σ_{ST}(hourly wind) is the standard deviation of wind forecast errors that is a function of the predicted wind portrayed in Figure 7.

$$\text{Reg req} = 3 * \sqrt{\left(\frac{1\% \text{ hourly load}}{3}\right)^2 + \sigma_{ST}(\text{hourly wind})^2}$$

A similar approach was used for the hour-ahead wind forecast error. However, in this case it was assumed that the errors that were not occurring often could be compensated for with off-line reserve. Therefore, one standard deviation of the hour-ahead forecast error was required to be spinning, and two standard deviations could be non-spinning. Also, since the reserves were used in the production cost simulations for the study, it was ensured that if the reserves had to be used for the hour-ahead forecast error of the hour in question, those reserves did not have to be kept in real time. In other words, if reserves were needed because less wind was available than forecast, the model would release that amount of reserves in real time since the reserves were used for the forecast error and not needed further. The total amounts of all reserves used in the study are shown in Figure 8. The reserve requirement was an hourly value that was a function of wind levels. This is an important evolution in our understanding of reserves for wind (sometimes called "flexibility reserves"). The level of flexibility reserve needed at any point in time is a function of the current level of wind output. In its simplest form, this would provide the equivalent of a look-up table that could be used to determine the level of flexibility reserve needed. Variations of this approach are discussed in the next section. There has been some work to refine and adapt the EWITS reserves method, dropping the percentage of errors to be covered from 99% to 95%, which corresponds to the NERC CPS2 requirements. Further improvements can be found at http://wind.nrel.gov/public/WWIS/Reserve.pdf

5.2 Western wind and solar integration study

The Western Wind and Solar Integration Study Phase 1 (WWSIS) focused on the WestConnect region but included the entire U.S. portion of the Western Interconnection with wind, load, generation, and

Reserve component	Spinning (MW)	Non-spinning (MW)
Regulation (variability and short-term wind forecast error)	$3 \cdot \sqrt{\left(\dfrac{1\% \cdot \text{hourly load}}{3}\right)^2 + \sigma_{ST}(\text{hourly wind})^2}$	0
Regulation (next-hour wind forecast error)	$1 \cdot \sigma_{\text{next hour error}}$ (previous hour wind)	0
Additional reserve		2 × (regulation for next hour wind forecast error)
Contingency	50% of 1.5 × SLH (or designated fraction)	50% of 1.5 × SLH (or designated fraction)
Total (used in production simulations)	Sum of above	Sum of above

FIGURE 8

Summary of reserve methodologies for Eastern Wind Integration and Transmission Study (EWITS) and Single Largest Hazard (SLH).

transmission data. [11] The study analyzed the integration impacts of up to 30% wind energy and 5% solar energy in the study region, with 20% wind energy and 1% solar energy in the rest of the Interconnection. A large part of the study analyzed the impacts on operating reserves that the increased penetrations of wind and solar would have on the system.

The study assumed that three times the standard deviation of the 10-min variability would be held for the *variability reserve* (Following Reserve). The study recommended that one standard deviation of wind variability be held for *regulation reserve* (Regulating Reserve) and have AGC capability. The report also discussed the need for a dynamic requirement that depends on both the wind and load levels.

Figure 9 shows the reserve requirements based on the 3 sigma rule for all the individual regions in the study footprint. The rule can be compared with using the 3% load, and then with the 3 sigma for load variability alone. In areas with high wind and relatively low load (i.e. Wyoming), some significant changes in the *variability reserves* can be seen. Figure 10 also shows the same methodology using one standard deviation applied to the *regulation reserve* need.

This method was then analyzed further so that simple rules that did not need detailed look-up tables could be used by system operators when determining the needs for these *variability reserves*. At first a rule of 3% load plus 5% wind was analyzed but proved to be problematic under certain conditions. The team then used a best fit using three degrees of freedom. These included the coefficient to load level, the coefficient to wind level, and an amount of wind capacity where further incremental increases in reserve were not needed—in other words, $X*$Load $+ Y*$Wind, until Wind $> Z$(nameplate capacity). It was determined that this rule was a good compromise between accuracy of capturing the 3 sigma need

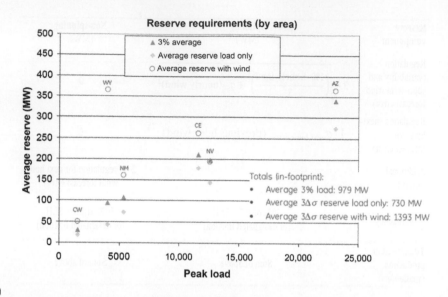

FIGURE 9

Reserve requirements using 3% or using 3 sigma rule for each area. Colorado West (CW), New Mexico (NM), Wyoming (WY), Colorado East (CE), Nevada (NV) and Arizona (AZ).

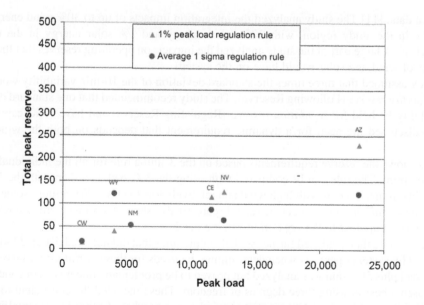

FIGURE 10

The 1% peak load *regulation* rule and average 1 sigma *regulation* rule for all regions. Colorado West (CW), New Mexico (NM), Wyoming (WY), Colorado East (CE), Nevada (NV) and Arizona (AZ).

Table 2 Western Wind and Solar Integration Study Phase 1 Reserve Rules for 30% Local Priority Scenario

	Load Only (% of Load)	30% Local Priority Scenario		
		Load Term (% of Load)	Wind Term (% of Wind Production)	Up to (% of Wind Nameplate)
Footprint	1.3	1.1	5	47
Arizona	2.2	2.2	5.6	36
Nevada	2.1	1	10.7	54
Colorado East	2.4	2	5.7	68
New Mexico	2	3.1	3.5	70
Wyoming	1.3	2.7	8.7	33
Colorado West	1.8	3.1	7.3	100

for variability reserve and simplicity. Table 2 shows the requirements for the study footprint and for each region inside the study footprint.

6. Discussion

Deciding the level of flexibility reserve that is needed for VERs is a topic that will continue to receive interest for many years to come. In the United States there are different opinions regarding the need for these reserves. For example, MISO (Mid-continent Independent System Operator) has not changed its reserve policy, in spite of the 12,000 MW of wind in the market. MISO has developed a tariff for dispatchable intermittent resources that would call upon wind or solar generators to reduce output when chosen from the economic stack and when needed. The New York ISO also allows wind energy to bid into the market, and can therefore be dispatched downward on an economic basis when needed.

Thus far ERCOT has increased its regulating reserve somewhat, but the fast energy market extracts significant flexibility from the generation fleet based on economic bids. The California ISO is pursuing a new ancillary service, flexi-ramp, which functions similarly to a flexibility reserve such as those used for WWSIS and EWITS. With a few exceptions, it is widely recognized that flexibility reserve need is dynamic: when wind generation is at or near its minimum output level (zero), it cannot fall much further. Likewise, when it is at or near maximum, it cannot increase much further. In between, at approximately 50% of its output, the variability and uncertainty per unit appear to be at or near maximum values.

Wind turbines can provide regulation services [7]. In principle this function could eliminate the need for any additional regulating reserve to be acquired when wind power is added to the system. However, there are policy collisions associated with this capability: at present, wind plants in the United States earn a production tax credit that does not recognize the benefit of providing regulation (or other ancillary services) that would reduce energy output. In some areas market rules may, at least at present, not allow wind turbines or solar plants to provide some ancillary services.

7. Summary

Some regions in the United States have begun to change their operating reserve policies due to the increased amount of VG, but there is not widespread agreement on the method used to evaluate these reserves. ERCOT has begun to consider the possibility of wind forecast errors and the increased variability of wind in its up- and downregulation reserve and its non-spinning reserve. Many NERC regions have made enhancements to allow newer resources like demand response and energy storage devices to provide different types of reserves. Few regions in North America are allowing VG to provide any type of reserve. However, ERCOT does require wind power to provide over-frequency frequency response. Hydro-Quebec is also requiring wind power to provide an emulated inertial response. In many European TSOs, many of the offshore wind power plants are beginning to be asked to provide active power control as well. Many of the high-penetration systems, including Denmark and Ireland, are looking at ways in which wind power can assist in reliability by providing different types of operating reserves. Overall, many of these policies have been around for a number of years, and significant changes are just starting to be enforced. With higher penetrations, we may see this trend continue.

Higher penetrations of VER are being studied with increasing accuracy. Optimal operating reserve determination has advanced from the way that the current standards have been developed. It is generally agreed that increased VER will create an increase in the total amount of operating reserves held in power systems. The main questions are how much and of what type. Most of the studies broke out the reserve categories in the ways that they are broken out today in the NERC, UCTE, or other jurisdictions. With these in mind, all of the studies have determined that no additional contingency reserves will be needed, as total wind power loss from the wind resource in the time frame of contingencies is not a possible scenario.

A key conclusion of many of the more recent studies is that these operating reserve requirements need to be dynamic. Rather than keeping the same requirements at all times of the day and all days of the year, system operators should be using the information available to them to understand when the system is at low or high risk, and then schedule operating reserves accordingly.

Current policies around system operations have been developed after many years of experience of reliably operating a particular system and of input of the stakeholders of that system. Many of these systems have very small amounts of VER and therefore have not seen the need to adjust these methods and policies due to the increased variability and uncertainty of VG. However, most are very aware of the impacts that they will likely see in the future as VER penetration increases.

Many of the rules and standards of current operations use a static reserve requirement. Either a single reserve value is used for all the time or simple time-of-day rules are built in to different reserve requirements. Some may use single variables as input to their reserve rules (e.g. load in UCTE secondary control reserve requirement). However, many of the newer studies are determining that the variables that cause variability and uncertainty should be used in an intelligent manner to determine a dynamic reserve requirement.

References

[1] Ela E, Milligan M, Kirby B. Operating reserves and variable generation; August 2011. NREL/TP-5500–51978.
[2] NERC. Reliability standards for the bulk electric systems of North America; June 21, 2013.

[3] Kirby B, Milligan M. A method and case study for estimating the ramping capability of a control area or balancing authority and implications for moderate or high wind penetration. American Wind Energy Association; 2005. WindPower 2005, May.

[4] Kirby B. Demand response for power system reliability: FAQ. ORNL/TM 2006/565. Oak Ridge National Laboratory; December 2006.

[5] Kirby B, Hirst E. Customer-specific metrics for the regulation and load-following ancillary services. ORNL/CON-474. Oak Ridge (TN): Oak Ridge National Laboratory; January 2000.

[6] Hirst E, Kirby B. Allocating costs of ancillary services: contingency reserves and regulation. ORNL/TM 2003/152. Oak Ridge (TN): Oak Ridge National Laboratory; June 2003.

[7] Kirby B, Milligan M, Ela E. Providing minute-to-minute regulation from wind plants. In: 9th international workshop on large-scale integration of wind power; October 2010.

[8] Dany G. Power reserve in interconnected systems with high wind power production. In: Proceedings of IEEE porto power tech conference, Porto, Portugal; September 2001.

[9] Ela E, Milligan M, Parsons B, Lew D, Corbus D. The evolution of wind power integration studies: past, present, and future. In: Proceedings of Power & Energy Society general meeting, Calgary, Canada; July 2009.

[10] Enernex Corporation. Eastern wind integration and transmission study. Prepared for the National Renewable Energy Laboratory; January 2010.

[11] Energy GE. Western wind and solar integration study. Prepared for the National Renewable Energy Laboratory; May 2010.

[12] Gooi HB, Mendes DP, Bell KRW, Kirschen DS. Optimal scheduling of spinning reserve. IEEE Trans Power Syst November 1999;14(4):1485–92.

[13] Milligan M, Ela E, Lew D, Corbus D, Wan Y. Advancing wind integration study methodologies: implications of higher levels of wind. In: Proceedings of American Wind Energy Association, Windpower 2010, Dallas, TX; May 2010.

[14] Ortega-Vazquez M, Kirschen D. Optimizing the spinning reserve requirements using a cost/benefit analysis. IEEE Trans Power Syst February 2007;22(1):24–33.

[15] Smith J, Milligan M, DeMeo E, Parsons B. Utility wind integration and operating impact state of the art. IEEE Trans Power Syst August 2007;22:900–8.

[16] Wang J, Wang X, Wu Y. Operating reserve model in the power market. IEEE Trans Power Syst February 2005;20(1):223–9.

Advances in Market Management Solutions for Variable Energy Resources Integration

Xing Wang

Network Management Systems, Alstom Grid Inc., Redmond, WA, USA

1. Introduction

Increasing integration of Variable Energy Resources (VER) into the system has introduced new challenges to grid and market operations. This chapter provides an overview of wholesale electricity markets and market management systems. The intermittent nature of VER's increases the need for system ramping capability in real-time balancing market and causes issues in long-term market pricing and resource adequacy. Enhancements are being made in both market design and market analytical tools in terms of managing operational uncertainties introduced by VER integration.

This chapter also focuses on two areas of market enhancement. The first is the idea to establish a ramp market in real-time balancing operation to create the right market incentives for resources to provide sufficient ramping energy to compensate VER volatility. The second topic is to manage short-term VER uncertainty by applying robust optimization to look-ahead unit commitment.

2. Wholesale electricity markets and market management systems overview

Over the past two decades, power industry deregulation has been adopted internationally to create more open and competitive electricity markets, although the success mainly has been observed on the wholesale market side rather than on the retail side. The history and trend of global power industry restructuring can be found in [1] and [2].

Different from other economic systems, the complexity of power grids and the fact that electricity still cannot be stored in large scale make the design and operation of an electricity market a difficult task. The main principles of electricity market design include the following [2]:

- Establish fair and effective trading mechanisms for market participants. A fundamental character of any market is its products, or commodities, that are traded in the market. For the electricity market, electric energy is the dominant product. Hence, the design of the electricity market must establish fair and effective trading mechanisms for the suppliers and the consumers to trade electricity. This includes trading in spot markets where physical delivery of the products can be reliably achieved, as well as allowing forward bilateral contracts that are essential for overall

Renewable Energy Integration. http://dx.doi.org/10.1016/B978-0-12-407910-6.00008-9

market liquidity. Reliability is always the number one priority of power systems; therefore, an electricity spot market must take into account the security of the transmission grid. In addition to energy trading, an electricity market should also support the competitive provision of ancillary services, such as regulation and various system reserves.

- Establish open access to transmission services. To support a robust spot market for generation and demand resources, it is necessary to ensure that all market participants have open access to transmission services, such that the cleared generation offers and demand bids do not exceed available transmission capacity. Open access to the transmission system involves determining transmission capacity (total and available) and the methodology for designating to market participants the rights to use the capacity. There are two basic approaches to designating transmission rights: physical transmission rights (PTRs) and financial transmission rights (FTRs). Holders of PTRs are assured access to transmission according to the PTR terms. Holders of FTRs, on the other hand, are entitled to receive congestion credits according to the FTR terms.
- Coordinate between system operation and market operation. Physical power system constraints must be modeled explicitly in market operation, from forward markets to real-time markets. To achieve this goal, both the business processes of system and market operations and the corresponding software systems must be streamlined and harmonized.

Despite different market designs in different countries, wholesale electricity markets share a number of ingredients. Using the Locational Marginal Price (LMP)-based wholesale market design in the United States as an example, the following general business processes, shown in Figure 1, can be found in the operation of most market designs.

- *Capacity market.* Capacity market is one market-based approach to resolve the long-term resource capacity adequacy issue. It typically looks forward 3–5 years and is open to both generating resources (both conventional and renewable) and demand-side responses.
- *Financial transmission rights (FTR) market.* The FTR market provides an auction mechanism and a secondary market for market participants to trade FTR, which is a financial instrument for market participants to hedge against their risks because of potential congestion charges in the day-ahead (DA) market. The product in an FTR market is the right of using transmission capacity. The time window of an FTR auction market ranges from 1 month to 3 years.
- *Multi-day-ahead unit commitment.* The commitment of generators with long notification time and minimum up- or downtime ($\geq 24\,h$), such as nuclear and hydro, will be determined in a multi-day-ahead unit commitment process. The determined commitment will be stored in the current operating plan (COP) and will be treated as must-run in the DA market.
- *Day-ahead market.* The DA market is a financial market with physical transmission security and resource-operating limit constraints. The reasons that make DA market a financial market are (1) bid-in demands (fixed or price sensitive) from market participants are used in the power-balancing constraint of DA market clearing; (2) a DA market allows virtual bids or offers that are financial instruments to hedge the LMP differences between the DA market and the real-time (RT) market. DA market clearing co-optimizes between energy and ancillary services. It produces resource commitment, energy dispatch, regulation or reserve assignments, energy LMP, and reserve market clearing prices (MCP). Another important DA process after DA market clearing is the DA reliability unit commitment (DA-RUC), which commits additional resources to satisfy

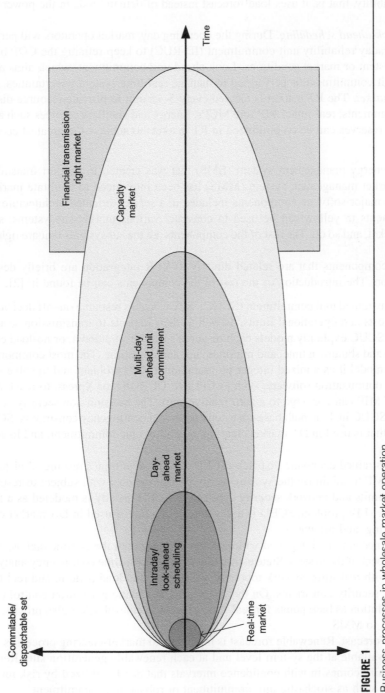

FIGURE 1

Business processes in wholesale market operation.

system reliability, that is, it uses load forecast instead of demand bids in the power-balancing constraints.

- *Intraday/look-ahead scheduling.* During the operating day, market operators will perform rolling forward intraday reliability unit commitment (ID-RUC) to keep refining the COP based on the latest system or market conditions. Look-ahead unit commitment, which aims at intrahour fast-start unit commitment, is performed to manage real-time system uncertainties.
- *Real-time market.* The RT market is cleared every 5–15 min to provide resource dispatch MW, reserve assignments, real-time LMP, and MCPs. Energy and ancillary services such as regulation and various reserves can be co-optimized in RT markets to achieve the greatest economic dispatch.

Similar to the energy management system (EMS) that was created to support transmission system operation, a market management system (MMS) has been introduced to facilitate market operation. Figure 2 shows major software components included in a service-oriented architecture (SOA)-based MMS. Components in yellow can be used to construct various market subsystems, such as a DA market, RT market, and so on. The rest of the components are the subsystems that are tightly integrated with MMS.

The MMS components that are related directly to VER integration are briefly described in the following section. The introduction of the rest of the components can be found in [2].

- Security-constrained unit commitment (SCUC). SCUC makes resource on–off decisions based on resources' costs and operational limits, as well as their impacts to transmission security constraints. SCUC explicitly models each resource's startup, shutdown, or no-load costs, energy cost, startup and shutdown time, and minimum up- and downtime. The most common method for SCUC is to model it as a mixed integer programming (MIP) problem and to solve it with commercial optimization software, such as CPLEX, GUROBI, or Xpress, to reach the global optimum (if MIP can converge to a zero relative gap). The transmission security constraints enforced in SCUC include both base-case constraints and contingency constraints. SCUC is a key application that is used in DA market clearing, reliability unit commitment, and look-ahead unit commitment.
- Security-constrained economic dispatch (SCED). SCED determines resources' dispatch MW as well as LMP to minimize the system energy production cost with subject to resources' operational limits and network security constraints. SCED usually is modeled as a linear programming (LP) problem. SCED is a key application that is used in DA market clearing, RT market clearing, and pricing.
- EMS. MMS needs to be integrated with EMS to consider both the commercial aspect and the reliability aspect of the system. State estimation (SE) and real-time contingency analysis (RTCA) feed MMS with real-time network topology, generation, and load pattern, and real-time transmission security constraints. On the other hand, automatic generation control (AGC) takes RT SCED solution as base points for real-time generation control. EMS also provides short-term load forecast to MMS.
- Renewable Forecast. Renewable forecast is input data to market-clearing processes. Renewable forecast is available at the system level and at each renewable generation site. Renewable forecast usually comes in with confidence intervals that can be utilized by risk management applications, such as stochastic unit commitment or robust unit commitment.

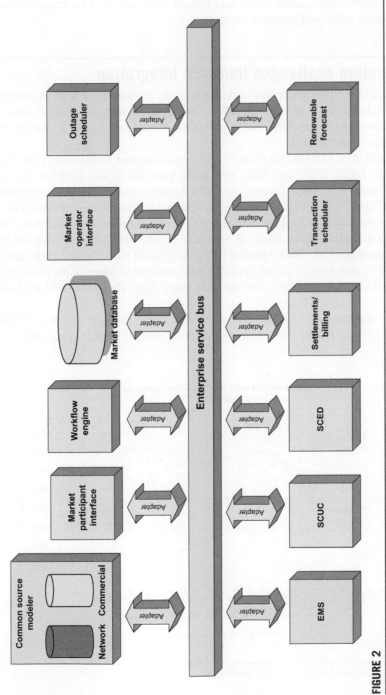

FIGURE 2

Market management system components and architecture. EMS, energy management system; SCUC, security-constrained unit commitment; SCED, security-constrained economic dispatch. (For interpretation of the references to color in this figure legend, the reader is referred to the online version of this book.)

Depending on the market design, some MMS may require only part of these components, whereas some MMS may require additional software components.

3. Market operation challenges from VER integration

Since the deregulation of the power industry started, electricity markets have been evolving at a rapid pace to address new challenges from both physical power systems and power economics. In recent years, increasing capacity of renewable generation has introduced new issues to market design and corresponding technical solutions [3].

Taking California ISO (CAISO) as an example, California adopted a renewable portfolio standard of 33% by 2020. In addition to large-scale renewable generation, the governor has called for 12,000 MW of distributed generation. Figure 3 provides the projected net load ramping requirements in January 2020 by CAISO. The net load curve after accounting for renewable generation shows a very different load pattern from today's load curve. A morning load pickup ramping-up event of 6700 MW in 2 h is followed by a ramping-down event of 7000 MW in 3 h, which then is followed by an even sharper evening ramping-up event. Such a pattern will require much more flexible generation capacity, which can be ramped up and down in a short period of time, which will increase the risks for CAISO to reliably operate its DA and RT markets. See the case study from California ISO by Loutan et al. (Chapter 6).

Another example of significant VER penetration can be found in the state of Texas. The integration of more than 10,000 MW of intermittent renewable resources into the market operated by the Electric

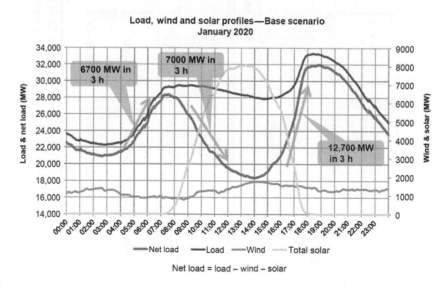

FIGURE 3

California ISO January 2020 load wind and solar profiles [4].

Reliability Council of Texas (ERCOT) has resulted in planning and operational difficulties, as well as economic impacts. Figure 4 was obtained from the ERCOT Website and shows the forecasted wind generation vs the actual wind generation on September 29, 2013. Although the latest shot-term wind power forecast (STWPF) matched the actual hourly average fairly closely, the DA wind power forecast can be way off, which introduces additional difficulties for ERCOT to operate its DA market and ensure system reliability via the DA-RUC process. See the case study from ERCOT by Dumas and Maggio in this book.

VER generation have caused many operational difficulties, including reactive power support, active power control, forecasting, scheduling, and determining the proper amount of reserves to ensure the reliability of grid operation. The main challenges from significant VER integration for the market design and operation are as follows:

- Impact on real-time system power balancing and pricing.
- System reliability issues with VER uncertainty in the DA market and intraday scheduling processes.
- Long-term resource capacity adequacy.

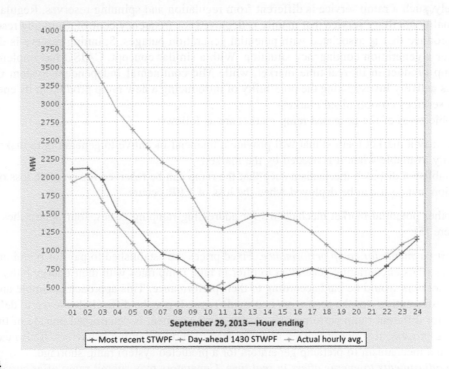

FIGURE 4

Electric Reliability Council of Texas (ERCOT) wind generation forecast and actual comparison on September 29, 2013. STWPF, shot-term wind power forecast.

4. Advances in market management solutions for VER integration

Advances in both market design and market management solution are needed to address the VER integration challenges discussed in the previous section. Although many recent market design enhancements for VER integration can be found in [1], we focus our discussion in real time and the look-ahead time frame with two topics. One topic is about establishing a real-time ramp market, and the other one is applying robust optimization to a look-ahead unit commitment process.

4.1 Establish a ramp market for VER integration

System ramp capacity is measured in today's system operation to cope with significant load fluctuation. Sufficient system ramp capability is the key to dispatch wind generation and other intermittent energy resources. In general, the generators try to avoid being dispatched up and down frequently, even if they can provide fast ramp energy, because of the concern of wear and tear. Therefore, there is a need to define a new ancillary service, called ramp service or load-following service, to provide economic incentive to the generators to offer their true ramp capability. Conversely, such a ramp service is different from regulation and spinning reserves. Regulation is a higher quality ancillary service that mainly is deployed for frequency control. Spinning reserve is a 10-min produce that is associated with potential generation outages. Spinning reserve is deployed only under a generation contingency situation. With a similar concept, CAISO has implemented a Flex Ramp product in its real-time market, while Mid-continental Independent System Operator (MISO) is actively investigating the possibility of introducing a new ramp product in its energy and ancillary services co-optimized markets.

The objectives of the proposed ramp market are as follows:

- To identify a market method that will provide a financial incentive to generators to make ramp capacity available when it is needed by the market.
- To establish a dispatch mechanism such that the ramp can be reserved for possible loss of VER situation and so it can be deployed when the loss of VER occurs.

Among other potential market mechanisms for addressing ramping issues, four approaches are discussed here [5]:

- *Ex post payments for ramp performance.* Fixed prices are established to pay for premium ramp services (thus encouraging generators to offer in high ramp rates with their energy bids), and the generators would be paid according to their performance. This approach is simple and provides incentives for the generators to provide ramp. It may be difficult, however, to define the prices for different ramp performances. As the ramp payment is determined outside of the clearing process, it is not optimized and may lead to certain gaming behaviors. It also cannot provide a mechanism to preramp generators for a predicted system ramp shortage.
- *Ramp adjustments to energy offers in real time.* Generators may submit ramp offer curves. The economic dispatch (ED) engine adjusts the energy offer curves in each interval based on the additional ramp cost required to move generators. With such a method, ramp procurement is optimized and can directly affect LMP. Although ramp is valued explicitly in the market-clearing

process, it does not consider load volatility beyond the dispatch target interval, that is, it does not help preposition units for upcoming ramp events. In addition, when there is no system-wide ramping shortage (which is the case for most intervals in a day), ramp offer prices may penalize the fast units during certain periods, for example, units being dispatched to relieve congestion.

- *Interval-coupled look-ahead dispatch.* This approach allows the market clearing engine to make better use of the available ramp capacity in the market over a longer look-ahead period of time than a single interval. By looking at multiple coupled intervals, however, the LMP in the dispatch target interval will be affected by future intervals for overall system optimality.
- *Adding the ramp product to energy and ancillary services co-optimization.* The procurement of energy and ramp service will be co-optimized along with the rest of the reserve products in market clearing to account for the lost opportunity cost (LOC). Ramp energy deployment will be part of the real-time dispatch process.

The co-optimization approach would serve the objectives of a ramp market if the system ramp requirement can be defined. Calculation of the ramp requirement will involve many things, including the following:

- short-term load forecast;
- net scheduled interchange (NSI);
- self-scheduled resource plan deviation;
- VER generation forecast;
- system demand adjustment entered by real-time operators.

Assuming that the system ramp requirement can be identified for the target dispatch interval, it would be possible to identify the LOC for resources whose energy dispatch must be held back to meet the ramp requirement.

Figure 5 shows that the ramp requirement changes from one interval to the next and is based on system load forecast, net scheduled interchange (NSI) forecast, and VER forecast.

Ramp requirement can be defined at reserve zone level as well as if there is a concern about the deliverability of ramp deployment due to transmission bottlenecks.

The shadow prices of the ramp requirement constraints are used to calculate the MCP for the ramp service, which include the LOC for the resources that are prepositioned to provide ramp. All resources that are dispatched out of merit to satisfy the ramp requirement will be paid with their LOC.

Deployment of reserved ramp service naturally will occur when there is a need for additional balancing energy in the target interval while the ramp requirements for future intervals decrease. For example, as shown in Figure 5, the ramping capability reserved for target interval $t + 5$ will be deployed for energy in target interval $t + 10$ because of the much-decreased ramping requirement for $t + 15$. Deployed ramping energy will be settled with the LMP at $t + 10$.

Further study shows that the single interval co-optimization of energy and ramp procurement and the multi-interval look-ahead economic dispatch essentially are related. In fact, it can be proven mathematically that if the system ramp requirement is defined as the difference between the load forecasts at two adjacent dispatch intervals, the single interval co-optimization of energy and ramp is equivalent to a multi-interval ED when its objective function includes only the target dispatch interval's production cost, that is, resources will be dispatched only out of merit in the target interval for reliability reasons.

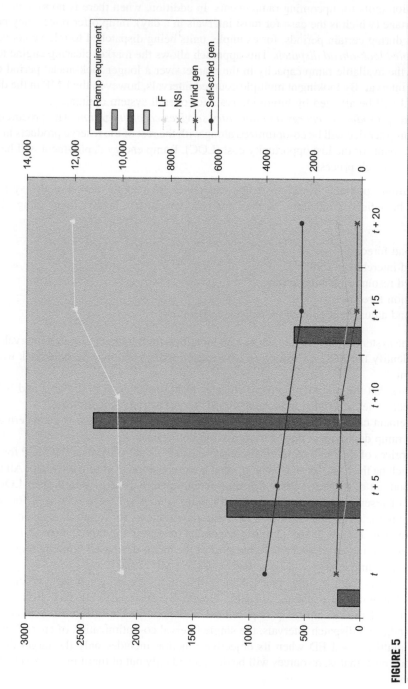

FIGURE 5

Illustration of the fluctuation of system ramping requirements. LF, load forecast; NSI, net scheduled interchange.

A simplified single-interval economic dispatch problem can be formulated as follows:
Objective function:

$$\text{Minimize} \sum_{r \in R} (P_{r,t} * C_{r,t} * \text{Duration}_t) \tag{1}$$

where r is the index for resources, R is the set of resources, $P_{r,t}$ is the MW output of resource r at interval t, and $C_{r,t}$ is the energy cost of resource r for interval t.

Subject to the system power-balancing constraint for distich target interval t:

$$\sum_{r \in R} P_{r,t} = \text{SysLoad}_t \tag{2}$$

System ramp requirement constraint:

$$\sum_{r \in R} (P_{r,t+1} - P_{r,t}) = \text{SysRampReq}_t \tag{3}$$

If we assume that

$$\text{SysRampReq}_t = \text{SysLoad}_{t+1} - \text{SysLoad}_t \tag{4}$$

we have:

$$\sum_{r \in R} (P_{r,t+1} - P_{r,t}) = \text{SysLoad}_{t+1} - \text{SysLoad}_t \tag{5}$$

With system power balancing equation at interval t, we have $\sum_{r \in R} P_{r,t} = \text{SysLoad}_t$. The ramp requirement constraint (3) then becomes

$$\sum_{r \in R} P_{r,t+1} = \text{SysLoad}_{t+1} \tag{6}$$

It is actually the power balance equation for interval $t+1$ that means the original single interval co-optimization problem for target interval t (Eqns (1)–(3)) has been transformed into a two-interval ED problem (with power-balancing equations for both interval t and interval $t+1$) that only minimizes the production cost of interval t (Eqns (1),(2),and (6)).

From the pricing perspective, the shadow price of constraint (6) represents the cost of an out-of-merit dispatch in the interval t because of the need to satisfy the system load at interval $t+1$, that is, the ramp capability required from interval t to interval $t+1$.

4.2 Managing VER uncertainty with robust and stochastic unit commitment

As introduced in Section 2, SCUC is a critical decision support tool for operators to ensure that the system is operating reliably and economically in both market and vertical utility environments. SCUC is used in various market-clearing and resource-scheduling processes, which all need to deal with operational uncertainties and risks, in particular when there is a significant amount of VER penetration. Four different approaches have been taken to address uncertainties in SCUC:

- Reserve adjustment approach, which handles uncertainties by applying conservative reserve requirements. Such an approach is used widely in control rooms; however, the solution is less economic, and it is difficult to determine and justify and right amount of reserve adjustment.

- Scenario-based deterministic approach, which allows operators to run several scenarios with various bias on load, VER forecast, and so on, and let operator make the final commitment decision. This approach also is used widely, but requires a lot of human experiences, which may lead to uneconomic, insecure, and suboptimal solution.
- Stochastic programming approach, which explicitly incorporates a probability distribution of each sample scenario for more economic solution when considering uncertainties. Practical limitations of this approach, however, include computational challenges and identifying the right set of sample scenarios as well as their accurate probability distributions.
- Robust optimization approach, which can provide an optimal solution that immunizes against all realizations of the uncertain data within a deterministic uncertainty set. Such an approach requires only limited information about the uncertainty, such as the mean and the range of the uncertain data. Although its solution is potentially conservative in terms of cost, robust optimization does ensure the reliability against system uncertainty [6].

This section presents an example of applying robust optimization to a look-ahead SCUC (LA-SCUC) problem [7]. LA-SCUC aims to develop with an optimal set of fast-start resource commitment decisions to help real-time operators prepare the system to deal with predicted events. LA-SCUC is a critical decision support tool for operators to manage real-time operational risks.

In the proposed two-stage robust optimization (TSRO) LA-SCUC formulation, the commitment decisions are the first-stage decision variables; the economic dispatch decisions are the second-stage variables and a function of the uncertain load. In the first stage, we enforce unit physical constraints (e.g. startup/shutdown, min up/downtime constraints, etc.). In the second stage, dispatch constraints (e.g. load balance, reserve limits, transmission line-flow limits, generation operating limits, ramping limits, etc.) are enforced.

The robust unit commitment problem can be represented by the following simplified formulation [7]:

$$\text{Min}_{x,y}\left(C^{T}x + \text{Max}_{d\in D}b^{T}y(d)\right) \tag{7}$$

Subject to the following:

$$Fx \le f \tag{8}$$

$$Hy(d) \le h(d), \forall d \in D \tag{9}$$

$$Ax + By(d) \le g, \forall d \in D \tag{10}$$

$$Iy(d) = d, \forall d \in D \tag{11}$$

where D represents the uncertainty set, and x and y are unit commitment decisions and economic dispatch decisions, respectively. To simplify the discussion, we only model load forecast uncertainty. VER forecast uncertainty can be modeled using the same approach.

Uncertain load forecasts are considered to be within certain ranges. Accordingly, a basic uncertainty set can be described as follows:

$$D_0 = \left\{ d_{it}: D_{it}^{l} \le d_{it} \le D_{it}^{u}, \forall t, \forall i \in N \right\} \tag{12}$$

where d_{it} are the uncertain load forecast values, and D_{it}^l and D_{it}^u are the corresponding lower and upper bounds.

A two-stage decomposition approach is used to solve the TSRO problem iteratively:

1. *Master problem:* unit commitment is considered to be the master problem in the decomposition framework. In the first iteration, we set the load at the nominal forecast level and solve a deterministic unit commitment to obtain the starting point for the whole algorithm. The solution from the master problem is used in the subproblem of the second stage. At the beginning of each iteration, the master problem is solved again with an additional set of cuts.
2. Subproblem: the subproblem aims to solve the economic dispatch problem under the worst-case load scenario with a fixed-unit commitment from the first stage. The solution of the subproblem discovers the worst-case scenarios, which are used to generate the cuts.

According to linear programming and duality theory, we can transform the subproblem from a max–min programming into a single-maximization problem. Readers are referred to (7)–(9) in [6] for this reformulating process. This maximization problem is a nonlinear programming problem composed of a bilinear objective function and linear constraints. To solve such a bilinear programming problem effectively for large-scale problems, we propose a bilinear heuristic algorithm as described in [8]. By fixing different sets of decision variables in the bilinear programming approach, we obtain the following two subproblems.

$$\text{SUB}^1: \max_{\varphi,\psi,\eta} \lambda^T (Ax - g) - \varphi^T h + \eta^T d^* \tag{13}$$

Subject to:

$$-\lambda^T B - \varphi^T H + \eta^T I = b^T \tag{14}$$
$$\varphi \geq 0, \ \lambda \geq 0, \ \eta \text{ free}$$

$$\text{SUB}^2: \max_d \lambda^{*T} (Ax - g) - \varphi^{*T} h + \eta^{*T} d \tag{15}$$

Subject to:

$$-\lambda^{*T} B - \varphi^{*T} H + \eta^{*T} I = b^T \tag{16}$$
$$d \in D$$

The heuristic algorithm to solve the bilinear programming approach is described as follows:

1. Pick an extreme point $d^* \in D$.
2. Solve SUB^1 with d^* and store the objective value as $\omega_1(y,d)$.
3. Solve SUB^2 with the dual variable value obtained from step 2, and store the objective value as $\omega_2(\varphi,\lambda,\eta)$.
4. If $\omega_2(\varphi,\lambda,\eta) > \omega_1(y,d)$, go to step 2, otherwise stop.

When applying this algorithm to solve real-world large-scale market unit commitment problems, a significant bottleneck is to build the dual-model SUB1. Such a dual model is vulnerable to any changes of the primal model with thousands of constraints and variables. To improve this bilinear heuristic algorithm, we design a modified algorithm to avoid formulating the dual model. Instead of solving

SUB[1], we consider the dual-model SUB[1], which turns out to be the original primal subproblem with fixed-load and first-stage decision variables:

$$DSUB^1: \quad \min_{y \in \Omega(x,d^*)} b^T y \tag{17}$$

where $\Omega(x, d^*) = \{y: Hy \leq h, Ax + By \leq g, Iy = d^*\}$.

Now, we can solve the DSUB[1] instead of SUB[1] in the second step of the algorithm. According to strong duality theorem, optimal objective values of DSUB[1] and SUB[1] should be equal. And it is easy to acquire the dual solutions (i.e. shadow prices of the constraints) from the optimization solver as the input for the third step. With this method, we avoid formulating the dual model, which is risky in a large-scale software system.

There is no guarantee that such a heuristic algorithm can obtain the optimal solution of the original bilinear programming. This algorithm, however, avoids solving the nonlinear programming and provides a near-optimal solution for large-scale problems.

We use a PJM (Pennsylvania-New Jersey-Maryland) LA-SCUC case to demonstrate the proposed TSRO approach. The case has four look-ahead commitment intervals and about 200 committable quick-start resources. In general, forecast (load, VER, etc.) becomes less accurate (i.e. deviation becomes larger) when looking further in the future. Thus, we assume that the positive–negative deviations of the nominal loads for four look-ahead intervals are 1%, 5%, 10%, and 15%, respectively. Figure 6 illustrates the forecasts and how uncertainties vary in four time intervals. "Norm" means the nominal loads; "high" and "low" are nominal loads with positive and negative deviations through all the intervals, respectively.

One practical strategy for the market operator to handle uncertainty in real time is to run multiple deterministic LA-SCUC with different scenarios and pick the commitment recommendations from one

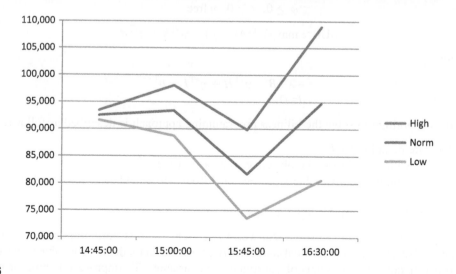

FIGURE 6

Illustration of the forecast deviation over commitment intervals.

Table 1 Comparison of Total Cost Between TSRO and Deterministic Scenarios	
Model Type	**Total Cost ($)**
Deterministic model (nominal demand)	7,110,998.91
Deterministic model (high demand)	8,115,738.09
Deterministic model (low demand)	6,363,717.40
Robust model	8,311,788.86

or more scenarios. On the basis of the uncertain load profile, the market operator usually prefers to look into "high" and "low" cases with the reliability concern, while the "norm" case is to consider system economics. We follow this deterministic process to run three scenarios (i.e. "norm", "low", and "high") and compare them with the TSRO solution (Table 1).

It can be observed that neither the high or the low are scenarious are the worst-case scenarios with this load profile. The robust optimization approach identifies the worst-case scenario (i.e. the worst ramping) and provides a more reliable solution to immunize against uncertainties. To better illustrate this finding, we compare the unit commitment solution of the robust optimization approach with those of "high" and "low" in Figure 7, where RO means the robust optimization; HL and LL mean high-load and low-load scenarios, respectively. One can observe that many units are shut down in the third period, and many of units are started up in the fourth period for the robust optimization approach. The reason behind this observation is that the worst-case scenario occurs when the system has the greatest ramping event between the third and fourth intervals. Such a dramatic ramp event calls for more startup and shutdown of the available fast-start resources and thus incurs the highest cost.

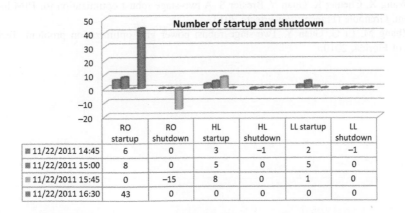

	RO startup	RO shutdown	HL startup	HL shutdown	LL startup	LL shutdown
■ 11/22/2011 14:45	6	0	3	−1	2	−1
■ 11/22/2011 15:00	8	0	5	0	5	0
■ 11/22/2011 15:45	0	−15	8	0	1	0
■ 11/22/2011 16:30	43	0	0	0	0	0

FIGURE 7

Comparison of number of startup and shutdown between two-stage robust optimization and high and low scenarios. RO, robust optimization; HL, high-load scenarios; LL, low-load scenarios.

5. Conclusion

Significant VER integration has presented tough but exciting challenges for market design and operations. One common effort is to improve decisions, ranging from real-time balancing decisions to long-term revenue and capacity adequacy decisions, to handle the much greater uncertainty from VER integration than in the past. The uncertainties always will be present, but they can be managed efficiently with improved market design and innovative decision support tools. It is expected that new market products, such as ramp product, and new optimization algorithms, such as robust unit commitment, will be adopted by the wholesale electricity markets to provide a reliable and competitive environment for VER integration.

References

[1] Sioshansi FP. Evolution of global electricity markets – new paradigms, new challenges, new approaches. Academic Press; 2013.

[2] Cheung K, Rosenwald G, Wang X, Sun D. Restructured electric power systems and electricity markets (Chapter 2). In: Restructured electric power systems: analysis of electricity markets with equilibrium models; 2010.

[3] Jones L. DOE report strategies and decision support systems for integrating variable energy resources in control centers for reliable grid operations – global best practices, examples of excellence and lessons learned; 2009.

[4] Rothleder M. Presentation titled flexible supply and renewable energy: solar and the impact on load curves. California ISO; August 2013.

[5] Wang X. Dispatch wind generation and demand response. Panel paper for IEEE PES general meeting; 2011. Detroit Michigan, USA.

[6] Bertsimas D, Litvinov E, Sun XA, Zhao J, Zheng T. Adaptive robust optimization for the security constrained unit commitment problem. Power Syst IEEE Trans 2013;28(1):52–63.

[7] Wang Q, Wang X, Cheung K, Guan Y, Bresler S. A two-stage robust optimization for PJM look-ahead unit commitment. Grenoble (France): Powertech; 2013.

[8] Jiang R, Zhang M, Li G, Guan Y. Two-stage robust power grid optimization problem. Technical report. University of Florida; 2010.

Electric Reliability Council of Texas Case Study: Reserve Management for Integrating Renewable Generation in Electricity Markets

John Dumas, David Maggio

ERCOT, Inc., Wholesale Market Operations, Taylor, Texas, Williamson

1. Introduction

The Electric Reliability Council of Texas (ERCOT) is the independent system operator that manages the electric power grid serving 85% of the load in the state of Texas. This corresponds to approximately 23 million electric customers. Within ERCOT's wholesale electricity market, there are more than 550 generation resources, a number of which use a renewable fuel source such as wind energy. At the end of 2012, the ERCOT system had more than 10.4 GW of installed wind-generation capacity. As a result, these resources accounted for 13% of the installed capacity and 9.2% of the energy used for that year. A summary of energy mix and generation capacity by fuel type for 2012 can be seen in Figure 1. Growth is expected to continue for wind generation and other renewable generation resources in the years to come.

As the independent system operator, a primary function of ERCOT is to balance electricity supply and demand. This task includes several time frames, including day-to-day, hour-to-hour, and even second-to-second. A key component in being able to do this is the procurement and deployment of ancillary services. Ancillary services are products that are provided by generation and demand-side resources within the market and are used to support the reliable transmission of energy across the system. System conditions requiring this type of support include errors in forecasted demand, unexpected loss of generation, and short-term changes in the amount of energy consumed. These ancillary services include regulation-up service, regulation-down service, and nonspinning reserve service, which will be a key focus of the case study.

Two attributes often associated with renewable generation resources are uncertainty and volatility. In the context discussed here, uncertainty is a lack of confidence in the amount of generation that is going to be available for some period of time in the future. Volatility is referring to short-term changes in a generator's fuel source, such as a decrease in wind speed. These characteristics can be concerning to system operators and make the job of balancing supply and demand more challenging. Several things can be done to reduce the potential reliability impacts of the additional uncertainty and volatility. This chapter will discuss steps that have been taken by the system operator and market participants in ERCOT that have aided in the integration of renewable generation resources with an emphasis on changes that have been made to the ancillary services.

Renewable Energy Integration. http://dx.doi.org/10.1016/B978-0-12-407910-6.00009-0

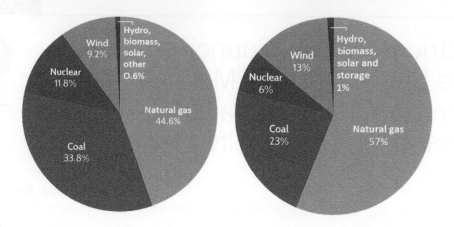

FIGURE 1

Percent of annual energy produced and installed generation capacity for 2012 by fuel type.

2. Study of the impacts of wind generation on ancillary service requirements

As an initial step in understanding the challenges of integrating renewable generation resources, ERCOT commissioned General Electric (GE) to perform a study of the impact of wind generation on ERCOT's ancillary service requirement.[1] AWS Truepower (AWST), LLC, formally AWS Truewind, was a large contributor to the work that was completed. This analysis primarily was performed in 2007, with a final report being provided to ERCOT and the stakeholders in March of 2008. During this time, there was a growing concern regarding what wind generation would mean to ancillary services in light of the rapid growth in installed renewable capacity created by the Competitive Renewable Energy Zone project.

The objectives of the analysis were to look at the current methodology being used to determine ancillary service needs, gauge the impacts of wind generation for various levels of penetration, and determine any enhancements to the methodology that may be appropriate. The scenarios studied included one case with five GW of installed wind capacity, two cases with 10 GW of installed wind capacity, and a more extreme case with 15 GW of installed wind capacity. To have a point of reference, there was a case with zero wind resources. The two 10 GW cases were selected to consider benefits that may have resulted from additional geographic diversity created by placing a larger percentage of the future-modeled wind generation resources in the Gulf Coast region, as opposed to the western portion of Texas.

Several conclusions and recommendations resulted from the analysis, and some key points were particularly noteworthy. One of these points was the concept of "net load" or the treatment of wind

---

[1]General Electric Company. Analysis of wind generation impact on ERCOT ancillary service requirements; March 2008 (On-line). Available: http://www.ercot.com/content/news/presentations/2008/Wind_Generation_Impact_on_Ancillary_Services_-_GE_Study.zip.

power output like negative load for the purpose of analyzing ancillary service needs. Another recommendation was for ERCOT to consider acquiring various types of wind power forecast products, including one that looks at potential variability with subhourly granularity. Section 3 will discuss the various wind power forecasts that ERCOT currently is using in system operations, due at least in part to these recommendations.

3. Wind power forecasts in ERCOT operations

Over the past several years, ERCOT has been working to incorporate wind power forecasting into system operations. Although there were several challenges in achieving this, the efforts have culminated into two distinct wind power forecasting products; an hourly forecast produced for a rolling 48 h period and the ERCOT Large Ramp Alert System (ELRAS).

3.1 Hourly wind power forecasts

The first step taken by ERCOT in regards to forecasting wind was the procurement of an hourly wind power forecast from AWST. The forecast provides hourly granularity and is updated each hour for a rolling 48 h window. The 48 h time frame is intended to line up with processes that occur in the day ahead, such as ERCOT's day-ahead market and reliability unit commitment studies. The forecast was first provided to the system operators in early 2008 and is still in production at this time.

Two separate values are provided for each hour; a value that represents a 50% probability of exceedance forecast (meaning that there should be a 50% chance that actual wind power will exceed the forecasted value) and an 80% probability of exceedance forecast. These forecasts are produced for the individual wind generation resources and at an aggregate level. The individual resource forecasts are provided to the companies that interact with ERCOT to schedule those resources and the aggregate forecasts are provided publically through ERCOT's Website. Figure 2 provides an example of the aggregate forecast for the rolling 48 h period.

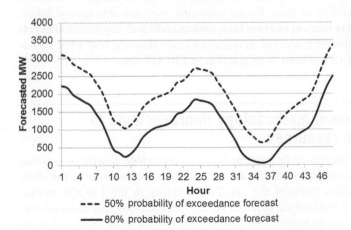

FIGURE 2

Example of the aggregate 48 h hourly wind power forecast provided by AWS Truepower in production at Electric Reliability Council of Texas.

A primary use for the hourly wind power forecast by ERCOT is the reliability unit commitment studies that are performed by the system operators. These studies are performed for each hour of the day both in the day ahead and continually throughout the actual day of operation and are designed to analyze concerns of having sufficient generator capacity to meet forecasted demand and whether projected transmission constraints can be managed with the resources expected to be available for dispatch. The software executing the process is using the 50% probability of exceedance forecast as the capability that can be expected from wind generation resource; however, the 80% probability of exceedance forecast is provided to offer additional situational awareness. To the degree that the forecast is wrong, this will create an inherent error in the studies being performed. This concern is addressed using ancillary services and is discussed in Section 4.

3.2 The ERCOT large ramp alert system

In addition to the hourly wind power forecast, it was determined that ERCOT should investigate the procurement of a forecast focused on providing system operators with information on the likelihood of a large event, where a large event is described as a significant change in wind power output over a relatively short period of time. Following the initial stages on integrating wind power forecasting into operations, ERCOT and many other forecast consumers began to recognize a weakness in the forecasts that they had asked vendors to produce. This weakness was that forecasts were tuned with the objective of minimizing error over some extended period of time, with the error typically being measured using mean absolute or root mean square error. The concern with a forecast tuned in this manner is that it will tend to "smooth out," or underforecast, extreme events to minimize the risk of forecasting an event that typically does not occur. As a result, ERCOT contracted AWST and worked together with them to develop a tool called ELRAS that could be used in concert with the hourly wind power forecast with the purpose of giving operators an improved global understanding of potential system conditions that could occur over the next few hours.

ELRAS was designed to look at the probability of a wind ramp event over three different time frames; the next 15 min, 60 min, and 180 min. For each of the three time frames, ramp events of various sizes were defined. For example, an event could be defined as the aggregate wind power output decreasing by 2000 MW or more during a 60 min period. Once these events are defined, the tool is designed to indicate to users the probability of these events occurring over the next several hours. Supplementary information is provided to users to increase their understanding of any potential events and conditions. This information includes such material as animated maps showing how wind speeds are expected to traverse the state. ELRAS has been in production in ERCOT since 2010.

4. Ancillary service requirement methodology improvements to integrate wind generation resources

As discussed, one expected outcome of increasing penetration levels of renewable generation was that it would increase the need for ancillary services. This likewise was seen as a conclusion of the study performed by GE. This section examines three of the ancillary services in the ERCOT market considering how the services are defined and how the requirement for those services has changed over time. Although other ancillary services are included in the ERCOT market, the three discussed in the following section have been affected most directly by the integration of renewable generation resources.

4.1 Regulation-up and -down reserve service

Regulation-up and -down reserve service is energy that is deployed by ERCOT in response to frequency deviations from the system's scheduled frequency. In the ERCOT market, these are treated as separate services and either are self-provided by market participants, traded bilaterally between market participants, or procured within a centralized market. Regulation-up reserve service refers to instructions that are at a level greater than a generator's economic set-point, whereas regulation-down reserve service deployments are instructions at a level less than a generator's economic set point. These services can be provided by demand-side resources.

Deployments are calculated every 4 s to restore frequency and to balance energy supply and demand between executions of the real-time security-constrained economic dispatch. As a result of the real-time market being run only every 5 min, system changes that occur over that period generally are made up using regulation service deployments. This concept is illustrated in Figure 3. The difference between expected and actual wind power output, signified by the grey shadowing, is what must be covered by regulation reserves, either up or down service depending on the direction in error. One additional point to take from Figure 3 is that other factors also are changing within the 5 min period. System demand is being shown in the plot, but these factors also include unexpected losses of generation or load.

Over the past several years, ERCOT has been making improvements to its methodology for determining the amount of regulation-up and -down reserve service that is required to maintain a

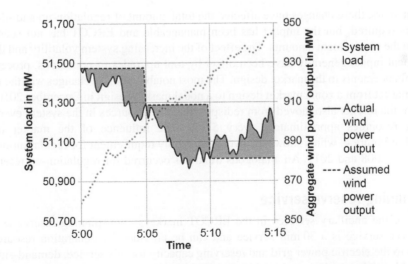

FIGURE 3

Example illustrating the need for regulation-up and -down reserve service in balancing changes in wind power output over a period of 5 min.

FIGURE 4

Average monthly regulation-down reserve requirement for the Electric Reliability Council of Texas market comparing years 2008–2012.

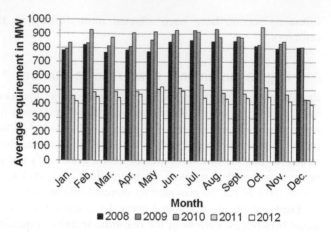

reasonable level of reliability considering potential impacts of renewable generation. These include the following:

- Historical data used for the analysis is adjusted to reflect the impacts of additional wind generation capacity that has been added to the system since the period of time from which the data are being taken. This process is being performed as a direct response to conclusions from the study performed by GE and uses the results of their analysis [1].
- Inputs for the methodology process have been increased to include historical 5 min net-load changes to determine regulation-up and -down reserve service. Net-load is defined in this context as system-wide demand minus aggregate wind power output.

During recent years, these changes have affected the total amount of regulation-up and -down reserve service that is required, but the impact has been manageable and ERCOT has not seen any major concerns with the requirement amounts. The effect of the increasing system volatility and the inclusion of the additional inputs generally have been offset by tool upgrades in operations, process improvements, and enhancements to the market design. The most notable of these changes was the transition of the ERCOT market from a zonal market design to a nodal market design in December 2010. As part of this transition, the market also moved from redispatching the resources in the system every 15 min to redispatching resources approximately every 5 min. The influence of the market transition is demonstrated in Figure 4, which shows the average monthly requirement for regulation-down reserve service between 2008 and 2012. An analogous trend has occurred for regulation-up reserve service.

4.2 Nonspinning reserve service

Another one of the ancillary services in the ERCOT market is nonspinning reserve service. Nonspinning reserve service is a 30 min service and can be provided by generation resources that are synchronized to the electric power grid and reserving capacity for this service, demand-side resources that are capable of being disrupted within the 30 min time frame, and generation resources that currently are not synchronized to the electric power grid but could be able to provide energy and follow dispatch instructions within 30 min. Like regulation-up and -down reserve service, nonspinning

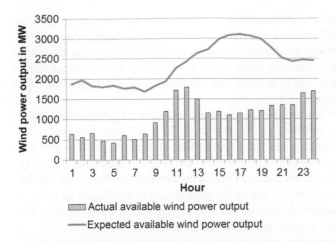

FIGURE 5

Example of wind power forecast error illustrating the need for nonspinning reserve service.

reserve service is either self-provided by market participants, traded bilaterally between market participants, or procured within a centralized ERCOT market for ancillary services.

Nonspinning reserve service is a product that can be thought of as resource capacity that can be used to replace other reserves or capacity that was expected to be available but is now absent. Reasons for deploying the service include the unexpected loss of a generation resource or uncertainty created by errors in the wind power forecast used by system operators. The potential impacts of the wind power forecast uncertainty is illustrated in Figure 5. This forecast error results in false assumptions for system operators performing reliability unit commitment studies and may mean that nonspinning reserve service must be used to ensure that supply and demand can remain reliably balanced.

Before having a significant amount of wind power generation on the system, the requirement for nonspinning primarily was driven by the potential loss of the single-largest generator on the power grid within ERCOT. After reconsidering the uses of the service and evaluating the uncertainty that can exist as a result of demand and wind power forecast errors, the methodology has been enhanced to directly use historical forecasts errors for demand and wind power as an input for the analysis performed to determine future requirements.

The most notable impact that the methodology changes have had on the nonspinning reserve requirement is the hours of the day in which the service is being procured. Under the old process, concerns largely were focused on the peak demand portions of the day, for example, the middle to late afternoon on a hot summer day in Texas. The observation, however, was made that this is not necessarily the time in which the forecast uncertainty is at its greatest.

5. Additional actions and future considerations

Although the steps taken to incorporate wind power forecasting products and improve the processes for determining ancillary service requirements have been highly successful, other changes and market design features have played important roles in integrating renewable generation resources. One example of such a change is the inclusion of renewable generation resource dispatch as part of the optimization for security-constrained economic dispatch. Developing the technology and rules to apply this concept took a considerable amount of time and effort; however, the approach creates a more efficient market result and reduces potential burden on other generation resources. A key component in accomplishing this was the ability of renewable generation resources to update information about the renewable generation resource on a second-to-second bases, such as maximum power output capability as a function of changing wind speeds, and to provide that information to system operators for consideration in the economic dispatch engine.

Significant benefits can be realized from having renewable generation resources participating in the market and providing services in a matter similar to other resources with differing fuel types. An example of this within ERCOT is a rule that requires wind generation resources to provide primary frequency response. Primary frequency response can be described as the near-instantaneous change in resource power output resulting in a deviation in frequency. This immediate reaction helps contain the frequency to give time to other systems to respond, including the deployment of regulation-up reserve service or the execution of economic dispatch. Resources are not required to reserve capacity for primary frequency response unless they also provide ancillary services.

ERCOT has been analyzing future resource mix scenarios to investigate any potential insufficiencies in its current levels and processes for determining ancillary service requirements. Using funding from the U.S. Department of Energy, ERCOT staff, in collaboration with KEMA, has initiated a study looking at these concerns.[2] Performance of the existing ancillary services is being measured, focusing on frequency control performance looking at both statistics in the study model's frequency deviations and North American Electric Reliability Corporation (NERC) Control Performance Standard 1 (CPS1) requirements. Preliminary results of this effort indicate that the ancillary requirements and processes need to be modified to manage the volatility growth introduced by increased levels of renewable generation penetration.

Development of renewable generation resources likely will not be the only source of additional system volatility. Electrical vehicle charging or retail market programs designed to encourage electricity consumers to change current behavior also may result in energy demand patterns and fluctuations that are quite different from what currently is being observed by system operators. The impacts of these factors may be relatively small at this point in their evolution; however, these are just a couple of the several factors that engineers will have to consider when planning for the future power grid.

The probable outcome of increased ancillary service needs raises the question of who is going to be providing these products. The types of generation resources currently providing ancillary services will continue to be important in ensuring that the market is able to meet the system reliability needs. As market conditions change, it may become economically feasible for renewable generation resources to begin being offered in the ancillary service markets. These changes may provide opportunities for emerging technologies to play a more significant role. Examples of this are energy storage technologies, such as battery storage units or compressed air energy storage (CAES) or demand-side resources, like aggregations of residential air conditioning units. ERCOT will continue to actively analyze all the potential concerns on the power grid and within the ERCOT market and pursue opportunities to address those risks through emerging technologies and new market products.

---

[2]Matevosyan J. Ancillary services adequacy study. Presented at the 2012 Great Lakes symposium at the Illinois Institute of Technology; September 2012 (Online). Available: http://www.iitmicrogrid.net/event/greatlake2012/publication/PPTs/9-Transmission%20Planning%20Issues%20for%20Variable%20Energy%20Resources/9-GreatLake2012_Matevosyan.pdf.

Case Study: Grid and Market Integration of Wind Generation in Tamil Nadu, India

10

Anish De, Puneet Chitkara
Mercados EMI Asia, New Delhi, India

1. Background

1.1 Penetration of variable renewable energy sources in India

Renewable energy in India has grown at a fast pace in the past few years, led by wind, with substantial contributions from biomass, small hydro, and, more recently, solar energy. Wind energy deployment is now two decades old in Tamil Nadu and other states in South India. When compared with the other Indian states, Tamil Nadu has considerable experience in operating a power system with variable renewable energy (VRE) resources. The installed capacity of wind in Tamil Nadu is around 7149 MW, compared with a total wind generating capacity in India of 17,365 MW [1]. Wind energy currently contributes 40% of Tamil Nadu's installed generating capacity.

Going forward, the Government of India (GoI) has set out an ambitious plan that proposes to rapidly expand installed capacity of VRE. GoI's aspirational targets for wind and solar capacity in India is 45,000 MW and 20,000 MW by 2022, respectively [2]. These aspirations, however, will be strongly challenged by realities on the ground that, unless addressed, likely will cause stagnation or even deceleration of the capacity addition pace that the country has witnessed in the past decade.

Wind generation in India exhibits strong seasonal traits and usually is concentrated between May and September. It is anticipated that significant accelerated growth in the deployment of VRE and, in particular, wind and solar energy will occur in the short to medium term as evidenced by escalating targets and the amount of activity in the market in on-shore and off-shore wind energy. The growth of VRE generation is not fueled solely by carbon concerns. State governments in India in wind resource–rich states have tended to set high renewable portfolio obligations (RPOs) for themselves, because in these states, the electricity utilities and the large industrial customers see a reliance on wind- and solar-based generation as "economically efficient" from a long-term perspective. It is, however, important to observe that the pace of development of wind-based generation in Tamil Nadu has slowed despite the state having the maximum untapped wind potential. Of late, even other states like Gujarat and Rajasthan that have lower VRE penetration levels than Tamil Nadu have pointed out the challenges in further expanding VRE-based generation. The raison d'être for increasing inhibitions of the states is rooted in the institutional, technical, and regulatory frameworks that govern the power sector in India, as discussed in this chapter.

Renewable Energy Integration. http://dx.doi.org/10.1016/B978-0-12-407910-6.00010-7

1.2 Institutional and regulatory framework for VRE

In India's federal constitutional structure, the states and national governments have specified legislative and administrative powers. Electricity is on the concurrent list of the constitution, that is, the state governments are entitled to enact their own statutes and govern the electricity sector operations in their respective states, while abiding by the applicable national statutes. The power sector in each state is managed largely by the state-owned utilities—in generation, transmission, and distribution—and is regulated by the State Electricity Regulatory Commission (SERC). Interstate assets and transactions are regulated by the Central Electricity Regulatory Commission (CERC). Within the control area[1] of a state, the responsibility of operating the power system in a secure and reliable manner rests with the State Load Dispatch Centre (SLDC). Each SLDC maintains a schedule of its interchange with other control areas through its Regional Load Dispatch Centre (RLDC), which is a central government entity responsible for operating the interstate transmission system (ISTS) within a region in a secure and reliable manner. The Indian power system, therefore, operates as a loose pool, where each SLDC schedules and dispatches its state's generation resources and demand and requisitions for withdrawal from or injection into the ISTS from the RLDC based on long-term, medium-term, or short-term contracts (which may be bilateral or through a power exchange) that the state utilities may have with interstate generation facilities, including independent power producers or traders. *Therefore, "balancing" power for all VRE-based generation in a state is required to be managed by the SLDCs. Currently, no commercial mechanism allows for recovery of the costs of balancing from either the intrastate generators causing such imbalance or the beneficiaries of energy from such power plants.* On the contrary, under the frequency linked unscheduled interchange (UI) rate for the settlement of deviations, the host utility becomes the risk bearer for such changes. The net UI impact on the utility is substantial.

The simplistic tariff arrangements and banking arrangements prevalent in the state are out of sync with the complexities of operations in the Indian power markets. In real time, when frequency is expected to increase and UI rates are expected to be less than the Feed-in Tariff (FiT) rates, the SLDCs tend to curtail wind generation. The UI rate, being a regulated rate, may not match the "market value" of energy at that time. Therefore, this conflict between UI mechanism, which essentially was designed for conventional systems, and the FiT mechanism causes considerable "spillage of wind" during high wind seasons.

1.3 Case study of wind integration in Tamil Nadu

AF-Mercados EMI conducted a survey of wind farms in Tamil Nadu to identify the micropractices in VRE management in the state. The diagram represents the grid network in the Udumalpet area of Tamil Nadu where the survey was conducted. Udumalpet is one of the wind generation hubs in Tamil Nadu (Figure 1).

The intrastate transmission network in Tamil Nadu is weak and hence balancing resources to manage wind variability, in most instances, cannot be procured from balancing resources at far off locations. This location allows balancing wind with limited pumped storage–based hydro generation and, to some extent, by cycling coal-based generation.

---

[1]An electrical system bounded by interconnections (tie lines), metering, and telemetry that controls its generation or load to maintain its interchange schedule with other control areas whenever required to do so and contributes to frequency regulation of the synchronously operating system.

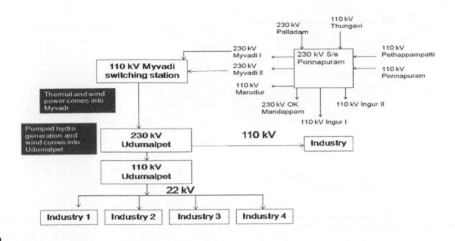

FIGURE 1

Network in Udumalpet.

The region is an industrial hub. Industrial consumers, with adequate incentives, can be useful providers of demand response (DR) reserves. DR ramp rates can match the ramp rates of wind generation.

2. Analysis: Implications of wind integration for the host state

2.1 Identification of various costs of grid integration of VRE-based electricity generation

The costs of grid integration of wind-based power in Tamil Nadu can be analyzed in terms of the following:

1. The cost of running tertiary resources, such as Kadamparai pumped hydro;
2. The cost of UI withdrawals from the ISTS network; and
3. The cost of load shedding.

The Tamil Nadu SLDC resorts to ramping of tertiary reserves, such as Pumped Hydro Power Plants (e.g. the Kadamparai near Udumalpet area), in the event of loss of wind generation. This is illustrated in Figure 2. The cost of such plants is borne by Tamil Nadu, and there is no mechanism for its recovery from either the wind generator or the beneficiary of the power generated or the green attribute.

There is considerable variation in generation profiles even on two consecutive days. To balance the state demand and supply, the SLDC also deviates from its schedule given to RLDC. This is illustrated in Figure 3, based on actual operations data for the state on two consecutive days.

Figures 2 and 3 demonstrate the following:

1. On August 17, 2011, when the wind was low, the actual withdrawal from the ISTS exceeded the schedule, and the state paid heavy unscheduled interchange charges.

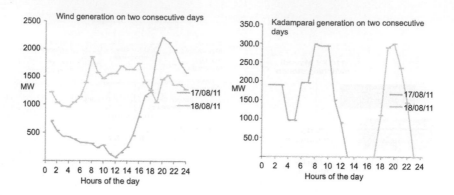

FIGURE 2

Wind generation on two consecutive days.

2. On August 18, 2011, in anticipation of low wind, the state seems to have given a higher schedule in the ISTS network. The wind generation on August 18, 2011, however, was more than expected, and hence the state had to underdraw with respect to the schedule. The UI regulations, in 2011, provisioned that if a state underdraws by more than 250 MW, the rate of compensation for the extent of the underdraw would be in the range of Rs 1.50/kWh, which was much lower than the FiT rate. The state, therefore, again has to suffer a loss.

As may be observed from these figures, the loss of generation due to wind cannot be sufficiently compensated by ramping up Kadamparai because the power plant is limited by its capacity or by

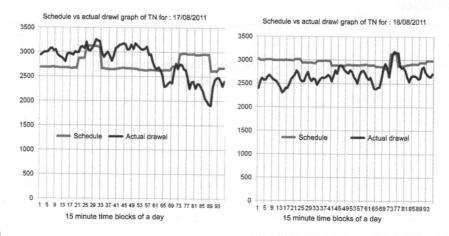

FIGURE 3

Actual withdrawal exceeded scheduled withdrawal considerably on August 17, 2011, and August 18, 2011.

overdrawing from the ISTS because this is a costly proposition. The state, therefore, resorts to load shedding (i.e. disconnection of consumers). This is illustrated in Figure 4.

The cost, in this case, is carried by the consumers of the state, and the loss of revenue from sale of power is carried by the state utility.

2.2 The problem of "discarded wind"

The host utility in which the wind generator is located has no incentive to continue to allow the wind generator to generate if the UI prices are below either the FiT rates or the regulated prices in the case of the other two commercial mechanisms. Thus, in some instances, the wind generator could be instructed to reduce generation even when the frequency is below 50 Hz, let alone instances when the frequency exceeds 50 Hz. Instances of such behavior were captured through an analysis of the Palladam S/S data for high wind months.

As frequency increases, and if at the same time wind generation also increases, the utility is faced with an operational decision that is influenced by financial considerations. When the frequency starts increasing but is still below 50.2 Hz, and the UI rates are lower than FiT rates or contract rates, then SLDCs are inclined to back down wind generators, because unscheduled off-take from the regional grid is commercially more advantageous. The feeders for which the following graphs are presented have significant load. The negative values indicate that the wind generation was curtailed, and there was a net load on the feeder. This is seen to be happening in most cases when frequency was greater than 49.5 Hz—where the corresponding UI rate starts approaching the FiT rate (Figure 5).

Such behavior could have been avoided by altering the UI mechanism or adopting an alternative institutional mechanism as proposed subsequently.

2.3 Why does neighboring SLDC not help when wind generation declines?

During high frequency periods, the home state pays FiT for wind power, but states that overdraw pay very little UI charge. During low-frequency periods, if wind power suddenly reduces, the burden of UI falls on the home state where the wind is integrated—other states who may have balancing resources will not help because they are negatively affected by lower off-take (which will happen under the UI framework if the other state with balancing resources generates more from its own sources and hence draws less from the ISTS grid). For example, when wind declines in Tamil Nadu and frequency is low,

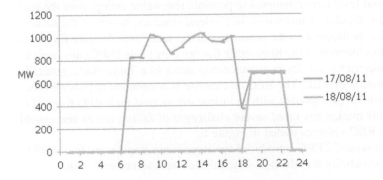

FIGURE 4

Lower wind generation on August 17, 2011, made up through load shedding.

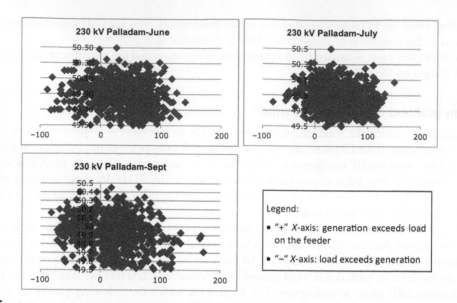

FIGURE 5

Generation–load vs frequency at the 230 kV Palladam substation.

and neighboring Andhra Pradesh is drawing as scheduled, it needs to be incentivized to ramp up its hydro- and gas-based power plants. Under the current UI mechanism, Andhra Pradesh is effectively penalized if an increase of such in-state generation results in lower off-take from the grid by more than 10% or 250 MW (whichever is less). Unless these incentive issues are addressed, the VRE resource-rich states would be disinclined naturally to carry the costs of propagation of VRE beyond a certain level, and other states capable of assisting the resource-rich states would avoid doing so.

3. Regulatory and policy measures to encourage grid and market integration of VRE generation

Policy and regulation at the national level have attempted to promote renewable energy over the past decade through a range of fiscal and regulatory incentives. In a federal structure, however, the success of India's renewable energy initiatives depends on the economics at the state level. Despite favorable FiT rates, incentives to state utilities inherent in the Renewable Energy Certificate (REC) mechanism, and other financial devolutions, electricity utilities in resource-rich states like Tamil Nadu generally have become opposed to the expansion of VRE. The REC mechanism was intended to provide states that are not endowed with their own VRE generators with an instrument to meet their RPO from REC supply in the national market. This market has faced severe challenges of falling prices and unsold REC stock, as illustrated through REC volumes traded in Figure 6.

To partly correct the incentive issues, CERC introduced the Renewable Regulatory Fund (RRF) mechanism to offset the costs of variability that the host state utilities have to contend with.

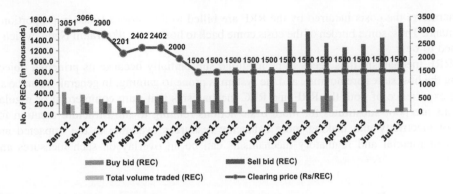

FIGURE 6

Past trend of the REC market at Indian Energy Exchange (IEX) (nonsolar).

3.1 The RRF mechanism

As discussed in elsewhere in this book, the RRF mechanism was implemented through the order of the CERC dated July 9, 2013, titled "Procedure for the Implementation of the Mechanism of Renewable Regulatory Fund" [3]. The mechanism was intended to insulate the host state from bearing the costs of variability through socialization (up to 30% deviation from schedules) and attribution (generators to pay the costs beyond 30% variation from schedule). The specific formulation introduced has its defects. An analysis of the regulations indicates that the wind generators benefit from incorrect schedules. This is observed in practice through the following data collected for the wind farms in Tamil Nadu, which indicates regular overscheduling [4] Figure 7.

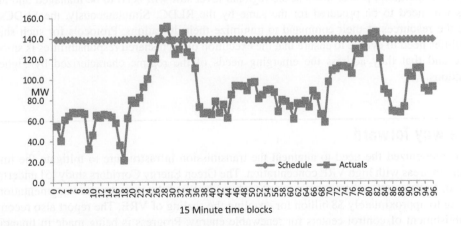

FIGURE 7

Snapshot (sample) of real-time wind scheduling data in Tamil Nadu.

Furthermore, the costs incurred by the RRF are billed to the various states in proportion to their peak demand. Thus, some burden of the costs come back to host states like Tamil Nadu, which they are disinclined to bear.

The RRF mechanism thus needs to be modified substantially because its principal objectives of insulating the host states are not met, and the system is prone to gaming. In general, it can be surmised from the experience of both the RRF and REC mechanisms that pure commercial or regulatory arrangements are poor substitutes for more basic actions relating to infrastructure creation for transmission of electricity and making more balancing resources available to the impacted areas and entities. Commercial and regulatory mechanisms can be an overlay on such measures and not a substitute.

3.2 Deviation settlement mechanism, ancillary services market, and forecasting requirements

The Indian electricity grid is already one of the largest synchronized grids in the world. The issues of VRE integration on a large scale must be addressed in conjunction with the other issues involved with management of the overall power system. To cater to the changing requirements of the grid and new mechanisms such as the deviation settlement mechanism, CERC intends to have an ancillary services (AS) market in which reserves for balancing the grid in real time can be procured. This, however, would require the states and wind generators alike to forecast demand (and net load) and generation, respectively. Forecasting codes, a prerequisite for grid and market integration of VRE, are yet to be developed by the CERC or the SERCs.

In addition to AS markets and forecasting codes to meet the requirements of the deviation settlement mechanism that the direction of ACE be reversed every six time blocks (1.5 h), it will be an imperative that various control areas (of different states) act in concert to manage the grid. Although in the federal structure of India grid management responsibilities are shared between central and state-level system operators, power flows at the regional level still will need to be managed and ancillary services will need to be procured for the same by the RLDC. Simultaneously, the SLDCs would manage the resources at their command to minimize their deviations. Protocols for such shared responsibilities need to evolve to ensure that the execution of such shared responsibilities is smooth and effective and that they address the emerging needs of the regime characterized by higher VRE contributions.

4. The way forward

India has recognized the need to augment the transmission infrastructure to mitigate the impact of variability in areas with high VRE concentration. The Green Energy Corridors study [5] undertaken by Powergrid, India's national transmission utility, has identified the infrastructure augmentation needs amounting to approximately $8 billion for enabling the scaling of VRE. The report also recommends the establishment of control centers for renewable energy. Progress is being made in financing and implementation of these recommendations.

In contrast, the provision of balancing power or mitigating the impact of variability through the provision of complementing energy supplies has received lesser attention in India. As this chapter has discussed, the traditional definition of control areas under the federal structure is a serious impediment to effective balancing resource sharing. This can be overcome through a variety of measures that can encompass physical sharing of quick-response balancing resources through bilateral agreements or markets, or through a combination of these means, essentially for enlarging the control areas.

The mechanisms must induce the physical control areas (defined by state boundaries at the lowest level) to act in a cohesive manner to provide balancing resources for intermittent sources of power like VRE. It is important to radically redesign or disband the UI mechanism, which, having laudably performed its role until now, has become a serious impediment for VRE integration and expansion.

India has been emphasizing decentralized forecasting and scheduling for management of VRE. As the RRF experiment has demonstrated, this is inadequate. For better integration of VRE-based generation in electricity grids and markets, it is important that such resources be scheduled and forecasting be done both at the centralized and decentralized levels. A framework for forecasting and planning needs to be established. This should allow for the following:

1. Ability to develop an accurate real-time production forecast for any particular wind generator strongly correlates to the availability of site-specific and precise real-time data;
2. RLDCs and SLDCs to obtain accurate forecasts of site-specific meteorological data and renewable energy production to maintain reliable and efficient system operation;
3. Commercially responsive forecasting to be done at the substation level by scheduling coordinators or aggregators of renewable energy; and
4. Centralized forecasting to be undertaken by the RLDCs and SLDCs along with decentralized forecasting at the wind farm or pooling station level.

In a federal structure like India, with electricity being often at the fault lines between central and state-level responsibilities and actions, the implementation of these measures will face inevitable challenges. India needs to evolve an acceptable framework for network management that allows for the renewable energy–related goals to be met without infringing on the division of responsibilities under the federal structure.

References

[1] http://www.inwea.org/ [accessed 30.08.13].
[2] Ministry of New and Renewable Energy, Government of India. Strategic plan for new and renewable energy sector for the period 2011–17. http://mnre.gov.in/file-manager/UserFiles/strategic_plan_mnre_2011_17.pdf, February 2011.
[3] Central Electricity Regulatory Commission, New Delhi. http://posoco.in:83/Flasher/CERC%20Order%20for%20implementation%20of%20RRF%20Mechanism%20from%2015.07.13.pdf, July 2013.
[4] Southern Regional Power Committee, Bangalore. http://www.srpc.kar.nic.in/html/index.html Data for July 15, 2013 [accessed 10.09.13].
[5] Power Grid Corporation of India Limited. Transmission plan for envisaged renewable capacity: a report, vol. 1; July 2012.

Forecasting Renewables

Forecasting Renewable Energy for Grid Operations

11

Audun Botterud

Decision and Information Sciences Division, Argonne National Laboratory, Argonne, USA

1. Introduction

Renewable resources create new challenges in the planning and operation of the electric power grid. In particular, the variability and uncertainty in the renewable resource availability must be properly accounted for in the complex decision-making processes required to balance supply and demand in the power system. It is becoming increasingly evident that forecasting is a key solution to efficiently handle renewable energy in power grid operations. In this chapter, we first discuss the potential applications of renewable forecasting for grid operators and electricity market participants. We then give a brief overview of the current methods used to forecast wind and solar power, with emphasis on recent developments to tailor forecasting products to the specific needs in the power grid. Finally, we take a visionary look ahead and discuss opportunities and challenges for forecasting and its grid applications on the road toward future power systems dominated by renewable energy.

2. Forecast applications in grid operations

In this section we discuss potential forecasting applications for grid system operators, renewable energy producers, and other market participants. The different entities are all seeking improved forecast accuracy, although what constitutes a good forecast oftentimes depends on the specific forecast user [1]. In the discussion below, we briefly describe current use of renewable forecasting and look at possible future innovations, particularly related to the use of probabilistic forecasts. Uncertainty management is a key challenge in grid operations, and it is our opinion that probabilistic forecasts will play an increasingly important role toward this end as the penetration of renewable energy continues to increase.

2.1 Grid system operators

System operators are responsible for balancing supply and demand in the power grid on a continuous basis, accounting for expected and unexpected situations. Forecasts of electric load have played an important role in grid operations for many years, but the introduction of forecasts for renewable energy is more recent. Today, system operators in regions with significant penetration of wind and solar power

Renewable Energy Integration. http://dx.doi.org/10.1016/B978-0-12-407910-6.00011-9

typically have their own forecasts, sometimes multiple forecasts for improved accuracy and information, to guide their decision-making processes. In the longer operational time frame, from days to weeks, system operators may want to schedule maintenance of transmission lines during periods of low-forecasted renewable energy generation. In the daily operations of electricity markets, from day-ahead (DA) scheduling to real-time (RT) dispatch, renewable energy forecasts can play multiple roles, as discussed below.

Operating reserves are an important measure for system operators to address the uncertainty and variability in grid resources. Traditionally, system operators have determined how much operating reserves are required to maintain system reliability based on deterministic heuristics and static rules (e.g., the N-1 rule). However, with the large-scale introduction of renewable energy, there is a shift toward more dynamic assessments of operating reserve requirements that consider the state of the renewable resources in the system. Forecasting can clearly play an important role in these assessments. As an example, in the United States, the Electricity Reliability Council of Texas (ERCOT) has for several years adjusted their operating reserve requirements on a monthly basis based on historical forecasting errors [2]. In the industry and research communities, a number of potential solutions have been proposed to properly account for the impact of renewables, their variability, and forecasting errors in the determination of operating reserves and flexibility needs in system operation [3,4]. In addition to dynamic reserves, probabilistic assessments of reserve requirements take on higher importance with renewable energy forecast uncertainty [5,6]. Moreover, the use of probabilistic demand curves for reserves has been proposed to address wind power forecast uncertainty and to better reflect the marginal value of operating reserves for system reliability in pricing and dispatch [7].

Bids and offers from consumers and suppliers of electricity, along with the operating reserve requirements, are considered in the clearing of the DA market to produce a DA schedule and corresponding prices for energy and reserves. Currently, the system operator's forecast is usually not used as an input to the market clearing, but it may influence the operating reserve requirements, as discussed above. The expected amounts of renewable generation should be reflected in the offers from wind and solar generators. However, it has also been suggested that the system operator's own forecast could be used directly in the DA market clearing. In this case, a probabilistic forecast, typically represented as scenarios, could be used to implicitly schedule reserves through stochastic programing [8]. An advantage is that the expected cost of operating the system can be minimized across a range of potential outcomes, which should lead to lower average costs in the long run. However, the challenges of computational complexity and how to adequately price energy and reserves under stochastic scheduling are still open questions in the research community. As such, the use of dynamic reserves based on probabilistic forecasts or historical forecasting errors offers a more straightforward, although less sophisticated, approach to handling renewable energy in market operations.

After the clearing of the DA market, there may still be a need to adjust schedules due to contingencies or when realized levels of load or renewable generation deviate from their DA schedules. This rescheduling process is typically called reliability commitment in U.S. markets, as the focus is now more on reliability than economics. At this stage, the system operator uses its own forecasts of renewable energy as well as load to make sure that sufficient resources are available to meet RT loads. System operators currently use point forecasts and deterministic optimization models with reserve requirements for this purpose. However, the use of stochastic unit commitment [9–11] or robust optimization [12] has been proposed to handle the uncertainty from renewable generation at the reliability stage. The latter approach may have advantages in terms of less computational burden and relaxed requirements for the forecasted probability distributions. However, robust optimization is very conservative strategy and not as well linked to fundamental axioms of rational decisions under uncertainty, as it plans for worst-case scenarios rather than an expected utility metric. In general, stochastic scheduling methods may be easier to implement at the reliability stage, since it occurs after the DA market clearing; system operators already use their own forecast at this stage, and the focus is on reliability. The resulting commitment will still influence RT dispatch and prices, and possibly the convergence between DA and RT prices.

As the RT dispatch approaches, forecasting plays an important role in balancing supply and demand in a cost-efficient manner while respecting transmission limits and other physical constraints of the power system. Whereas some countries and system operators give renewable generation priority in the dispatch, the trend in the United States is to treat wind and solar generation more like other dispatchable generators. Several system operators request that wind power submits their forecasted quantity along with a corresponding price offer into the RT market [13]. This way, wind power may be dispatched below their available capacity during periods of surplus supply, which typically happens during low load or transmission congestion. Although "free energy" is being wasted, economic curtailment of renewables can be the most cost-efficient solution in surplus situations to avoid incurring additional shutdowns and start-ups with associated costs for other generators. Still, to make accurate dispatch decisions, including potential curtailment of renewables, it is critical that the short-term forecast of renewable generation is accurate. Moreover, as system operators are implementing look-ahead dispatch schemes [14], the forecast takes on additional importance because the system dispatch and price in the current RT period will depend on the expected conditions in the next few time periods. Visualization of weather observations and forecast data can also contribute to improved situational awareness in the control room.

A more detailed discussion of system operators' use of wind power forecasting in U.S. electricity markets is provided in Botterud et al. [15] and Rogers and Porter [13]. Potential impacts of solar PV on system operation are discussed in Mills et al. [16].

2.2 Wind and solar power producers

The main objective for most producers of renewable energy is to maximize their profits. Renewables are oftentimes sold on long-term contracts, which reduce the risk exposure for the wind and solar power plants. However, renewables still need to be scheduled into the market along with other generation and demand resources, either by the producers themselves or by the purchasers of the long-term contracts. Forecasting is therefore important for operational decision making and may also be required to meet the interconnection standards set by the system operator. Moreover, the producers of renewable energy want to schedule maintenance during periods with limited revenue losses, i.e., typically during low wind and solar resource conditions. Longer-term forecasts (days to weeks ahead) and seasonal trends can therefore be very useful for deciding the timing of maintenance, along with other factors such as expected prices in the electricity market, equipment conditions, and availability of maintenance technicians.

Wind and solar power producers, or the purchasers of their generation, also have to decide on the best strategy for selling their energy into the electricity market. There are typically rules that penalize RT deviations from scheduled generation, and such deviation penalties are also becoming commonplace for renewable generation. At the same time, RT prices tend to be much more volatile than DA prices, leaving producers more exposed to price uncertainty if they do not sell their generation in the DA market. Several algorithms have been proposed for trading of wind power in electricity markets (e.g., Ref. [17–19]). In general, basing DA offers on the best possible point forecast will contribute to minimizing deviation penalties. However, the point forecast is not necessarily the best approach when forecast errors for both renewable energy generation and electricity market prices are considered. Risk preferences will also influence the optimal bidding strategy, as well as the specific rules of the electricity market. In U.S. markets with locational marginal prices, generators can benefit from deviating from their DA schedule if the difference between the DA and RT prices is larger than the deviation penalty. In contrast, most European markets are designed with a two-price system for RT balancing to penalize any deviations from the schedule, although the penalty is not always symmetric. All these factors must be considered to derive the best offer strategy at different stages of market operations. Probabilistic forecasts of wind and solar power will be critical for renewable generators to control the tradeoff between risk and return.

2.3 Other electricity market participants

Renewable energy forecasts are also becoming increasingly important for other market participants such as generation companies, load serving entities, and traders. These entities need to continuously monitor and analyze the market in support of their scheduling and trading activities. Renewable energy is already making a significant impact on market prices in regions with large penetration levels. As an example, Figure 1 shows DA prices for a node in the MISO market, where a new wind farm was built between 2008 and 2012. When comparing the DA prices for the two years, there is a distinct drop in prices and a much higher frequency of negative prices in 2012. Whereas the overall price reduction is largely driven by lower fuel prices (in particular for natural gas), the large increase in negative prices is

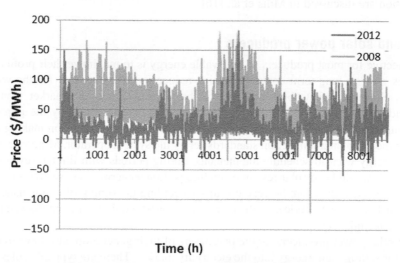

FIGURE 1 DA prices in selected Minnesota node within the MISO market in 2008 and 2012.

very likely caused by the new wind farm, which creates frequent occasions of over-supply at this location, thereby driving prices down below zero. Moreover, wind farms may have an incentive to produce even during negative prices due to support schemes such as tax and renewable energy credits, also contributing to explain the negative prices. Negative prices have been occurring more frequently in markets in both Europe and the United States in recent years, creating profitability challenges for other generation resources that may not be able to shut down their plants during such periods because of operational constraints. Overall, to analyze and understand the complex price dynamics in electricity markets, it is becoming crucial to consider the impact of renewable generation, and the use of forecasting is a key solution toward this end.

3. Forecasting wind and solar energy: the basics

This section gives a brief overview of the general approaches for forecasting weather-driven electricity. The overall approach for forecasting wind and solar power is similar, but there are also some important differences when it comes to data and algorithms, as further discussed below. Whereas wind power forecasting has reached a relatively mature stage, forecasting of solar power is still at a relative infancy.

3.1 Wind power forecasting

A wind power forecasting system usually consists of a complex configuration with multiple interacting components, as illustrated in Figure 2. The amount of wind power generation from a site or region is mainly driven by the weather conditions and particularly the wind speed. Hence, numerical weather prediction (NWP) models form the foundation of a good wind power forecast. Results from NWPs, oftentimes nested and downscaled from large geographical areas to the specific region of interest, are merged with weather observations (e.g., wind speed, wind direction, temperature, pressure) and data for the current and most recent wind farm generation as input to a statistical model whose objective is to produce the most accurate wind power forecast. Computational learning algorithms such as neural networks or support vector machines are typically used for

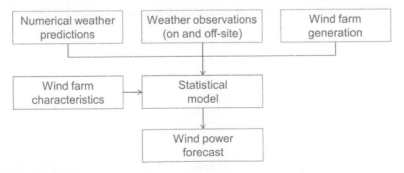

FIGURE 2 Schematic for wind power forecasting.

this purpose. The importance of the different types of inputs to the statistical model depends on the forecast horizon. Longer forecasting horizons (several hours to days) depend heavily on NWP, whereas observational data from weather stations and wind farms are more important for the very short-term forecasts. The average forecasting error depends on the complexity of the local weather conditions and the terrain surrounding the wind farms, and it increases with the forecast horizon. Moreover, the total wind generation in a region typically has a lower forecasting error, since there is an aggregation effect and forecasting errors from individual wind farms have tendency to cancel out. For an extensive overview of the state of the art in wind power forecasting, we refer to Refs [20] and [21].

3.2 Solar power forecasting

The main components of a solar power forecasting system (Figure 3) are similar to the ones for wind power forecasting. However, there are some important differences. A solar photovoltaic (PV) forecasting system typically does an intermediate step of forecasting solar irradiance, which is the main determinant of solar power generation. Irradiance depends heavily on the level of cloudiness. Hence, forecasting of cloud patterns is critical to produce a good forecast. Sky cameras and satellite observations therefore play an important role for forecasting irradiance and solar power in the short term (minutes to a few hours). For longer forecast horizons, it is again NWP that is the main contributor to a good forecast. A majority of the literature on solar forecasting focuses on irradiance forecasts (e.g., Refs [22] and [39]). However, the final step of converting irradiance to power output is also important, and this is done either by considering the physical and technical characteristics of the solar PV system [23] or by using statistical algorithms [24].

The same accuracy metrics (e.g., mean square error, absolute error, bias) apply to solar and wind power forecasts, but solar power forecasts metrics are sometimes adjusted to only account for hours with sunlight. Moreover, although the output from individual solar PV farms can experience extreme

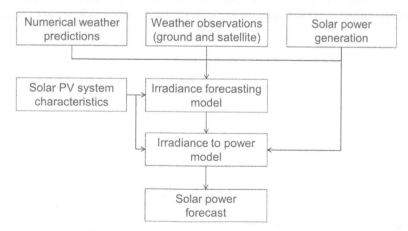

FIGURE 3 Schematic for solar PV forecasting.

variability in a short amount of time and therefore may be very hard to forecast, forecasts for a region typically have much smaller errors due to the aggregation effect [16]. For an overview of the state of the art in solar power forecasting, we refer to Kleissl [25].

4. Emerging forecasting products

Renewable energy forecasting is a rapidly evolving field, and there is a continuous effort to adapt products to the needs of the forecast users. Three directions with relevance to the discussion of forecasting applications in section 2 are briefly discussed below.

4.1 Probabilistic forecasts

Since so many different aspects of grid operations deal with the same general challenge of managing uncertainty in a cost-efficient and reliable manner, the potential value of using probabilistic forecasting is increasingly recognized among system operators and market participants. Computational learning methods such as quantile regression, fuzzy inference, or kernel density estimation can be used to make conditional statistical estimates of the uncertainty around the point forecast based on historical forecasting errors (e.g., Refs [26–28]). Another approach for quantification of forecast uncertainty is to run an ensemble of NWP simulations [28,29]. A probabilistic forecast can be represented with a set of forecast quantiles for each individual forecast horizon. However, for decision problems with inter-temporal constraints, such as the unit commitment problem, accounting for temporal correlations is important and a scenario representation (e.g., Ref. [30]) is oftentimes required. An advantage of the NWP approach is that the information in the NWP ensemble can be used to account for both spatial and temporal correlations. However, statistical models have also been proposed for spatiotemporal forecasts [14]. The next chapter of this book provides a more detailed discussion of probabilistic forecasts.

4.2 Ramp forecasts

A recent trend in forecasting of renewable energy is the interest in understanding and predicting rapid changes in the power output [31,32]. Such ramp events, particularly in the downward direction, may create challenges for reliable grid operations, and a ramp forecast is therefore seen as an important operational tool among some system operators. Others, however, argue that there is no need for a specific ramp forecast beyond the information that is already embedded in the regular forecasts of renewable energy. A ramp event can be characterized according to three features: direction, magnitude, and duration. However, there is no universal definition of a ramp event. In reality, what constitutes a significant ramp will depend on the specific system, such as its size and the availability and flexibility of other resources balancing the grid. Since the task for a ramp forecast is to predict specific events, the metrics used to evaluate such forecasts (e.g., precision, recall, and critical success index) are also event based. Several models for ramp forecasting have been proposed (e.g., Refs [33–36]). Since forecasting of individual events is very difficult, a common approach is to provide information about the likelihood of a ramp to take place. Ramp forecasts are most relevant for the very short-term time horizon (0–6 h). Hence, the use of NWP has been of limited value for this purpose so far. However, this may change as efforts are currently under way in the United States to improve short-term

NWP through higher resolution models with more rapid updates, also using a wider network of hub-height weather measurements as inputs [37].

4.3 Net load forecasts

Load forecasting in the electric power system is a mature field and has been a topic of interest for a long time [38]. Electric loads tend to follow stable and predictable patterns, giving rise to rather low load forecasting errors in relative terms. However, with the advent of distributed generation such as rooftop solar PV, load forecasting takes on a new dimension. In some areas, distributed PV already meets a significant fraction of the total load. This generation is connected to the distribution network, behind the meter, and can therefore typically not be directly measured or controlled by the utility or system operator. Hence, to schedule and dispatch the resources in the bulk power system, system operators need a forecast of the net load, accounting for the impact of distributed generation. Although the distributed nature of solar PV installations creates aggregation effects that reduce forecasting errors, it is still likely that net load forecasts will have larger forecasting errors than traditional forecasts of gross loads, with potential implications for the overall balancing of supply and demand in the grid. Therefore, developing net load forecasts with high accuracy, accompanied by good predictions of the forecast uncertainty, will take on increasing importance as more rooftop PV and other distributed generation resources are added to the power system.

5. Looking ahead

Wind power forecasting has received more attention so far and is therefore at the most mature state, but forecasting of solar energy is also developing fast given the rapid expansion of solar PV in many regions. NWP is the fundamental starting point for forecasting of both wind and solar power. Hence, forecasting of renewable energy will benefit as NWP models move toward higher resolution and more frequent update cycles, enabled in part by better computing resources. More measurements of wind at hub-height and solar irradiance at ground level can also contribute to enhance NWP and ultimately the forecasts of renewable energy. Moreover, the development of more sophisticated statistical algorithms will contribute to higher forecast accuracy and to improvements in emerging products such as probabilistic and ramp forecasts. With the increasing amounts of available data, the key challenge is to filter out the important information and convert it to formats that can be used effectively by decision makers in the power grid.

System operators and electricity market participants are rapidly building experience in the use of renewable energy forecasting. However, in most countries the power industry is still at an early stage of what may evolve into a low-carbon renewable power grid, with fundamental changes in planning and operational procedures as well as electricity market design. Hence, more work is needed to develop tools and methods to enable the best use of forecast information in operational decisions. The core challenge for system operators is to manage the grid in a cost-efficient and reliable manner, accounting for increasing levels of uncertainty and variability from renewable resources. Although several promising directions are emerging at the research stage (e.g., stochastic unit commitment or dynamic reserves based on probabilistic forecasts), there is a long way to go in developing and implementing the best forecasting solutions, considering all the complexities in grid operations and the many different needs and objectives of the various stakeholders in the grid. Producers of renewable energy have a narrower objective of maximizing profits while keeping risk under control, and it is obvious that forecasting will play an important role toward this end. Likewise, forecasting of renewable energy will become critical for all market participants as renewable resources influence the overall scheduling, dispatch, and pricing in the electricity market.

The expansion of renewable energy also triggers the need to revisit the design of electricity markets to create the right incentives for maintaining reliability while keeping costs to a minimum as the share of low-marginal-cost variable resources increases. From a forecasting perspective, it is important that the users of renewable energy forecasts have the incentives to invest in and help develop the best possible forecasting products. The costs of introducing advanced forecasting and decision support tools need to be measured against the potential for substantial benefits to the system. Moreover, markets must be designed so that wind and solar producers have incentives to bid their most accurate forecasts into DA and RT markets. The prices for energy and ancillary services products also need to reflect the expected variability in renewable generation as well as its forecast uncertainty. Furthermore, market participants must be able to hedge against the possibility of more volatility in short-term prices and to recover both their fixed and variable costs from the revenue streams offered in the electricity market.

Some concepts of good market designs for integration of renewable energy are generally agreed upon, such as the importance of more frequent scheduling over larger balancing areas based on the best possible forecasts. At the same time, the rapid expansion of renewables does not happen in isolation. The impacts of other important trends, such as the current shale gas revolution and the development of more demand response, smart buildings, electric vehicles, and energy storage, must also be considered. The devil is in the details when it comes to creating market rules for a commodity with such a complex delivery system as electricity, and there are still a number of open electricity market design questions that will need to be addressed. One thing is certain, however: forecasting of renewable energy is an important part of the solution to enable a clean, reliability, and cost-efficient future electricity grid.

Acknowledgments

The author is grateful for fruitful collaborations with multiple researchers in the area of wind power forecasting and electricity market operations in recent years, including Zhi Zhou, Jianhui Wang, Emil Constantinescu, and Edwin Campos (Argonne National Laboratory), and Vladimiro Miranda, Jean Sumaili, Ricardo Bessa, Hrvoje Keko, Joana Mendes, Joao Gama, Carlos Ferreira, and Claudio Monteiro (INESC TEC, Porto, Portugal). This chapter draws heavily on those collaborations.

The author also acknowledges the U.S. Department of Energy (DOE) and its Wind and Water Power Program for funding underlying research for this chapter. The submitted manuscript has been created by UChicago Argonne, LLC, Operator of Argonne National Laboratory ("Argonne"). Argonne, a U.S. DOE Office of Science laboratory, is operated under Contract No. DE AC02-06CH11357.

References

[1] Bessa RJ, Miranda V, Botterud A, Wang J. 'Good' or 'Bad' wind power forecasts: a relative concept. Wind Energy 2011;14(5):625–36.

[2] Electricity Reliability Council of Texas (ERCOT). ERCOT methodologies for determining ancillary service requirements. ERCOT; December 2009.

[3] Ela E, Milligan M, Kirby B. Operating reserves and variable generation. Technical Report NREL/TP-5500-51978. National Renewable Energy Laboratory; August 2011.

[4] Navid N, Rosenwald G. Market solutions for managing ramp flexibility with high penetration of renewable resources. IEEE Trans Sustain Energy 2012;3(4):784–90.

[5] Matos MA, Bessa R. Setting the operating reserve using probabilistic wind power forecasts. IEEE Trans Power Syst 2011;26(2):594–603.

[6] Ortega-Vazquez MA, Kirschen DS. Estimating spinning reserve requirements in systems with significant wind power generation penetration. IEEE Trans Power Syst 2009;24(1):114–24.

[7] Zhou Z, Botterud A. Dynamic scheduling of operating reserves in co-optimized electricity markets with wind power. IEEE Trans Power Syst 2014;29(1):160–71.

[8] Morales JM, Conejo AJ, Liu K, Zhong J. Pricing electricity in pools with wind producers. IEEE Trans Power Syst 2012;27(3):1366–76.

[9] Papavasiliou A, Oren SS. Multi-area stochastic unit commitment for high wind penetration in a transmission constrained network. Operations Res 2013;61(3):578–92.

[10] Tuohy A, Meibom P, Denny E, O'Malley M. Unit commitment for systems with significant wind penetration. IEEE Trans Power Syst 2009;24(2):592–601.

[11] Wang J, Botterud A, Bessa R, Keko H, Carvalho L, Issicaba D, et al. Wind power forecasting uncertainty and unit commitment. Appl Energy 2011;88(11):4014–23.

[12] Bertsimas D, Litvinov E, Sun XA, Zhao J, Zheng T. Adaptive robust optimization for the security constrained unit commitment problem. IEEE Trans Power Syst 2013;28(1):52–64.

[13] Rogers J, Porter K. Wind power and electricity markets. Utility Wind Integration Group; 2011.

[14] Xie L, Gu Y, Zhu X, Genton MG. Short-term spatio-temporal wind power forecast in robust look-ahead power system dispatch. IEEE Trans Smart Grid 2014;5(1):511–20.

[15] Botterud A, Wang J, Miranda V, Bessa RJ. Wind power forecasting in US electricity markets. Elect J 2010; 23(3):71–82.

[16] Mills A, Ahlstrom M, Brower M, Ellis A, George R, Hoff T, et al. Dark shadows: understanding variability and uncertainty of photovoltaics for integration with the electric power system. IEEE Power Energy Mag 2011;9(3):33–41.

[17] Matevosyan J, Söder L. Minimization of imbalance costs trading wind power on the short-term power market. IEEE Trans Power Syst 2006;21(3):1396–404.

[18] Pinson P, Chevallier C, Kariniotakis GN. Trading wind generation from short-term probabilistic forecasts of wind power. IEEE Trans Power Syst 2007;22(3):1148–56.

[19] Botterud A, Zhou Z, Wang J, Bessa RJ, Keko H, Sumaili J, et al. Wind power trading under uncertainty in LMP markets. IEEE Trans Power Syst 2012;27(2):894–903.

[20] Giebel G, Brownsword R, Kariniotakis G, Denhard M, Draxl C. The state-of-the-art in short-term prediction of wind power: a literature overview. 2nd ed 2011. Anemos.Plus, Deliverable D-1.2.

[21] Monteiro C, Bessa RJ, Miranda V, Botterud A, Wang J, Conzelmann G. Wind power forecasting: state-of-the-art 2009. Report ANL/DIS-10–1. Argonne National Laboratory; Nov.2009.

[22] Lorenz E, Hurka J, Heinemann D, Beyer HG. Irradiance forecasting for the power prediction of grid-connected photovoltaic systems. IEEE J Sel Top Appl Earth Observations Remote Sens 2009;2(1):2–10.

[23] King DL, Boyson WE, Kratochvil JA. Photovoltaic array performance model. Report SAND 2004-3535. Sandia National Laboratory; August 2004.

[24] Bacher P, Madsen H, Nielsen HA. Online short-term solar power forecasting. Solar Energy 2009;83(10): 1772–83.

[25] Kleissl J. Current state of the art in solar forecasting. Report. California Institute for Energy and Environment; 2010.

[26] Bessa RJ, Miranda V, Botterud A, Wang J, Constantinescu EM. Time adaptive conditional kernel density estimation for wind power forecasting. IEEE Trans Sustain Energy Oct. 2012;3(4):660–9.

[27] Nielsen HA, Madsen H, Nielsen TS. Using quantile regression to extend an existing wind power forecasting system with probabilistic forecasts. Wind Energy 2006;9(1–2):95–108.

[28] Pinson P. Estimation of the uncertainty in wind power forecasting [Ph.D. Dissertation]. École des Mines de Paris; 2006.

[29] Constantinescu EM, Zavala VM, Rocklin M, Lee S, Anitescu M. A computational framework for uncertainty quantification and stochastic optimization in unit commitment with wind power generation. IEEE Trans Power Syst 2011;26(1):431–41.

[30] Pinson P, Papaefthymiou G, Klockl B, Nielsen HA, Madsen H. From probabilistic forecasts to statistical scenarios of short-term wind power production. Wind Energy 2009;12(1):51–62.

[31] Ferreira C, Gama J, Matias L, Botterud A, Wang J. A survey on wind power ramp forecasting. Report ANL/DIS 10-13. Argonne National Laboratory; Dec. 2010.

[32] Kamath C. Understanding wind ramp events through analysis of historical data. In: Proceedings of the IEEE PES transmission and distribution conference and expo, New Orleans, LA, USA; 2010.

[33] Cutler NJ, Outhred HR, MacGill IF, Kay MJ, Kepert JD. Characterizing future large, rapid changes in aggregated wind power using numerical weather prediction spatial fields. Wind Energy 2009;12(6):542–55.

[34] Ferreira C, Gama J, Miranda V, Botterud A. Probabilistic ramp detection and forecasting for wind power prediction. In: Billinton, Karki, Verma, editors. Reliability and risk evaluation of wind integrated power systems. Springer; 2013. pp. 29–44.

[35] Potter CW, Grimit E, Nijssen B. Potential benefits of a dedicated probabilistic rapid ramp event forecast tool. In: Proceedings of the IEEE power systems conference and exposition, Seattle, Wash., USA; March 2009.

[36] Zack JW, Young S, Nocera J, Aymami J, Vidal J. Development and testing of an innovative short-term large wind ramp forecasting system. In: Proceedings of the European wind energy conference & exhibition (EWEC), Warsaw, Poland; April 2010.

[37] Orwig K, Clark C, Cline J, Benjamin S, Wilczak J, Marquis M, et al. Enhanced short-term wind power forecasting and value to grid operations. In: Proceedings of 11th annual international workshop on large-scale integration of wind power into power systems, Lisbon, Portugal; Nov. 2012.

[38] Gross G, Galiana FD. Short-term load forecasting. Proc IEEE 1987;75(12):1558–73.

[39] Reikard G. Predicting solar radiation at high resolutions: a comparison of time series forecasts. Solar Energy 2009;83(3):342–9.

Probabilistic Wind and Solar Power Predictions

12

Luca Delle Monache, Stefano Alessandrini

National Center for Atmospheric Research, Boulder, CO, USA

1. Introduction

Among the limiting factors to the penetration of wind and solar energy is the variable nature and limit to predictability of wind speed and solar irradiance. To maintain grid stability, it is fundamental to know a priori the renewable energy production that can be combined with other less variable and more predictable sources (e.g. coal, natural gas) to satisfy the energy demand [1]. In a deregulated market, it is crucial for the utility companies that provide electricity to be able to deliver the committed energy to the grid. In this sense this is a challenge for wind and solar energy when compared to other types of energy sources because of their intrinsic limited predictability. Accurate wind and solar power predictions and reliable estimates of their uncertainty are key aspects to mitigate these disadvantages.

Wind and solar power predictions can be categorized into two main groups: deterministic and probabilistic. A deterministic forecast consists of a single value for each time in the future for the variable to be predicted. Different methods for deterministic predictions can be found in the literature and a thorough review of the state-of-the-art is reported in [2]. Probabilistic forecasting provides probability density functions (PDF) from which probabilities of future outcomes of events can be estimated (e.g. [3], and references therein). This implies also providing information about uncertainty in addition to the commonly provided single-valued power prediction. A deterministic prediction can provide useful information for decision making. Its utility, however, is fundamentally limited as it represents only a single plausible future state of the atmosphere from a continuum of possible states, which result from imperfect initial conditions and model deficiencies that lead to nonlinear error growth during model integration. Accurate knowledge of that continuum, the forecast PDF, provides considerably more utility to decision making (e.g. Ref. [4]).

In the following sections a review of several state-of-the-science techniques to produce probabilistic power predictions is presented (Section 2), followed by an introduction on how these predictions can be verified and evaluated (Section 3), and conclusions (Section 4).

2. Probabilistic power predictions methods

2.1 Numerical weather/power ensemble predictions

One way to generate probabilistic power predictions is to produce a forecast PDF of the hub-height winds and then convert these winds to probabilistic power estimates by mean of power curves. Leith [5] proposed a Monte Carlo approximation to stochastic dynamic forecasting, referred to here as a numerical weather predictions (NWP) ensemble, where the deterministic NWP model is run multiple times (referred to as ensemble members) over the valid period with plausible variations to each separate run. The NWP ensembles have been created using different model initial conditions (e.g. Ref. [6]), parameterizations within a single model (e.g. Ref. [7]), stochastic approaches (e.g. Ref. [8]), numerical schemes (e.g. Ref. [9]), models (e.g. Ref. [10]), and coupled to ocean and land surface ensembles (e.g. Ref. [11]).

2.2 Calibration/post-processing methods

Ensemble forecasts allow the quantification of forecast uncertainty, which could lead to cost-effective decision making while integrating wind power generation in the energy market. However, the accuracy of the PDF resulting from ensemble predictions of hub-height wind speed based on the direct NWP output is affected by forecast errors (e.g. Ref. [12]). Numerous approaches can be found in the literature to address these deficiencies after the NWP ensemble is generated, and as such they are referred to as calibration or postprocessing methods. Among the methods that focus on the univariate correction of wind speed there are the ensemble model output statistics (EMOS [12]), Bayesian model averaging (BMA [13]), and variance deficit calibration [3]. Other methods focus instead on a bivariate calibration approach of wind speed and direction as the adaptive wind vector calibration (AUV [14]), the bivariate extensions to EMOS [15] and to BMA [16], and ensemble copula coupling (ECC) combined with univariate EMOS (ECCEMOS [17]) or univariate BMA [18].

   Junk et al. [19] compared the performance of state-of-the-science postprocessing methods for the calibration of 100-m wind speed and wind vector ensemble forecasts at distinct off- and onshore sites in Central Europe. The authors found that the bivariate recursive and adaptive wind vector calibration developed by [14] outperforms univariate and bivariate EMOS among others at onshore sites while yielding comparable performance at offshore sites.

2.3 Probabilistic power predictions based on a historical data set and a deterministic numerical prediction

There is a class of methods that generates probabilistic predictions based on an historical data set including observations of the quantity to be predicted (i.e. wind speed or power) and a single deterministic model prediction. These methods are particularly suited for wind and solar energy applications because they can produce probabilistic power predictions at specific locations (i.e. wind farms) without the need to generate in real-time multiple NWP-based predictions (see Section 2.1), which are computationally expensive and need to be calibrated (see Section 2.2). These techniques include the analog ensemble (AnEn [3,20,21]), quantile regression (QR [22]), and logistic regression (LR [23]).

   In the AnEn, for each forecast lead time and location, the ensemble prediction of a given variable is constituted by a set of measurements of the past. These measurements are those concurrent to past deterministic NWPs for the same lead time and location, chosen based on their similarity to the current forecast. While the mechanics of LR are quite different from AnEn, both approaches consider the past relationship between predictor variable(s) and the predictand to produce a forecast of the predictand given the predictors' values in the current forecast cycle. One difference with LR is that the predictand is the probability of an event, such as the probability of 10-m wind speed greater than 5 m/s, rather than the value (or PDF) of 10-m wind speed itself. In LR a nonlinear function is fit to past pairs of the predictor(s) and the predictand, which as an observed value takes on a probability of either 1.0 (event occurred) or 0.0 (event did not occur). The AnEn has been found to have similar skills and characteristics in term of probabilistic predictions for common events as LR [21] and QR [3], but it performs better when predicting rare events of both wind speed and power.

3. Verification and value of probabilistic predictions

The quality of probabilistic predictions can be rigorously and thoroughly assessed by evaluating key attributes, including statistical consistency, reliability, sharpness, resolution, and value. While these attributes and their associated metrics are briefly reviewed below, thorough descriptions can be found in [23].

3.1 Statistical consistency

An ensemble is statistically consistent when its members are indistinguishable from the truth (i.e. the PDF from which the members are drawn is consistent with the PDF from which the truth is drawn). If so, an observation ranked among the corresponding ordered ensemble members is equally likely to take any rank in the range $1, 2, ..., n + 1$, where n is the number of ensemble members. Collecting the rank of the observation over a number of cases and plotting the results generates a rank histogram, which tests as flat (i.e. uniform rank probability of $1/(n + 1)$) for a statistically consistent ensemble.

Figure 1, adapted from [21]; shows an example of a rank histogram for a 9-h forecast of 10-m wind speed, where the AnEn is compared to the Environment Canada Regional Ensemble Prediction System (REPS). It also shows the missing rate error (MRE), which is the fraction of observations lower (higher) than the lowest (highest) ranked prediction above or below the expected missing rate, $2/(n + 1)$. The REPS rank histogram (Figure 1(a)) reveals a severe lack of statistical consistency, with a notable negative forecast bias (observed wind speed often greater than all REPS members) as well as an underspread condition (highest probabilities in the two outer ranks). The AnEn has much better statistical consistency, displayed by a nearly uniform rank histogram (Figure 1(b)) with a slightly overspread condition (MRE equal to -1.1%), as is more evident in Figure 1(c) where a tighter vertical axis range is used. Similar results were obtained for solar probabilistic power prediction (not shown).

Examination of statistical consistency over all forecast lead times is accomplished following the general definition that the mean square error of the ensemble mean should match the average ensemble variance over a large number of verifications. Comparing the square root of those two statistics (to display results with the predicted variables' natural unit) over all forecast lead times produces a dispersion diagram that reveals whether an ensemble is properly dispersive (i.e. able to simulate average forecast error growth). See Figure 5 in Ref. [21] for an example of a dispersion diagram.

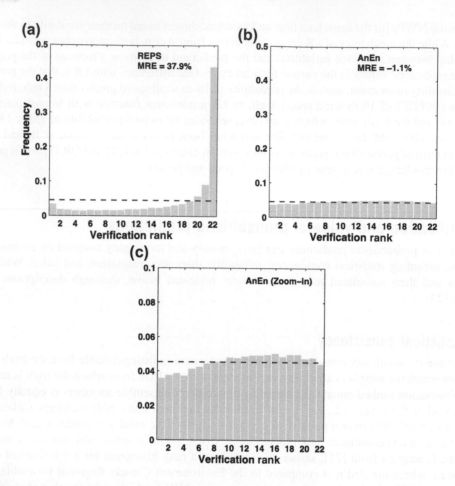

FIGURE 1

Rank histogram for probabilistic prediction of 10-m wind speed for (a) the raw REPS, (b) the AnEn, and (c) AnEn but with a tighter vertical axis range, and with inset MRE results for (a) and (b). Gray histogram bars show the frequency of the observation occurring in each rank. The dashed black line indicates perfect, uniform probability.

Adapted from Ref. [21].

A more in-depth assessment of statistical consistency at a particular forecast lead time is possible with a binned spread-skill plot (Figure 2), which compares ensemble spread to the root-mean-square-error (RMSE) of the ensemble mean over small class intervals of spread rather than just considering the overall average spread as in the dispersion diagram. Good statistical consistency now requires the two metrics to match at all values of ensemble spread (i.e. results along the plot's 1:1 diagonal).

In the example shown in Figure 2 adapted from [3], the European Center for Medium-Range Weather Forecasts (ECMWF) Ensemble Prediction System (ECMWF-EPS), and the Limited-area Ensemble Prediction System (LEPS) developed within the COnsortium for Small-scale MOdeling

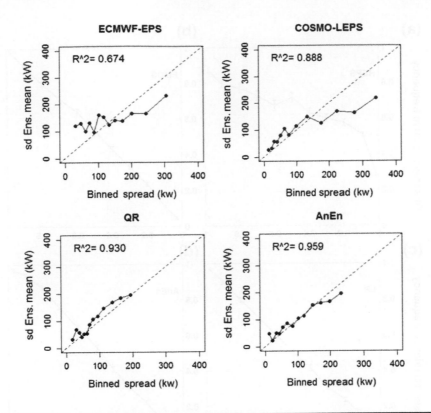

FIGURE 2

Binned ensemble spread versus standard deviation of the ensemble mean error for 9–15 h ahead power forecasts. Shown forecast include ECMWF-EPS (upper left), COSMO-LEPS (upper right), QR (bottom left) and AnEn (bottom right). The R^2 values reported indicate the square power of correlation indexes of the series of points.

Adapted from Ref. [3].

(COSMO-LEPS) show an underdispersive condition for smaller spread values and overspread at higher values. The QR generally shows a slightly underdispersive condition, while AnEn is adequately able to represent forecast uncertainty, since its spread dependably reflects the deterministic error variance.

3.2 Reliability

Ideally, a large set of 30% probability forecasts will verify with a 30% occurrence rate of the event (called the observed relative frequency). In perfectly reliable (or calibrated) forecasts, the observed relative frequency equals the forecasted probability for any given level of probability, resulting in the 1:1 diagonal line on a reliability diagram that plots class intervals of forecast probability against observed relative frequency [23].

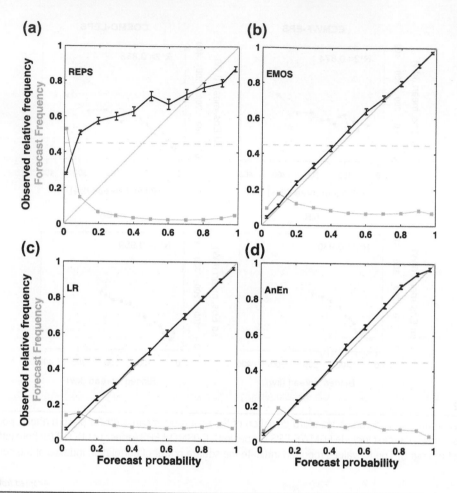

FIGURE 3

Reliability (black lines with vertical error bars) and sharpness (gray lines with square marks) for (a) REPS, (b) EMOS, (c) LR, and (d) AnEn. Results are shown for forecast hour 9 and 10-m wind speed greater than 5 m/s. The horizontal dashed line represents the event's observed frequency over the verification period (i.e. sample climatology), while the diagonal 1:1 line represents the perfect reliability. The error bars indicate the 95% bootstrap confidence interval.

Adapted from Ref. [21].

Figure 3 adapted from [21], shows an example of a reliability diagram (black lines with error bars) for the event of 10-m wind speed greater than 5 m/s at forecast hour 9. The REPS forecasts are the least reliable, with notable underestimation (overestimation) of the observed relative frequency below (above) approximately 0.7 (Figure 3(a)). The EMOS forecasts have rather good reliability (Figure 3(b)), while both LR and AnEn (Figure 3(c) and (d), respectively) forecasts are also imperfect but exhibit roughly the same degree of good reliability as EMOS.

3.3 Sharpness

A sharper (more narrow) forecast PDF has a greater concentration of probability density and produces probability values more toward the extremes (i.e. close to 0% or 100%) for any given event threshold. Sharpness, which is a property of the forecasts only, can be diagnosed in a reliability diagram by plotting how often (relative frequency) each class interval of probability is used. A sharper forecast leads to better resolution (see next subsection) if the forecasts are reliable [12].

Figure 3 shows an example of a sharpness diagram (gray lines with square marks) for forecast hour 9 and 10-m wind speed greater than 5 m/s; the REPS forecasts (Figure 3(a)) are very sharp with the majority occurring in the 0–10% range, but this is due to overconfidence as indicated by the poor reliability. The EMOS forecasts (Figure 3(b)) have a lower yet trustworthy sharpness resulting from the calibration's correction of REPS overconfidence. AnEn sharpness (Figure 3(d)) is comparable to EMOS (Figure 3(b)) and LR forecasts (Figure 3(c)). See Ref. [21] for details.

3.4 Resolution

Resolution measures the forecasts' ability to a priori sort out when an event occurs or not. Probability forecasts with perfect resolution forecast 100% on occasions when the event occurs and forecast 0% when the event does not occur.

The relative operating characteristic (ROC) skill score (ROCSS) is based on the ROC curve, which plots the hit rate (correct forecasts divided by total occurrences of the event) against the false alarm rate (false alarms divided by total nonoccurrences of the event) to show the forecast's ability to discriminate between events and nonevents. The ROC curve (as well as the ROCSS) thus depends upon resolution and not reliability, and the area under the ROC curve, known as the ROC score, conveys overall forecast value. The ROCSS translates the ROC score into a standard skill score so that a ROCSS equal to 1 comes from perfect forecasts and a ROCSS lower than 0 indicates lower performance than climatological forecasts. An Example of a ROCSS plot for wind power predictions can be found in Figure 3 of Ref. [3].

3.5 Economic value of probabilistic wind power forecasts

There are several methods for assessing the value of probabilistic wind power forecasts, but those are mostly independent from the user needs. In this section the economic impact of a probabilistic wind power forecast is evaluated with a simple model of an electricity market. This model aims to simulate the real situation where penalties must be paid by the producers of the energy market on *unbalancing*, which is computed as the difference between the hourly promised (a day ahead) energy and the delivered one. The penalties can be related to the day-ahead and spot market energy prices; the day-ahead market occurs once a day usually in the early morning and defines the hourly energy production for the day after, while the spot market occurs every hour to balance the energy production with the real demand. These rules aim to protect, for example, the balancing authority that must buy energy at a higher price at the spot market to satisfy the demand, if the producers are not able to deliver the promised energy.

The following is an example with a simplified market model and wind power data from a real wind farm located in Italy. Probabilistic wind power forecasts are generated with AnEn applied to meteorological forecasts produced by a limited area model with 4-km horizontal grid resolution. The market model simulates the mechanism by which producers promise to deliver energy to the grid in terms of a contract to provide a target amount of electricity, E_c, for a particular hour of the day ahead. The producer is paid a fixed unit price, P_c, for this promised electricity. If at that particular hour wind production is not sufficient to satisfy the contract, then the generator must satisfy the contract by purchasing the shortfall on the spot market at a cost of P_s per unit greater than P_c. If, in the opposite case, the real production exceeds the promised energy, the excess of production is paid at a price P_{s1} lower than P_c. In this latter case there will be some other energy producers (usually from conventional sources) who must be reimbursed by the amount of energy they are not able to deliver to the grid.

A model where the penalties faced by the producers are asymmetric can be implemented. This encourages the producers participating to the market to adopt a specific strategy in their approach to bidding. In fact, it can be shown (e.g. Ref. [24]) that the optimal day-ahead market bid for a wind energy producer is a certain quantile of the distribution of wind power generation forecast. The optimal value of E_c is such that the marginal income from an additional unit of promised electricity is exactly balanced by the expected penalty from failing to meet E_c for that unit.

The optimal quantile changes in time since P_s, P_c, and P_{s1} are not known a priori at the time of the bid and change every hour. The optimal bid E_c might significantly differ from the point forecasts of wind power production, usually considered as the value of the energy with the expected highest probability of occurrence.

In Figure 4 the cumulative income of a real wind farm (located in Italy) is plotted as a function of the hours over a one-year period. Each hourly income is computed with P_s, P_{s1}, and P_c being the occurred buy and selling prices of the spot market and the spot price of the day-ahead market, respectively. The black line refers to the income obtainable with 1 MW of installed capacity and a perfect forecast (the promised energy in the day-ahead market is exactly equal to the produced energy).

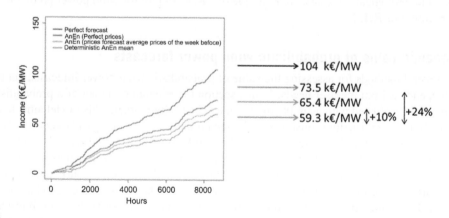

FIGURE 4

Cumulative incomes obtainable with 1 MW of installed capacity by a wind farm located in Italy as a function of the hours of the year 2011. The different lines refer to different approaches in setting the amount of energy to be sold in the day-ahead market.

The red line refers to the income achievable by selling during the day-ahead market an amount of energy obtained by a deterministic (or spot) wind power forecast that doesn't consider any economic optimization model. Note that in this exercise the deterministic forecast is based on the AnEn mean, i.e. is still an ensemble-based prediction. The blue and green lines indicate the cases where the amount of energy sold in the day-ahead market is set by using the optimal bid E_c. Since P_s, P_c, and P_{s1} are not known a priori at the time of the bid, they need to be forecasted. The AnEn power forecasts are used in both cases (blue and green line). In case of the green line, the prices are forecasted with a persistence approach (the average price occurred during the week before at the same hour of the day). In the blue line case, a perfect price forecast is used (setting the prices occurred "a posteriori"). In this experiment, the economic benefit of a probabilistic approach (green line) compared to a deterministic one (red line) is roughly a 10% of increment on the total annual income. This increment can reach the amount equal to 24% by improving the accuracy of the price forecast (blue line, perfect price forecast used). In the case of the green line, the efficacy of a probabilistic approach is partially limited by the inaccuracy of the price forecast method used.

4. Conclusion

Different methods to generate wind and solar power probabilistic predictions have been presented, as well as a range of metrics to assess the performance of probabilistic forecasts. The analog ensemble technique provides reliable and calibrated probabilistic wind and solar power forecasts, with a better performance when compared to model-based ensemble forecast. The potential economical benefit of probabilistic predictions versus deterministic forecasts was also shown with a real case example. As probabilistic approaches are further developed and refined, such benefit will likely increase, and the usefulness of reliable uncertainty quantification will positively impact also the optimization of daily operations of on- and offshore renewable energy parks.

References

[1] Mahoney WP, et al. A wind power forecasting system to optimize grid integration, special issue of IEEE transactions on sustainable energy on applications of wind energy to power Systems, vol. 3; 2012. pp. 670–682.

[2] Giebel G, et al. The state-of-the-art in short-term prediction of wind power: a literature overview. 2nd ed 2011 [ANEMOS.plus].

[3] Alessandrini S, et al. A novel application of an analog ensemble for short-term wind power forecasting. Submitted to renewable Energy; 2013.

[4] Hirschberg PA, et al. A weather and climate enterprise strategic implementation plan for generating and communicating forecast uncertainty information. Bull Am Meteorol Soc 2011;92:1651–66.

[5] Leith CE. Theoretical skill of Monte Carlo forecasts. Mon Weather Rev 1974;102:409–18.

[6] Molteni F, et al. The new ECMWF ensemble prediction system: method- ology and validation. Quart. J Roy Meteorol Soc 1996;122:73–119.

[7] Hacker J, et al. The U.S. Air Force Weather Agency's mesoscale ensemble: scientific description and performance results. Tellus 2011;63A:625–41.

[8] Buizza R, et al. Stochastic representation of model uncertainties in the ECMWF Ensemble Prediction System. Q J R Meteorol Soc 1999;125:2887–908.

[9] Thomas SJ, et al. An ensemble analysis of forecast errors related to floating point performance. Weather Forecast 2002;17:898–906.

[10] Krishnamurti TN, et al. Multimodel ensemble forecasts for weather and seasonal climate. J Clim 2000;13: 4196–216.

[11] Holt T, et al. Urban and ocean ensembles for improved meteorological and dispersion model- ling of the coastal zone. Tellus 2009;61A:232–49.

[12] Gneiting T, et al. Calibrated probabilistic forecasting using ensemble model output statistics and minimum CRPS estimation. Mon Weather Rev 2005;133(5):1098–118.

[13] Raftery A, et al. Using Bayesian model averaging to calibrate forecast ensembles. Mon Weather Rev 2005; 133:1155–74.

[14] Pinson P. Adaptive calibration of (u, v)-wind ensemble forecasts. Q J Roy Meteorol Soc 2012;138:1273–84.

[15] Schuhen N, et al. Ensemble model output statistics for wind vectors. Mon Weather Rev 2012;140:3204–19.

[16] Sloughter J, Gneiting T, Raftery A. Probabilistic wind speed forecasting using ensembles and Bayesian model averaging. J. Amer. Stat. Assoc. 2010;105(489):25–35.

[17] Schefzik R, et al. Uncertainty quantification in complex simulation models using ensemble copula coupling; 2013. arXiv:1302.7149.

[18] Möller A, et al. Multivariate probabilistic forecasting using ensemble Bayesian model averaging and copulas. Q J Roy Meteorol Soc; 2012.

[19] Junk C, von Bremen L, Heinemann D, Spaeth S, Kühn M. Comparison of post-processing methods for the calibration of 100 m wind ensemble forecasts at off- and onshore sites. J Appl Meteorol Climatol; 2013. Accepted to appear.

[20] Delle Monache L, et al. Kalman filter and analog schemes to post-process numerical weather predictions. Mon Weather Rev 2011;139:3554–70. 2011.

[21] Delle Monache L, et al. Probabilistic weather predictions with an analog ensemble. Mon Weather Rev 2013; 141:3498–516.

[22] Möller JK, et al. Time-adaptive quantile regression. Comput Statistics Data Analysis 2006;52:1292–303.

[23] Wilks DS. Statistical methods in the atmospheric sciences. 2nd ed. Academic Press; 2006. 627 pp.

[24] Roulston MS, et al. Using medium-range weather forecasts to improve the value of wind power production. Renew Energy 2002;28:585–602.

Incorporating Forecast Uncertainty in Utility Control Center

13

Yuri V. Makarov[1], Pavel V. Etingov[1], Jian Ma[2]

[1] *Pacific Northwest National Laboratory,* [2] *Burns & McDonnell*

1. Introduction

Uncertainties in forecasting the output of intermittent resources such as wind and solar generation, as well as system loads, are not reflected adequately in existing industry-grade tools used for transmission system management, generation commitment, dispatch, and market operation. Other sources of uncertainty include uninstructed deviations of conventional generators from their dispatch set-points, generator-forced outages and failures to start up, load drops, losses of major transmission facilities, and frequency variation. These uncertainties can cause deviations from the system balance, which sometimes require inefficient and costly last-minute solutions in the near real-time time frame. Major unexpected variations in wind power, unfavorably combined with load forecast errors and other factors, can cause significant power mismatches, which essentially would be unmanageable without knowing about the possibility of these variations ahead of time. In extreme cases, dispatch decisions could not be found because of the generators' start up, ramping, and capacity constraints.

With the growing penetration of variable resources, uncertainties could pose serious risks to control performance and operational characteristics, as well as the reliability of a power grid. Without knowing the risks caused by uncertainties (i.e. the probability, timing, and magnitude) of potential system imbalances, system operators have limited means to evaluate the potential problems and find solutions to mitigate their adverse impacts. Some important questions need to be addressed in counteracting the impact of uncertainties. For instance, when should one start more units to balance against possible fast ramps in the future over a given time horizon? Would the available online capacity be sufficient to balance against variations of uncertain parameters on the intrahour and minute-to-minute basis?

The need to evaluate uncertainties associated with wind generation and solar and to incorporate the knowledge into the algorithms and operating practices is well understood. Some wind forecast service providers offer uncertainty information for their forecasts. Several U.S. companies have developed wind generation forecasting tools with built-in capability to assess wind generation uncertainty [1,2]. Similar tools have been built in Europe. For example, in the European Union project, ANEMOS created a tool for online wind generation uncertainty estimation based on adaptive resampling or quantile regression [3]. A German company has proposed a tool for wind generation forecasting, assessing the uncertainty ranges associated with wind forecast, and predicting extreme ramping events [4]. Reference [5] discusses a wind generation interval forecast approach using the quantile method. Reference [6] used statistical analysis based on standard deviation to predict wind generation forecast errors. Work is under way to incorporate these uncertainties into power system operations [7,8].

Renewable Energy Integration. http://dx.doi.org/10.1016/B978-0-12-407910-6.00013-2

Unfortunately, in many cases, these efforts are limited to wind or solar generation uncertainties; they ignore the additional sources mentioned at the beginning of the chapter.[1] Moreover, these approaches, while considering the megawatt imbalances, do not address such essential characteristics as ramp (megawatts per minute) and ramp duration uncertainties (minutes), which are required from the generators participating in the balancing process.

This chapter considers sources of uncertainty and variability, overall system uncertainty model, a possible plan for transition from deterministic to probabilistic methods in planning and operations, and one example of uncertainty-based tools for grid operations.

This chapter is based on work conducted at the Pacific Northwest National Laboratory (PNNL).[2,3]

2. Sources of uncertainty and variability

Variable generation and system load are far from being the only sources of uncertainty. Additional uncertainty is created by uninstructed deviations of conventional generators from their set-points, forced outages of conventional generation and load drops, major intermittent loads, loss of major supply transmission facilities, frequency variations, and other sources influencing power balance in a control area. All these sources must be accounted for.

2.1 Load forecast errors

The load forecast errors, ΔL, contribute significantly to the overall uncertainty of the system's balancing requirement. In the operational environment, the load forecasts usually are provided for the next operating day (hourly block energy schedules), next operating hour (hourly block energy schedules or schedules for average load for smaller dispatch intervals whenever they are in use), and in real-time forecast (average load for within-hour dispatch intervals, e.g. 5, 10, or 15 ;min). The day-ahead error mean absolute percent error (MAPE) usually stays within several percentage points of the maximum load—see Figure 1(a). Instantaneous error values can significantly exceed MAPE from time to time, contributing to so-called tail events [10].

Load forecast errors depend on multiple factors including temperature and humidity forecast errors. This sensitivity can change with the air temperature—Figure 1(b). This forecast bias could create a significant problem (i.e. systematic overestimation and underestimation of the system load).

2.2 Wind power forecast errors

Issues related to the wind forecast error statistical characteristics attract a significant interest from researchers and practicing engineers. A review of the state-of-the-art practices in this area is given in [12].

---

[1]One of known exceptions is the comprehensive tool developed by Red Eléctrica de España (REE), the Spanish Transmission System Operator [9].

[2]The work and results included in this chapter were funded by the California Energy Commission. Additional funding came from the U.S. Department of Energy (DOE) Offices of Energy Efficiency and Renewable Energy (EERE), and Electricity Delivery and Energy Reliability (OE).

[3]Significant contributions to the work reflected in this chapter were made by the following PNNL engineers: Nader Samaan, Ning Lu, Pengwei Du, Ryan Hafen, Zhangshuan (Jason) Hou, Zhenyu (Henry) Huang, Krishnappa Subbarao, and others.

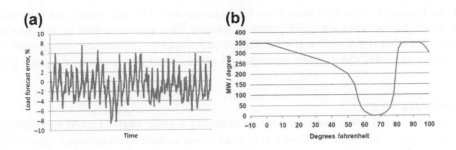

FIGURE 1

(a) Example of load error in a real-time system [10]; (b) impact of temperature forecast error on load [11].

Wind power forecasting errors sometimes are simulated using truncated normal distribution, whose characteristics are determined by curve fitting [13]. The truncation process is applied to reflect the natural limits posed by minimum (zero) and maximum (installed capacity minus the current capacity derate due to offline units) wind generation. This model cannot be accepted without caution because of indications that the wind power forecast error does not necessarily follow the normal distribution. Efforts are in place to propose better approximations for the error. For instance, beta distribution has been used in [14]. In [15], a different approach based on experimental probability density functions (PDFs) was proposed instead. The approach is suitable for handling nonparametric distributions, so that no hypothesis is needed regarding the wind generation error distribution law.

A continuing effort is in place aimed at improving the accuracy of wind generation forecasting algorithms. For example, in Germany, the day-ahead wind generation forecast error has been reduced to 4.5% [16]. Nevertheless, significant challenges remain with the very short-term forecasts. The persistence forecast model frequently demonstrates better performance than more scientific approaches. Prediction of wind generation ramps remain a challenging problem.

Wind generation forecast errors are sensitive to multiple external factors. In [17], a Bayesian model was developed that reflected the influence of external factors on the wind and loaf forecast errors.

2.3 Solar generation forecast errors

Characteristics of the solar power forecast errors have not been sufficiently well studied. In [18], a new model was proposed. This model has been used in 20–33% of California ISO's renewable generation penetration studies.

2.3.1 Variation of solar generation

Solar radiation (or solar irradiance) that determines the level of solar energy production at any specific location is neither completely random nor completely deterministic. Extraterrestrial (above clouds) solar radiation can be predicted confidently for any place and time interval. Solar radiation shows both yearly and daily variation. The area's atmospheric conditions (clouds, dust storms, etc.) determine the randomness of solar radiation at the ground level (also called global solar radiation). The ranges of yearly variation can be described by monthly maximum solar radiation for a sunny day, and the minimum solar radiation for a total cloudy day. The maximum and minimum solar radiation levels can be used for solar power generation forecasting as well as for forecast error simulations. The effect of clouds and other factors (e.g. water or ice concentration, types of water particles or ice crystals, water vapor amount, and aerosol type and amount) on solar generation varies.

2.3.2 Clearness index

The Clearness Index (CI) shows what percentage of the sky is clear. High CI could mean higher global solar radiation (i.e. global solar radiation levels being closer to their extraterrestrial values) and lower forecast errors. CI is used for solar power generation forecasting.

If the sky is clear, solar radiation and solar power production are more predictable based on the annual and daily extraterrestrial pattern. Thus, solar forecast errors are relatively small.

2.3.3 Simulating the solar generation forecast errors

Statistical characteristics of hour-ahead and real-time solar generation forecast errors are complex and depend on various factors, including the CI–solar radiation, types of solar generators (photovoltaic, concentrated thermal, etc.), geographic location, and spatial distribution of solar power plants, and other factors.

2.3.4 Standard deviation of the forecast error evaluated using CI

Different solar generation patterns during the day and night need to be considered in solar forecast errors. At night, solar irradiance is zero, and thus solar forecast errors are zero. Sunrise and sunset time are different in different seasons at different regions. Previous years' information regarding this matter can be categorized and used for solar forecast error evaluation.

The standard distribution of solar forecast errors can be described as a function of CI. Figure 2 shows the possible distribution of the standard deviation of solar forecast errors depending on the CI.

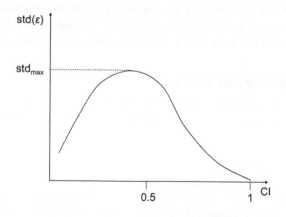

FIGURE 2

Distribution of the standard deviation of solar forecast errors depending on the Clearness Index (CI).

2.4 **Forced generation outages**

Generator outages are addressed by specially procured contingency reserves. The imbalances caused by forced generator outages initially are mitigated by the system governor response and automatic generation control (AGC) system, and then by committing and dispatching generation resources suitable for the intrahour balancing purposes, or by applying load reduction schemes. It could take 5–15 min or more to activate these systems. As a result, the system imbalances caused by forced generation outages could last for about 5–15 min. A schematic model for the balancing requirement is shown in Figure 3.

After a forced outage, the corresponding control area is subjected to a sudden imbalance, which depends on the level of generation on the tripped unit, system inertia, and available frequency (governor) response. The part of the initial imbalance addressed by the governor response does not contribute to the balancing requirement because of the fast recovery process involved (seconds). The AGC system in the affected control area starts to move regulating units to cover another part of the system imbalance (minutes). The job done by regulating units is addressing a part of the overall balancing requirement. Frequently, the available regulating reserve is not sufficient to completely restore the system balance. The imbalance stays in the system for about 10 min until the contingency reserve units are started, synchronized with the system (nonspinning reserve), and dispatched to the desired level (both nonspinning and spinning reserves). Then, the AGC units are moved back to completely restore the balance.

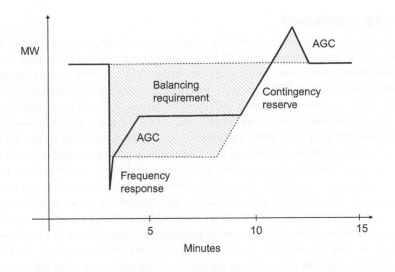

FIGURE 3

Balancing requirement caused by forced generation outage. AGC, automatic generation control.

2.5 Uninstructed deviation errors

The impacts of uncertainties caused by uninstructed deviations of conventional generators on the system balancing requirements frequently are neglected. Nevertheless, the total uninstructed deviations resulting from their inability to follow the set-points precisely could reach hundreds of megawatts and may have a profound impact on the system balancing requirements.

2.6 Discretization errors

Discretization errors are caused by the difference between scheduled values of the system net load (the balancing requirement) within a dispatch interval and the actual minute-by-minute variations of the net load. Unlike the forecast errors, the discretization errors are functions of the variability of balancing requirement and the size of dispatch intervals.

3. Overall uncertainty characteristics

The sources of uncertainty interact in a complicated statistical way, so that most of the time, their combined impact is reduced when compared with the sum of impacts of individual sources. The resulting statistical model of the uncertainties consists of continuous and discrete variations of multiple uncertain and variable parameters. It can form nonparametric distributions, include nonstationary stochastic processes; incorporate autocorrelation and cross-correlation moments between the particular parameters; and depend on other influencing external factors.

3.1 Balancing requirement

Power system balancing processes, including scheduling, real-time dispatch (load following), and regulation processes, traditionally are based on deterministic models. These models work relatively well with predictable traditional generation resources. The ever-increasing penetration of the renewable generations, such as wind and solar generation, causes significant challenges with handling the uncertainties and variability associated with these resources, and with incorporating the uncertainty information into grid operation and market functions. At the moment, possible random deviations of system generation and load from their estimated values are not reflected in the existing balancing processes and market applications, such as unit commitment and economic dispatch tools. With the increasing presence of variable resources, the deviations could result in unexpected power balancing problems and could cause serious risks to system reliability and control performance. Without a look-ahead assessment of the deviations and associated risks, system operators would have limited information about the likelihood and magnitude of the problems. Furthermore, these deviations could require procuring additional reserves and costly balancing services. There is a need to know whether the system will be able to meet the balancing requirements within the look-ahead horizon, what additional balancing capacity and ramping capability are needed, and what additional costs will be incurred by those needs.

An interconnected power system usually consists of one or multiple balancing authorities (BAs). Each BA must maintain a balance between its generation, load, interchange, and losses. The system's conventional generation is committed and dispatched to meet the total balancing requirement, BR

(MW). BAs' performance is judged based on control performance standards (CPS), requiring a certain degree of success in keeping the system imbalance within certain (sometimes statistically defined) bounds. The system balancing requirement, BR, which is the same as net load, and which is the balancing job required from conventional and renewable generators, power exports, and energy storage facilities, is expressed by the following:

$$BR = \Delta L - \Delta WG - \Delta SG - FO - UD + VDE - 10 \cdot B \cdot \Delta F - TE \qquad (1)$$

where Δ denotes the difference between the actual and the prescheduled values for dispatch intervals; L is the system load; WG is wind generation; SG is solar generation; FO reflects imbalances caused by forced generation outages; UD is the total uninstructed deviation of conventional generation units (including failure to start up); VDE is the variability errors within a dispatch interval (or scheduling discretization error caused by the difference between the block energy schedules and the continuous actual variation of BR); $10 \cdot B \cdot \Delta F$ is a frequency-dependent term; and TE is the time error correction term. Depending on the context and information availability, the meaning and the actual presence of the terms in Eqn (1) can change. For instance, the uninstructed deviation term, frequency-dependent term, and time error term are omitted and the actual values are replaced by their forecasts in the scheduling and real-time dispatch procedures.

Besides the capacity requirement in megawatts, BR uncertainty quantification should include such additional parameters as BR ramps (megawatts per minute), BR ramp duration (minutes), and cycling characteristics required from generators and energy storage facilities. These BR requirements all must be met to successfully balance the system.

3.2 Nonparametric nature of balancing requirement

Nonparametric distributions cannot be described using a standard probability distribution (e.g. the normal distribution), although sometimes they can be approximated using a standard distribution or a combination of distributions with certain limited accuracy. For instance, in [19], the truncated normal distribution was used to approximate distributions of wind and load forecast errors. The balancing requirement distribution could be a more difficult case. Sometimes its shape becomes essentially nonparametric. Figure 4 contains an example of real-life histograms of a nonparametric distribution of the regulation requirement in a large BA with wind and solar variable energy resources.

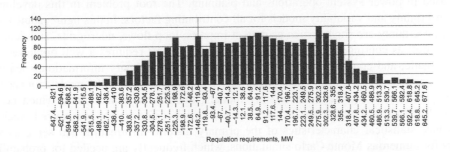

FIGURE 4

Hourly regulation requirement for a real balancing authority.

Frequently, the BR distribution exhibits long "heavy" tails indicating certain limited probability of major imbalances caused by unfortunate combinations of random factors contributing to BR in Eqn (1). Normally, the central part of the BR distribution could be or should be balanced using the existing balancing reserves, whereas the tail could create infrequent but significant problems because it requires balancing reserves that normally are not procured in the system.

3.3 Autocorrelation, cross-correlation, and nonstationary nature of forecast errors

The load and wind forecast errors usually exhibit strong autocorrelation between the subsequent forecasts and cross-correlation between different forecasts. Autocorrelation means that, for example, if for certain operating hours, a large positive forecast error is observed, it is likely that a similar error would be observed for the next hour [15]. Cross-correlation means statistical dependence. In nonstationary processes, statistical characteristics of forecast errors change over time.

4. Probabilistic operations and planning

The increasing randomness and variability impacts on the modern power systems are changing—and will be changing dramatically—the patterns of system behavior, how it is dispatched, and exchanges of energy. The existing deterministic approaches in utility control rooms are based on established dispatches and flow patterns, a few "typical" stresses, and known congested paths. Because of this, they are becoming increasingly inadequate for the uncertainty problem. A new generation of probabilistic methods, reliability and control performance criteria, tools, and business practices is very much needed. Existing probabilistic approaches, however, sometimes are addressing only the surface of the actual uncertainty problem because they represent just incremental or partial solutions, without recognizing the fundamental fact that modern power systems gradually are transforming into a completely different object—a huge stochastic machine.

There is a huge need to develop key elements of a new grid methodology based on probabilistic system models and performance and reliability criteria. The new methodology will address the fundamental new ways of system modeling, prediction, stressing, analysis, and control.

Figure 5 shows one conceptual view of the transition from deterministic to probabilistic technologies used in power system operations and planning. The root problem in this development is building uncertainty and variability models for all contributing factors. In this figure, some additional sources of uncertainty are shown, such as market impacts, loop flows, microgrids, and demand-side controls, as well as contingencies. In many instances, this effort requires a well-organized, coordinated, systematic, and continuous collection and processing of primary information, such as transmission lines and generator outage rates. The next step is building uncertainty models for each source, as well as the overall large-scale uncertainty model (statistical system model), which becomes a geographically disperse model when the transmission aspect of uncertainty is analyzed. Such models can reproduce statistical characteristics of the primary historic information and serve as scenario generators for numerous Monte Carlo simulations, which frequently are needed for probabilistic analyses. The high-performance computational framework becomes important because of the enormous computational challenge of probabilistic methods. Branches of the "tree" represent some other tasks that can be implemented using the new probabilistic framework. Together, with the addition of many

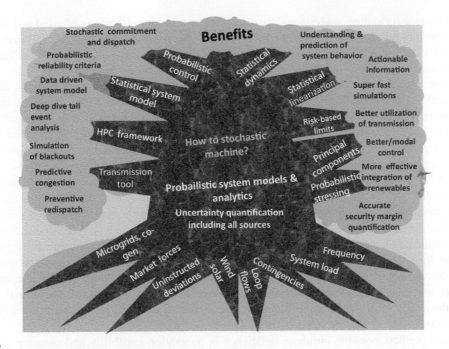

FIGURE 5

HPC = High-performance computing.

other tasks, they will produce a new understanding and a new look on power system planning and operation, and decision support tools for grid operators. Some new approaches will be developed, for instance, statistical linearization or probabilistic definition of system stresses. The global uncertainty quantification is linked closely with better prediction of system parameters and behavior.

5. Three levels of uncertainty integration in operations

Three possible modes of integration uncertainty information into system operation have been proposed: passive, active, and proactive [10,20]. Passive integration is the first level of integration, which brings awareness of uncertainties into control center software tools through information visualization and alarming. Active integration uses the uncertainty information to modify existing grid operation functions, such as unit commitment. Proactive integration develops new grid operation functions enabled by the uncertainty information.

6. Example: California ISO ramp uncertainty prediction tool

To assist the U.S. national objective of wider penetration of renewable resources into the existing generation mix without compromising system reliability and control performance, PNNL has developed a tool for online analysis and visualization of operational impacts of wind and solar generation. This tool, intended for use by the BAs, predicts and displays additional capacity and ramping requirements caused by the variability and uncertainties in forecasting loads and variable generation. The prediction is made for the next operating hours as well as for the next day [20].

The tool dynamically and adaptively correlates changing system conditions with the additional ramping and regulation needs triggered by the interplay between forecasted and actual load and output of variable resources. The approach includes three steps: forecast data acquisition, statistical analysis of retrospective information, and prediction of grid-balancing requirements for a specified time horizon and a given confidence level. An assessment of the capacity and ramping requirements is performed using a specially developed probabilistic algorithm based on histogram analysis, which is capable of incorporating multiple sources of uncertainty—both continuous (wind and load forecast errors) and discrete (forced generator outages and startup failures). A new method called the "flying-brick" technique has been developed to evaluate the look-ahead required generation performance envelope for the worst-case scenario within a user-specified confidence level. A self-validation process is used to validate the accuracy of the confidence intervals. These displays are created using the most up-to-date load and wind forecast along with the associated uncertainties, resources committed through the various market runs, generator-forced outage information, transmission line outage information, and related stochastic relationships between the input data sets. These visual displays forecast the ability of dispatchable resources to meet expected load and intermittent generation ramp requirements.

The tool predicts possible intrahour deficiency in generation capability and ramping capability. This deficiency of balancing resources can cause price spikes in a real-time market. An actual instance of the prediction and subsequent events are shown in Figure 6 and Figure 7.

The orange bands in the first graph show the generation requirements predicted by the tool at 90% confidence level (lighter) and 95% confidence level (darker). The gray band represents online

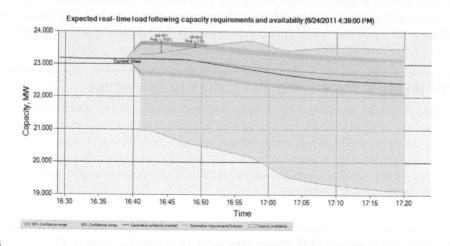

FIGURE 6

Predictions by the ramping tool of the deficiency of intrahour balancing capacity.

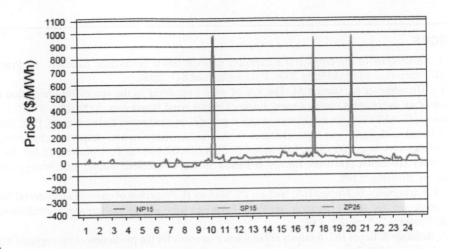

FIGURE 7

NP15, SP15, ZP26 are California ISO market pracing zones (June 24, 2011, Mountain Time) [21].

generation capacity availability in the system for both upward and downward needs. The blue line shows the generation dispatch, and the orange line shows the forecasted generation requirements. One can see that the deficiency of about 340 MW of the system generation upward capacity is predicted (red error bars). Figure 7 shows the real-time prices for corresponding times. As predicted by the tool, the deficiency occurred between 16:30 and 17:00 (Pacific Time), leading to a price spike. In response, with the tool fully deployed, corrective actions (additional unit commitment) could be taken by real-time dispatchers when the tool is fully deployed.

The tool has been enhanced to predict CAISO's regulation reserve requirements for the next operating day. This prediction has led to a demonstrable cost reduction by minimizing the regulation capacity procurement without compromising grid reliability. Further enhancements and integration with CAISO's real-time operations are under consideration.

7. Conclusion

The increasing presence of variable energy resources in the modern power systems, along with the other uncertainties and sources of variability, such as system load, create a significant random impact on system operation and utility control room functions, including the generation commitment and dispatch processes. Deterministic approaches and criteria are becoming increasingly inadequate to the current and especially future grid operator needs. A new generation of probabilistic reliability standards, operating procedures, and control room tools must be developed. Although this time-consuming effort is continued, some aspects of system operation can be modified based on probabilistic approaches right now. This chapter suggests a possible vision for the future immanent changes as well as some specific initial steps and tools to start these changes immediately.

References

[1] Zack J. An analysis of the errors and uncertainty in wind power production forecasts. In: Proc. WIND-POWER conference and exhibition 2006, Pittsburgh; June 4–7, 2006.

[2] Lerner J, Grundmeyer M, Garvert M. The role of wind forecasting in the successful integration and management of an intermittent energy source. Energy Central Wind Power July 2009;3(8).

[3] Kariniotakis G. ANEMOS, leading European Union research on wind power forecasting. In: Proc. international wind forecast techniques and methodologies workshop. Available online: http://www.bpa.gov/corporate/business/innovation/docs/2008/BPA_California%20ISO%20ANEMOS%20Presentation.pdf; July 24–25, 2008.

[4] Wind power prediction previento. Energy & Meteo Systems. Available online: http://energymeteo.de/de/media/e_m_Broschuere.pdf.

[5] Pinson P, Kariniotakis G, Nielsen HA, Nielsen TS, Madsen H. Properties of quantile and interval forecasts of wind generation and their evaluation. In: Proc. European wind energy conference & exhibition, Athens, Greece; February 2–March 2, 2006.

[6] Luig A, Bofinger S, Beyer HG. Analysis of confidence intervals for the prediction of the regional wind power output. In: Proc. 2001 European wind energy conference, Copenhagen, Denmark; July 2–6, 2001.

[7] Kehler J, Hu M, McMullen M, Blatchford J. ISO perspective and experience with integrating wind power forecasts into operations. In: Proc. IEEE general meeting, Minneapolis; July 25–29, 2010.

[8] Maggio D, D'Annunzio C, Huang S-H, Thompson C. Utilization of forecasts for wind-powered generation resources in ERCOT operations. In: Proc. IEEE PES general meeting, Minneapolis; July 25–29, 2010.

[9] Red Eléctrica de España (REE). Wind development, integration issues and solutions – TSO Spain. Presentation in the Northwest Wind Integration Forum, Portland, Oregon (Online) Available: http://www.nwcouncil.org/energy/wind/meetings/2010/07/; July 29 and 30, 2010.

[10] Makarov YV, Huang Z, Etingov PV, Ma J, Guttromson RT, Subbarao K, et al. Wind Energy management system EMS integration project: incorporating wind generation and load forecast uncertainties into power grid operations. PNNL-19189. Richland (WA): Pacific Northwest National Laboratory; 2010 (Online) Available: http://www.pnl.gov/main/publications/external/technical_reports/PNNL-19189.pdf.

[11] Walshe P. Temperature and demand forecasting in a large utility. TVA Presentation. (Online) Available: http://www.isse.ucar.edu/electricity/workshop/Presentations/pdf/Walshe.pdf.

[12] Monteiro C, Bessa R, Miranda V, Botterud A, Wang J, Conzelmann G. Wind power forecasting: state-of-the-art 2009. Report ANL/DIS-10-1. Illinois: Argonne National Laboratory; November 6, 2009 (Online) Available: http://www.osti.gov/energycitations/product.biblio.jsp?osti_id=968212.

[13] Constantinescu EM, Zavala VM, Rocklin M, Lee S, Anitescu M. A computational framework for uncertainty quantification and stochastic optimization in unit commitment with wind power generation. IEEE Trans Power Syst February 2011;26(1):431–41.

[14] Bludszuweit H, Dominguez-Navarro JA, Llombart A. Statistical analysis of wind power forecast error. IEEE Trans Power Syst August 2008;23(3):983–91.

[15] Makarov YV, Reyes-Spindola JF, Samaan N, Diao R, Hafen RP. Wind and load forecast error model for multiple geographically distributed forecasts. In: Proc. 9th international workshop on large-scale integration of wind power into power systems as well as on transmission networks for offshore wind power plants, Québec City, Québec, Canada; October 18–19, 2010.

[16] Ernst B, Schreirer U, Berster F, Scholz C, Erbring HP, Schlunke S, et al. Large scale wind and solar integration in German. PNNL-19225. Richland (WA): Pacific Northwest National Laboratory. (Online) Available: http://www.pnl.gov/main/publications/external/technical_reports/PNNL-19225.pdf.

[17] Lu S, Makarov YV, Brothers AJ, McKinstry CA, Jin S, Pease J. Prediction of power system balancing requirement and tail event. In: IEEE PES transmission and distribution conference; 2010. pp. 1–7.

[18] Makarov YV, Ma J, Loutan C, Rosenblum G. Solar forecast error simulation methodology for CAISO 33% renewables study. PNNL technical report, prepared for CAISO; January 2010.

[19] Makarov YV, Loutan C, Ma J, de Mello P. Operational impacts of wind generation in California. IEEE Trans Power Syst May 2009;24(2).

[20] Makarov Y, Etingov P, Ma J, Huang Z, Subbarao K. Incorporating uncertainty of wind power generation forecast into power system operation, dispatch, and unit commitment procedures. IEEE Transaction Sustainable Energy October 2011;2(4):433–42.

[21] California ISO. Real-time daily market watch for operating day of 06/24/11. (Online) Available: http://www.caiso.com/Documents/June%202011/DailyMarketWatch_Real-Time_Jun_24_2011.pdf.

Connecting Renewable Energy to Power Grids

PART

5

Connecting
Renewable Energy
to Power Grids

Global Power Grids for Harnessing World Renewable Energy

14

Spyros Chatzivasileiadis[1], Damien Ernst[2], Göran Andersson[1]

[1] *Power Systems Laboratory, ETH Zurich, Zurich, Switzerland,* [2] *Institut Montefiore, University of Liège, Liège, Belgium*

1. Introduction

Increased environmental awareness has led to concrete actions in the energy sector in recent years. Examples are the European Commission's target of 20% participation of renewable energy sources (RESs) in the EU energy mix by 2020 [1] and California's decision to increase renewable energy in the state's electricity mix to 33% of retail sales, again by 2020 [2]. At the same time, several studies have been carried out investigating the possibilities of a higher share of RESs in the energy supply system of the future. For instance, the German Energy Agency assumes 39% RES participation by 2020 [3], whereas a detailed study from the National Renewable Energy Laboratory suggests that meeting the US electricity demand in 2050 with 80% RES supply is a feasible option [4]. Czisch [5] discusses a 100% renewable energy supply system in Europe with interconnections in North Africa and West Asia. Jacobson and Delucchi [6] more recently investigated "the feasibility of providing worldwide energy for all purposes (electric power, transportation, heating/cooling, etc.) from wind, water, and sunlight." The authors made a detailed analysis and proposed a plan for implementation. They found that the barriers to the deployment of this plan are not technological or economic but rather social and political.

All these studies suggest that the development of the electricity network will play a crucial role in the efficient integration of increasing shares of RESs. Two reasons are most often mentioned. First, interconnecting RESs increases the reliability of their supply. Second, long transmission lines can help harvest renewable energy from remote locations with abundant potential and very low production costs. To exploit the benefits of such interconnections, concrete actions have been taken that will lead to the creation of regional supergrids. EU guidelines already encourage transmission projects, such as the Baltic Ring [7]. Projects such as Medgrid (www.medgrid-psm.com) and OffshoreGrid (www.offshoregrid.eu/) have been launched to interconnect Mediterranean states with Europe and transfer renewable energy to the major load centers. At the same time, initiatives such as Gobitec (www.gobitec.org/) in Asia and Atlantic Wind Connection (www.atlanticwindconnection.com) in the USA aim to interconnect the Asian power grids or transmit offshore wind energy to the east coast of the USA.

In Ref. [8] we suggested the next natural step of the electricity network: the Global Grid. With growing electricity demand, the need for green energy resources also will increase. The electricity networks will expand to harvest the renewable potential abundant in remote locations, forming supergrids of increasing size. The Global Grid aims at interconnecting the regional supergrids into one global

Renewable Energy Integration. http://dx.doi.org/10.1016/B978-0-12-407910-6.00014-4

175

electricity network. High-capacity long transmission lines will interconnect wind farms and solar power plants, supplying load centers with green power over long distances. Besides introducing the concept, in Ref. [8] we further highlighted the multiple opportunities emerging from it. We supported our analysis with studies of the economic feasibility of such a concept, and we further discussed possible investment mechanisms and operating schemes.

In this chapter, we introduce and elaborate on four possible stages that could gradually lead to the development of a globally interconnected power network. We extend our analysis with additional studies of the economic competitiveness of long transmission lines in different world regions, and we show that substantial profits can arise from intercontinental electricity trade.

Section 2 introduces the four possible development stages followed a brief illustration of the concept as we have envisioned it in Section 3. Sections 4–6 describe in detail all four development stages and provide examples, with quantitative analyses for each stage. In Section 7 we briefly discuss the additional opportunities emerging from a Global Grid. We conclude this chapter with Section 8.

2. Stages toward a global power grid

We expect that three main reasons will act as the major incentives toward the creation of a globally interconnected network. First, the need to harvest remote renewable energy resources–located either further off-shore or in deserts–will lead to continuously expanding regional supergrids. Second, taking advantage of the shift in peak demand periods between continents, remote RES plants located at similar distances from two regions can connect and sell their power always at peak price. Third, the time zone diversity between continents creates opportunities for electricity arbitrage, which can lead to a profitable electricity trade. Based on these reasons, in this section we present the four main stages we envision as leading to the development of a Global Grid environment.

The main driving force behind a Global Grid will be the harvesting of remote renewable resources. In the search for green electricity, new sites located even further away from the existing load centers and the current power grids, will be exploited. Deserts such as the Sahara or the Gobi serve as examples for solar power plants, and on-shore or off-shore locations with high winds such as the shores of Greenland or the Indian Ocean serve as examples for wind farms. This constitutes the first stage toward a Global Grid. Our cost–benefit analysis in this chapter will focus on the wind potential of such remote locations.

By building wind farms and solar parks in remote locations, a point will be reached at which an RES power plant will be equidistant from two power systems on different continents. A wind farm in Greenland, for instance, would be a realistic example of such a situation. Our analysis in Ref. [8] showed that connecting such a wind farm to both Europe and North America is a profitable solution. In this second development stage of the Global Grid, remote RES power plants can take advantage of the time difference between the continents to sell their power always at peak prices. For example, the wind farm in Greenland can sell its produced power 50% of the time during the peak demand in Europe and 50% of the time during the peak demand in North America. From there, an interconnected global power grid can start to form.

With an electrical connection between two continents, opportunities for electricity trade emerge, signaling the progression to the third stage. Wind (and solar) production is intermittent and most of the time does not use the total cable capacity. In addition, peak demand between the continents is shifted in time, which leaves room for electricity arbitrage. Therefore, electricity can be bought at lower prices in one area and sold at higher prices during peak demand in other areas. In Section 5.2 we explore the electricity trade potential for the Europe–North America connection over Greenland, using the remaining cable capacity.

As we will see in Section 5.2, the transmission corridor enabling electricity trade between the continents can result in substantial profits. Based on that, in the fourth stage of the development towards a Global Grid, direct interconnections between countries or continents can start to be built, independent of their connection to remote areas with high RES potential. Such an analysis for a cable between Europe and North America has already been carried out in Ref. [8], showing that, except for the most expensive RES generators, it would be more economical for the USA to import RES power from Europe than operate its own fossil fuel power plants. Here, we extend this analysis by considering real electricity prices in Germany and the USA and by examining the amortization period of such an investment.

Introducing the four development stages helps create a path leading to a Global Grid environment. It does not imply, nevertheless, that global interconnections will move through the four stages in a sequential manner. For example, the direct interconnections in the fourth stage can directly occur after the first stage; or the second and third development stage can occur simultaneously, for example, as we will see in Section 5, an off-shore wind farm can connect to both continents in order to sell at peak prices and, at the same time, benefit from the electricity trade.

3. The global grid: an illustration

The current section is devoted to a brief description of the Global Grid as we envision it. We hope this will produce a better understanding of this concept, first proposed in Ref. [8]. Figure 1 illustrates a possible global grid. We envision that the power supply of the Global Grid will depend on RESs. Large

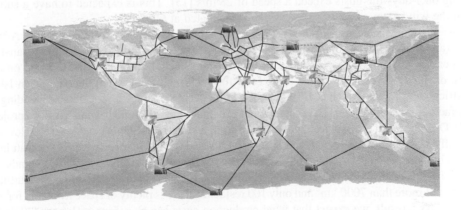

FIGURE 1 Illustration of a possible Global Grid.

The dotted lines indicate the high-voltage direct current (HVDC) lines with a length of more than 500 km that are already in operation; the lines currently in the building/planning phase are indicated by dashed lines (the list of the illustrated HVDC lines is not exhaustive). The locations of the renewable energy source power plants are based on solar radiation maps, average wind speeds, and sea depths [8].

renewable potential exists in remote locations, such as in deserts or offshore. Long high-voltage direct current (HVDC) lines will constitute the main arteries in a Global Grid environment, transmitting bulk quantities of power over long distances. HVDC lines are superior to alternating current technologies for long-distance transmission because they exhibit lower losses, provide active and reactive power support, and can connect nonsynchronous grids. Issues pertaining to the power generation and transmission by the Global Grid are described in more detail in Ref. [8].

4. Harvesting RESs from remote locations

Tapping the renewable potential in Greenland, as mentioned in Section 2, would be a realistic example of how we could progress to global interconnections. Greenland was selected in Ref. [8] as a representative example for three reasons. First, it has a significant wind and hydro potential [6,9]. Second, it is close to Iceland; Ref. [10] has already shown that the Iceland–UK interconnection is a viable option, and the two governments are currently discussing its possible realization [11]. Third, all interconnecting sections along this route have lengths or sea depths that are comparable to currently existing projects (see Ref. [8] for more details). Also important is that Greenland lies at an equal distance from both Europe and North America. Later in this chapter we extend our analysis of Greenland by exploring the possibilities for intercontinental trade.

In this section, we move to the southern hemisphere and study in more detail a wind farm on the Kerguelen Islands. These are a group of islands in the southern Indian Ocean that belong to France. Their climate is similar to that of Iceland and the Falkland Islands, with an average temperature between 0 and 10 °C [12]. The Kerguelen Islands are characterized by high continuous winds, with an average speed of about 9.7 m/s. For about 312.9 days a year they experience wind gusts above 16 m/s, and during 68.1 days the gusts exceed a speed of 28 m/s [13]. This is expected to have a substantial effect on the wind farm capacity factor, which is estimated at 60–70%.[1]

The Kerguelen Islands are located approximately an equal distance between South Africa and Australia (about 4000 km to South Africa and about 4150 km to Western Australia). In this analysis we assume a cable length of 4150 km and we focus on the connection to South Africa. Based on our calculations, detailed in Appendix A, the cable costs for a 4150-km route from the Kerguelen Islands to South Africa would lie between 0.019 and 0.054 USD per delivered kilowatt hour, depending on the capacity factor of the wind farm. The cost for wind generation for 2020 and beyond has been projected in Ref. [14]. Onshore wind costs are estimated to start from less than 0.04 USD per delivered kilowatt hour, whereas offshore wind is projected to cost between 0.08 and 0.13 USD per delivered kilowatt hour. All cost projections in Ref. [14], assumed a wind capacity factor of 40%. Given the strong winds that the Kerguelen Islands receive, the wind capacity factor in that region is likely be around 60%. Furthermore, with an area of more than 7000 km[2] and only 100 residents, there is plenty of space available for onshore wind farms. As a result, we expect that wind production costs can start from as low as 0.02 USD per kilowatt hour for onshore and should not exceed 0.09 USD per kilowatt hour for offshore wind farms.[2]

---

[1]Wind turbines are typically designed to reach their rated output power for wind speeds in the range between 14 and 25 m/s. Because of a lack of additional wind data for the Kerguelen Islands, we can assume that the wind turbines to be installed there will be designed such that the cut-out speed will be around 28 m/s.

[2]Kerguelen islands have resulted from volcanic formations. In such cases the sea depth increases quickly a few miles off the coast. Therefore, we expect that the majority of the wind farms installed on the Kerguelen islands would be onshore.

Figure 2 presents the total cost per delivered kilowatt hour, including both production and transmission to South Africa, plotted against varying wind production costs. In the graph we account for three different capacity factors of the wind farm—40%, 50%, and 60%—as well two cable cost projections: a high-cost cable and a low-cost cable. Each colored area represents a different capacity factor, with the lower costs per delivered kilowatt hour corresponding to a capacity factor of 60% and the higher costs corresponding to a capacity factor of 40%. The bottom border of each colored area represents the low-cost cable projections, whereas the top border stands for the high-cost cable projections. As a result, the total wind production and transmission costs will lie somewhere within a colored area, depending on the wind energy production costs and the cable costs. Besides the total costs for wind, in Figure 2 we also plot the expected revenues per kilowatt hour for wind energy production as determined by the South African government in 2011. In August 2011, South Africa launched a competitive bidding process for RESs. Two rounds of bids took place. The selected projects are expected to be commissioned by June 2014 (June 2015 for concentrated solar power plants) [15]. The average indexed bid prices for wind, expressed in US dollars, were 0.11 USD/kWh in the first round and 0.09 USD/kWh in the second round. These prices also are plotted as lines in Figure 2.

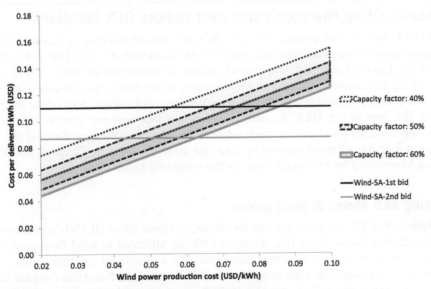

FIGURE 2 A wind farm on the Kerguelen Islands supplying South Africa.

Total cost per delivered kilowatt hour (production and transmission) for varying wind energy production costs, wind capacity factors, and cable costs. Each colored area corresponds to a different capacity factor and presents the cost spectrum depending on varying cost projections for the cable transmission and the wind power production. With Wind-SA-Bid are shown the constant prices per kilowatt hour that wind power producers in South Africa will receive, as determined in two rounds of a competitive bidding process.

As can be observed, for a wind capacity factor of 60%, the wind farm on the Kerguelen Islands can compete with local South African wind farms if the wind energy production costs are less than 0.085 USD/kWh for the low-cost cable projection. In the case of higher cable installation costs, the wind energy production costs should fall below 0.075 USD/kWh. Given the high wind potential of this region, and the fact that wind farms can be built on the island, the wind energy production costs, even by today's standards, are not expected to surpass these values. With decreasing capacity factors, as we observe in Figure 2, the wind energy production costs should decrease so that wind farms can remain competitive with those in South Africa. With a 40% capacity factor, the wind production costs should not exceed 0.033–0.055 USD/kWh in the worst case. Taking into account that Ref. [14] projects the production costs for onshore wind farms to start at less than 0.040 USD/kWh, values such as 0.033 USD/kWh are highly probable by 2020 and beyond.

By being already connected to South Africa, at a later stage the wind farm can also connect to Australia to take advantage of the time zone diversity and facilitate electricity trade. As the distance from the Kerguelen Islands to Australia is similar to the distance from the islands to South Africa, the cost projections, as shown in Figure 2, also are valid for the cable connection to Australia.

5. Interconnecting two continents over remote RES locations

In our analysis in Ref. [8], we assumed that a 3-GW wind farm off the eastern shores of Greenland is feasible; some investors have decided to connect a wind farm with a 3-GW line to Europe through Iceland and the Faroe Islands. We investigated whether a connection to North America would be profitable, taking into account two effects. First, because of the time zone difference, the wind farm will be able to sell its produced power always at peak prices, for example, 50% of the time to Europe and 50% of the time to the USA. Second, by creating a link between Europe and the USA via Greenland, opportunities for electricity trade between the continents emerge. Because the wind farm can produce power for a limited amount of time (we assume a capacity factor of 40%), the cable capacity can be reserved for electricity trade for the remaining hours.

5.1 Offering RES power at peak prices

In our analysis in Ref. [8], we estimated that the cable can deliver about 20 TWh/year after taking into account transmission losses. From this, about 10 TWh are allocated to wind farm production. This means that for about 50% of the time the cable capacity is available for electricity trade.

By building the transmission route from Greenland to the USA, we found that the wind farm's costs per delivered kilowatt hour would increase by 21–25%, assuming production costs of 0.06 USD/kWh.[3] If off-peak prices are half of peak prices, the revenues will increase by 31–33%, which results in additional profits of 7–12% for the wind farm, as shown in Table 1.

---

[3] As already mentioned, Ref. [14] projects production costs of less than 0.04 USD/kWh for onshore and 0.08–0.13 USD/kWh for offshore wind farms by 2020. By assuming higher production costs, variations in transmission costs affect the final cost per delivered kilowatt hour less. To account for the less favorable case, we assumed lower production costs, allowing the increase in transmission cost to play a more significant role in the final cost.

Table 1 Wind Farm in Greenland: Summary of the Cost-benefit Analysis Results for Connecting the Wind Farm to the USA

Transmission Route: Europe – USA via Greenland			
(Total Cable Energy Capacity: 20 TWh)			
	Wind Farm Production	**Electricity Trade**	
Utilization (% of total time)[a]	~10 TWh (40)	~6 TWh (30)	~10 TWh (50)
Profit increase (%)	7–12	24–27	39–42

[a]*The wind farm is located in the middle of the path from Europe to the USA. As a result, it incurs only half of the transmission losses. That means that the same amount of power, e.g. 10 TWh, can be delivered in less time, resulting in a lower utilization factor of the transmission path.*

5.2 Intercontinental electricity trade

In this section, we focus on the electricity trade opportunities that emerge from the connection of the wind farm to both continents. Our analysis is based on real price data for the year 2012. We obtained the hourly spot prices from the European Power Exchange in Germany [16] and the PJM Interconnection in the USA [17]. Because of the time zone difference between the two continents, the two electricity markets experience their peak and lowest prices at different times. Our analysis is detailed in Appendix B.2. We mainly examine two levels of utilization: 30% and 50%. As already mentioned, 50% corresponds to the maximum utilization rate of the cable for electricity trade; the rest is used to transfer the power produced by the wind farm. We also investigate a lower utilization rate of 30% to account for a less favorable case (e.g. reduced availability of the cable or of excess renewable energy). Our investigations show that through the revenues generated from the electricity trade, the route between Greenland and North America can be amortized within 10–12 years with a 50% utilization and within 14–17 years if the cable is used for electricity trade only 30% of the time. Translating these results into profits, we find that by exercising electricity trading 30% of the time, the net profits[4] will increase by 24–27%; this is in comparison with the case where the wind farm sells its wind power only to the UK and earns a *profit* of 0.06 USD for each delivered kilowatt hour.[5] When exercising electricity trade 50% of the time, the profits can increase up to 42%. Table 1 summarizes our results.

To conclude, it seems that being connected to both continents would be a profitable solution for the wind farm in Greenland. In the last two sections, we investigate two possible sources of income created by building a transmission route from Europe to the USA via a wind farm in Greenland: first, selling the produced wind power always at peak prices, either in Europe or in the USA, and second, by trading

---

[4]The investment costs of the additional cable have been deducted from the electricity trade revenues.

[5]The profits from the intercontinental electricity trade are positive, that is, the revenues surpass substantially the investment costs for the additional line in all cases. However, the *increase* in the final profit depends on the wind farm profits. Selling wind energy at higher prices to the UK results in lower variation of the total profit from the electricity trade revenues. In accordance with Section 5, profits of 0.06 USD/kWh imply a sell price of 0.12 USD/kWh. Here, we account again for the less favorable case in our calculations, assuming that the wind farm generates substantial profits by selling the produced power at a price twice its marginal cost.

electricity between the continents. Both options independently result in a profitable operation for the wind farm.

6. Intercontinental interconnections by direct lines

In Ref. [8], we analyzed the expected transmission costs per delivered kilowatt hour. We estimated the cost of a 5500-km, 3-GW submarine cable to be in the range between 0.0166 and 0.0251 € per delivered kilowatt hour, and we found that, except for the most expensive RES generators, it seems that it would be more economical for the USA to import RES power from Europe than operate its own fossil fuel power plants.[6]

Here, we extend our analysis to estimate whether the cost for a long submarine cable could be amortized through the revenues arising by the electricity trade between the two continents. Again, we used the hourly spot prices for 2012 provided by PJM Interconnection in the USA and the European Power Exchange in Germany [16,17]. In our calculations we accounted for the transmission losses incurred by an 8000-km-long corridor connecting the USA with Germany. We further assumed that before the realization of a direct submarine cable between the two continents, there will already exist long HVDC lines on land in both Europe and the USA. Therefore, we considered that the investment costs of this project correspond to the direct submarine cable between Europe and the USA, which has a length of 5500 km and a capacity of 3 GW. In our analysis, presented in Appendix B.1, we estimate that for an 80% utilization of the cable (i.e. the cable is used only 80% of the time), the amortization period ranges between 18 and 28 years. For the less favorable case in which the cable utilization is 50%, the amortization period increases to about 23–35 years, depending on the cost projections. Although such amortization periods might not be most attractive for private investors, these results highlight that, from the point of view of social welfare, such a cable is beneficial for society.

7. Discussion

In the previous sections, we saw how intercontinental interconnections can take advantage of the time zone difference and smooth out electricity supply and demand. In this way, excess electricity production will not be irrevocably lost but transmitted to where it is needed most. Besides this, there are several additional opportunities that emerge from such a concept. In the following we provide a brief overview of them. For a more detailed analysis that refers to both the benefits and challenges that arise, the interested reader can refer to Ref. [8].

7.1 Minimizing power reserves

With the increasing penetration of RESs, the necessary control reserve capacity is expected to increase [18]. Global interconnections can offer such services. Taking advantage of time zone diversity, they

---

[6]The study is based on electricity generation from conventional sources estimated in Ref. [14] with fuel cost projections based on Ref. [18]. Unconventional sources of oil and gas such as oil sands and natural gas shales are also considered in the projections of Ref. [18].

can reserve control capacity in areas with lower electricity demand and offer it at locations that experience their peak demand at the same time. By offering an additional source of control power that is supplementary to the control reserve options available in each control area, significant cost savings could emerge, as building additional "peaking" gas power plants for balancing renewable energy could be avoided. Aboumahboub et al. [19] investigated this, comparing the necessary conventional power plants in the presence (or not) of interconnecting lines between regions. Their results for both the European and a potential Global Grid showed that through interconnections the need for dispatchable conventional power plants could be reduced by two to eight times.

7.2 Alleviating the storage problem

Bulk quantities of storage will be necessary for absorbing nontransmissible power and relieving congestion (e.g. Ref. [3]). The HVDC links of the Global Grid have the potential to alleviate the storage problem in future power systems by absorbing excess power (i.e. at a low cost) and injecting it into regions where it is needed more. In terms of efficiency, the losses of an ultra-HVDC line (e.g. ±800 kV) amount to about 3% for every 1000 km [20]. This would imply that a 6000-km HVDC line using current technology has better efficiency than pump-hydro or compressed-air energy storage.

7.3 Additional benefits

Global Grid interconnections present additional opportunities that can prove beneficial to power systems. For example, they have the potential to reduce the volatility of electricity prices. They also allow the transmission of bulk quantities of power with less loss of power. Power system security can also be enhanced through such interconnections in two ways. First, they provide additional pathways for the power to flow. Second, because of HVDC technology, they offer independent active and reactive power control and can act as a firewall between the systems they interconnect, not allowing disturbances to spread.

8. Conclusions

The Global Grid advocates the connection of all regional power systems into one electricity transmission system spanning the whole globe [8]. Power systems are currently forming larger and larger interconnections, while ongoing projects plan to supply, for example, Europe with "green power" from the North Sea. Environmental awareness and increased electricity consumption will lead more investments toward RESs, which are abundant in remote locations (offshore or in deserts). The Global Grid will facilitate the transmission of this "green" electricity to the load centers, serving as a backbone.

The Global Grid concept was already introduced in Ref. [8], but this chapter presented four possible stages that could gradually lead to the development of a globally interconnected power network. Building long transmission lines to harvest remote renewable resources will be the main driving force in the first stage of the Global Grid development. From there, a point will be reached where an RES power plant will be equidistant from two power systems on different continents. Connecting the RES power plant to both continents to sell the produced power always at peak prices—taking advantage of the time zone diversity—will mark the second developmental stage of the Global Grid. Because RES power plants usually have a capacity factor below 50%, the remaining cable capacity can be used for intercontinental electricity trade. This signals the progression into the third stage of development. As long as the profits from electricity trade are substantial, in the fourth developmental stage direct interconnections between countries or continents can start to be built. Introducing the four development stages helps to lay out a path leading towards a global grid environment. Nevertheless, the progression into these stages does not need to happen in a sequential manner, e.g. the fourth stage can follow directly the first, or the second and third stage can occur simultaneously.

Quantitative analyses of all four development stages have been provided in this chapter, showing that global interconnections can be both technically feasible and economically competitive. By exploring the possibility of a wind farm in the Kerguelen Islands in the Indian Ocean, we showed that it could provide renewable energy to South Africa at a cost competitive to that charged by local wind farms. At a later stage, this wind farm could connect to Australia, leading to intercontinental interconnections. We further estimated that connecting a wind farm in Greenland to both Europe and North America results in a 7–12% increase in profit. At the same time, the remaining cable capacity leaves room for intercontinental electricity trade. Based on real 2012 prices from Germany and PJM Interconnection in the USA, we found that electricity trade results in additional profits of 24–42%. We concluded our calculations by conducting a cost-benefit analysis for a direct submarine cable between Europe and the USA. The revenues from the intercontinental electricity trade, again based on real prices, result in an amortization period of 18–35 years for the cable investments, depending on the cable utilization factor. This highlights that, from a social welfare point of view, such a cable is beneficial for society. Based on the results detailed in this chapter, additional studies are necessary on a technical, economic, and societal level. The research community and the industry should also be encouraged to actively participate in identifying challenges and developing solutions that could lead to a Global Grid.

Appendix A. Cable cost projections

In Ref. [8], we carried out a detailed analysis on the projected costs of a long HVDC submarine cable. In this work we adopt the same cost considerations. We will assume a 3000-MW, ±800-kV submarine cable. We selected the ±800-kV option because we believe that higher voltage levels will be adopted for long-distance transmission. We distinguish between two cost alternatives for the submarine cables. As a high-cost case, we assume a cost of €1.8 million/km for our 3000-MW line, the same as what was suggested in Ref. [21] for a 5000-MW sea cable. As a low-cost case, we assume the maximum cost of the completed HVDC projects (up to 2012 as presented in Ref. [8]). This is €1.15 million/km. The rest of the cost assumptions are the same for both cases. Concerning the voltage source converters, because of the higher voltage and the large capacity of the line, we assume the cost of each terminal converter to be €300 million. For additional details, the reader can refer to Ref. [8].

Table A.1 Kerguelen Islands to South Africa: Transmission Cost per Delivered Kilowatt Hour (in USD)

Cable Costs	Wind Farm Capacity Factor		
	40%	50%	60%
Low cost	0.036	0.029	0.024
High cost	0.054	0.043	0.036

Kerguelen Islands. For the connection between the Kerguelen Islands and South Africa we assume a 3000-MW, ±800-kV submarine cable with a length of 4150 km. Based on the cost assumptions above and in Ref. [8], Table A.1 presents the transmission costs per delivered kilowatt hour for different wind farm capacity factors.

Europe—North America over Greenland. For the route over Greenland, besides HVDC cables, building HVDC overhead lines will also be necessary. Thus, for HVDC overhead lines, we assume a cost of €600 million/1000 km, as also suggested by Weigt [22] and Delucchi and Jacobson [14]. The detailed cost analysis can be found in Ref. [8].

Europe—North America through a direct submarine cable. We assume a 3000-MW, ±800-kV submarine cable with a length of 5500 km. Note that the distance from Halifax, Canada, to Oporto, Portugal, is 4338 km, whereas the distance from New York City to Oporto is 5334 km.

Appendix B. Electricity trade between Europe and the USA: detailed analysis

Appendix B.1. Direct submarine cable

In this section we detail the analysis we carried out to investigate the possibilities for electricity trade between Europe and the USA. Because of the time zone difference, the peak demand, and thus the peak prices, are shifted in time. Opportunities for electricity trade emerge. Our analysis is based on real hourly price data for the year 2012 obtained from Germany in Europe and the PJM Interconnection in the USA. For Germany we took the hourly spot prices for 2012 from the European Power Exchange [16]. For PJM we took the real-time prices, specifically the system energy price, that is, the price component that is the same over the whole PJM area, ignoring cost of congestion and losses [17]. We assume that the two power systems are connected through an 8000-km line, from which 5500 km correspond to the submarine cable between Oporto and New York and the remaining 2500 km correspond to an HVDC corridor between Oporto and Germany. We further assume that by the time the investment for the intercontinental cable will take place, there will already exist several HVDC interconnections within Europe and the USA. These could be used for the transfer of power between Oporto and Germany. Therefore, as investment costs we assume the cost of the submarine cable between Oporto and New York. Still, our calculations take into account the incurred losses along the total length of the corridor, that is, the 8000 km.

We assume a time difference of 6 h between Germany and the US east coast, where the PJM Interconnection is located. The hourly prices in Euros are transformed to US dollars through average monthly exchange rates for 2012.[7] In our analysis we assume that the investment takes place in 2012, but the revenues are generated during the next 40 years since this is the expected lifetime of a cable [10, 24, 25]. Because data were available only for 2012, we assume that the prices over the next 40 years will be similar to those in 2012, considering a discount rate of 3%, as suggested by the National Renewable Energy Laboratory ([26], p. 9), and an inflation rate of 2.5%. We distinguish between three different utilization rates for the cable: 30%, 50%, and 80%. These rates reflect the equivalent amount of hours per year during which the cable is used. For example, 30% utilization means that the cable transmits power up to its full capacity 30% of the time. For a 30% utilization,

---

[7]Source: http://www.x-rates.com/average/?from=EUR&to=USD&amount=1&year=2012.

Table B.1 Amortization Period (in years) for a High-Cost and a Low-Cost Projection of the Cable

Cable Costs	Utilization		
	30%	50%	80%
Low cost	31	23	18
High cost	>40	35	28

electricity is traded when the price difference is more than US$35; for 50% utilization the trade takes place for prices more than US$23, and for 80% utilization the minimum price difference is US$10.

Table B.1 presents the amortization period for the submarine cable. The cable generates revenues by buying electricity with a low marginal price on one continent and selling it at higher system marginal price to the other continent. As can be observed, for 50% and 80% utilization of the cable the payback time is less than the minimum cable lifetime for both the high-cost and the low-cost scenario. For a utilization of about 30%, the cable is amortized only in the low-cost scenario. Here we stress that building a cable between the USA and Europe is not primarily a for-profit investment. The goal is to create an investment that will be beneficial society as a whole. Thus, any amortization period that is less than the lifetime of the project is considered positive because it is expected to benefit society.

Appendix B.2. Connecting Europe with the USA through a wind farm in Greenland

The connection of a wind farm in Greenland to both continents facilitates electricity trade between Europe and the USA. In this appendix we investigate whether the costs for building the line between Greenland and the USA could be covered from the revenues generated from electricity trade. In our analysis of Greenland in Ref. [8], we calculated that over a year, 50% of the cable energy capacity can be used for electricity trade. Two effects should be considered here. First, in reality the wind farm will often operate at capacities lower than its maximum. Therefore, on the one hand, only a part of the transmission capacity often will be available electricity trade; on the other hand, this capacity will be available for more than 50% of the time. Second, it may occur that during periods when there is substantial price difference between the two continents, the wind farm will be producing power at the same time. As a result, the transmission capacity cannot be used for profitable arbitrage. Because of these two effects, we assume that the transmission capacity factor for electricity trade will be about 30%, whereas in the best case it will not exceed 50%. Table B.2 presents the amortization period for

Table B.2 Amortization Period (in years) for the Transmission Path from Tasiilaq to New York for a High-Cost and a Low-Cost Projection of the Cable

Cable Costs	Utilization		
	30%	50%	80%
Low cost	14	10	8
High cost	17	12	10

three utilization levels of the cable: 30%, 50%, and 80%. (The 80% level is presented only for comparison with Table B.1.) As we can observe, for a utilization of 30–50%, the costs of the route from Greenland to North America can be recouped within 10–17 years. It should be noted that the payback period here does not take into account the additional profits the wind farm will make by always selling its produced wind power at peak prices. It is also interesting to point out that the route from Germany to New York via Greenland is a similar distance to that of the route from Germany to New York through a direct submarine cable in Oporto, Portugal.

References

[1] EC. Impact assessment on the EU's objectives on climate change and renewable energy. European Commission; 2008.

[2] State of California, senate bill X1-2 (sbx1 2); 2011.

[3] dena. Dena grid study II — integration of renewable energy sources in the german power supply system from 2015–2020 with an outlook to 2025. German Energy Agency; 2010. Final Report.

[4] National Renewable Energy Laboratory. NREL/TP-6A20-52409. In: Hand MM, Baldwin S, DeMeo E, Reilly JM, Mai T, Arent D, et al., editors. Renewable energy futures study, 4 vols. Golden, CO: National Renewable Energy Laboratory; 2012 [Online]: http://www.nrel.gov/analysis/re_futures/ [Last accessed 20.06.12].

[5] Czisch G. Scenarios for a future electricity supply: cost-optimized variations on supplying Europe and its neighbours with electricity from renewable energies. Institution of Engineering and Technology (IET); 2011.

[6] Jacobson MZ, Delucchi MA. Providing all global energy with wind, water, and solar power, part I: technologies, energy resources, quantities and areas of infrastructure, and materials. Energy Policy 2011; 39(3):1154–69.

[7] Boute A, Willems P. RUSTEC: greening Europe's energy supply by developing Russia's renewable energy potential. Energy Policy 2012;51(0):618–29.

[8] Chatzivasileiadis S, Ernst D, Andersson G. The global grid. Renewable Energy 2013;57(0):372–83. URL, http://www.sciencedirect.com/science/article/pii/S0960148113000700.

[9] U.S. Energy Information Administration (EIA). Greenland energy statistics. ca [Online]: http://www.eia.gov/countries/country-data.cfm?fips=GL; 2012 [Last accessed 20.06.12].

[10] Hammons T, Olsen A, Kacejko P, Leung C. Proposed Iceland/United Kingdom power link — An indepth analysis of issues and returns. IEEE Trans Energy Convers 1993;8(3):566–75.

[11] The Guardian. Iceland's volcanoes may power UK [Online]: http://www.guardian.co.uk/environment/2012/apr/11/iceland-volcano-green-power; 2012 [Last accessed 20.06.12].

[12] Wikipedia. Kerguelen islands [Online]: http://en.wikipedia.org/wiki/Kerguelen_Islands; 2013 [Last accessed 17.12.13].

[13] Lafayne C. Updated information on france's antarctic and sub-antarctic "weather-forecasting" interests for the international antarctic weather forecasting handbook [Online]: http://www.antarctica.ac.uk/met/momu/International_Antarctic_Weather_Forecasting_Handbook/update\%20France.php; 2008 [Last accessed 17.12.13].

[14] Delucchi MA, Jacobson MZ. Providing all global energy with wind, water, and solar power, part II: reliability, system and transmission costs, and policies. Energy Policy 2011;39(3):1170–90.

[15] Wikipedia. Energy in South Africa. [Online]: http://en.wikipedia.org/wiki/Energy_in_South_Africa; 2013 [Last accessed 17.12.13].

[16] EPEX. European power exchange www.epexspot.com/en/; 2013.

[17] PJM. PJM interconnection LLC www.pjm.com; 2013.

[18] U S Energy Information Administration (EIA). Annual energy outlook DOE/EIA-0383(2009). Washington, DC: US Department of Energy; 2009 [online]: http://www.eia.gov/countries/country-datacfm?Fips=GL [Last accessed 20.06.12].

[19] Milligan M, Donohoo P, Lew D, Ela E, Kirby B, Holttinen H, et al. Operating reserves and wind power integration: an international comparison. In: The 9th annual international workshop on large-scale integration of wind power into power systems as well as on transmission networks for offshore wind power plants conference; 2010.

[20] Aboumahboub T, Schaber K, Tzscheutschler P, Hamacher T. Optimization of the utilization of renewable energy sources in the electricity sector. In: Proceedings of the 5th IASME/WSEAS international conference on energy & environment (EE'10); 2010.

[21] Siemens. Ultra HVDC transmission system. ca [Online]: http://www.energy.siemens.com/hq/en/power-transmission/hvdc/hvdc-ultra/#content=Benefits; 2011 [Last accessed 20.06.12].

[22] DLR. Trans-mediterranean interconnection for concentrating solar power. Germany: German Aerospace Center, Institute of Technical Thermodynamics, Section Systems Analysis and Technology Assessment; 2006. Study commisioned by the Federal Ministry for the Environment, Nature Conservation and Nuclear Safety.

[23] Weigt H, Jeske T, Leuthold F, von Hirschhausen C. Take the long way down: integration of large-scale North Sea wind using HVDC transmission. Energy Policy 2010;38(7):3164–73.

[24] Skog JE. HVDC transmission and life expectancy. Memo Statnett-TenneT [Online]: http://www.tennet.org/english/images/19-UK-B7-HVDC_Transmission_and_Lifetime_Expectancy_tcm43-12302.pdf; 2004 [Last accessed 20.06.12].

[25] Wikipedia. NorGer. ca. [Online]: http://de.wikipedia.org/wiki/NorGer; 2011 [Last accessed 20.06.12].

[26] Short W, Packey DJ, Holt T. A manual for the economic evaluation of energy efficiency and renewable energy technologies. National Renewable Energy Laboratory; 1995. NREL/TP-462–5173.

Practical Management of Variable and Distributed Resources in Power Grids

15

Carl Barker
ALSTOM Grid

1. Preface

In the latter part of the nineteenth century, electrical power distribution originally covered small geographical areas to provide electric lighting. The choices of electrical power was Alternating Current (AC) or Direct Current (DC). By the early part of the twentieth century, AC electrical power transmission was taking a lead over DC, mainly because of the simplicity of conversion between voltage levels using transformers. The transformer made it possible to generate AC at low voltage and high current, transmit at high voltage and low current (and hence low loss), and then utilize the power at a low voltage and high current. Nevertheless, it was always noted that AC transmission was inferior or even impractical in some applications when compared to DC transmission. DC transmission, therefore, never really went away.

With developments in early "high-voltage" mercury-arc valves in the early 1950s it became possible to reap the benefits of DC transmission embedded within the AC network for specific applications where DC is more suitable, such as cable transmission; long-distance bulk power transmission on overhead lines; and interconnecting, unsynchronized AC networks.

Following the early development of DC transmission, the mercury-arc converters were replaced by power electronic devices with higher ratings, improving both performance and transmission ratings.

In the early 2000s a new technology was introduced: the voltage source converter. This technology further increased the flexibility of DC transmission. With this additional flexibility comes the realistic prospect of DC transmission networks, which may replace AC transmission applications in the future or may act as a "backbone," reinforcing the existing AC grid. This new infrastructure may be critical for allowing the power systems of tomorrow to adapt to the increased use of renewable sources and the global challenge to manage energy networks to make best use of the available sources and changing load distributions.

2. The early history of high-voltage direct current transmission

In the early 1880s a new technology became available that permitted then commonly used gas lighting to be replaced with a safer and cleaner solution. This technology was electric lighting, which had by then matured into a reliable technology suitable for production at an industrial level.

Renewable Energy Integration. http://dx.doi.org/10.1016/B978-0-12-407910-6.00015-6

A leading pioneer of this new technology was Thomas Edison [1], who gained early success with his direct current (DC) generation, distribution, and lighting systems operating at voltages of around 110 V. Early successes proved the value of the market, leading others to enter the market, principally George Westinghouse, whose alternating current (AC) system, originally based on an 1882 patented concept by the Frenchman Lucien Gaulard and the Englishman John Dixon Gibbs, competed with Edison's DC system. Westinghouse's first system, installed in 1886, was a 133-Hz single-phase system in Great Barrington, Massachusetts, USA [2].

Using AC transformers, George Westinghouse was able to transmit power at a higher voltage than that at which it was generated or consumed. In this way power could be transmitted with lower transmission losses and therefore transmitted economically over longer distances.

In 1888 Nikola Tesla presented his patented invention of a multiphase AC system at an Institute of Electrical and Electronics Engineers (IEEE) lecture. This lecture was attended by George Westinghouse, who saw how he could use Tesla's work to develop AC transmission systems to compete with Edison's DC systems. The battle of the currents, DC versus AC, had started.

The battle between the currents continued for many years, with many projects around the world being developed using DC and others using AC. However, AC transmission continued to have advantages, where generation and utilization had significant physical separation. Various attempts to overcome the perceived limitation of DC with regard to transmission distance were made. An early example of DC transmission was the Thury method [3], which used a number of series-connected DC generators, electrically insulated from the ground and each other, driven by a common prime mover. An example of such a system was built in 1905 between Moutiers and Lyon in France, covering a distance of 180 km and rated at 57.6 kVdc and 75 A [4].

In 1924 the English Electric Co., Ltd, reported a mechanical rectifier designed for high-voltage DC (HVDC) applications around 100 kVdc, called the Highfield-Calverley transverter [5]. However, by the mid-1920s many parts of the world had started to standardize systems to AC, and consequently the use of DC power generation, transmission, distribution, and utilization diminished. Nevertheless, the advantages of DC transmission over AC transmission were understood, and therefore research into methods of DC transmission continued.

The method of HVDC transmission that would be recognizable as an element in modern power systems was developed in the 1940s and 1950s [6]. This method converted power from the existing AC transmission to DC using a static electrical conversion system. It also used AC transformers to convert the existing AC power system voltage to an appropriate level to give a desired DC voltage transmission level.

Following the early applications of HVDC using static electrical conversion, there was initially a slow increase in the number of HVDC projects built. Those projects that were constructed were justified by special conditions that made HVDC more economical than AC transmission, such as transmission through cables or over long distances. These applications are discussed in the following sections. In more recent times the number of HVDC schemes introduced into AC transmission systems, either to interconnect independent AC systems or to increase the power throughput of existing AC systems, has significantly increased (Figure 1). This trend is only expected to continue.

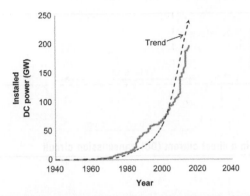

FIGURE 1 Number of high-voltage direct current (DC) schemes planned. GW, gigawatt

3. **HVDC for cable transmission**

From the early days of electrical power transmission, moving AC power via cable as opposed to overhead transmission lines was not cost-effective over relatively short distances and unpractical over slightly longer distances.

The problem with AC transmission can be understood if the electrical equivalent of a cable is considered. A cable, in its simplest form, is composed of a conductor, surrounded by an insulator, surrounded by a screen. The conductor will have resistance and inductance along its length. The insulation will form the dielectric of a capacitor, with the central conductor acting as one plate of the capacitor and the screen the other plate. A cable can, therefore, in its simplest form, be modeled as a simple circuit, such as that shown in Figure 2.

In an AC system the voltage reverses polarity each half-cycle, that is, it goes from peak positive voltage to peak negative voltage and back to peak positive voltage repeatedly during a time period defined by the AC frequency. Applying this voltage to one end of a cable will cause a current to flow into the cable's stray capacitance, known as a charging current. This charging current will flow through the resistance of the cable conductor and generate losses in the form of heat. The cable will have the

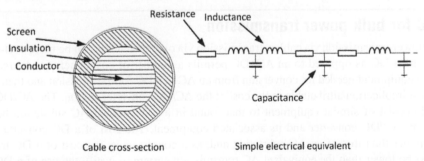

FIGURE 2 A simple cable model

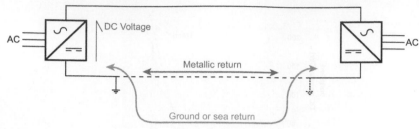

FIGURE 3 Return current flow in a direct current (DC) transmission circuit

AC, alternating current.

capability of operating stably up to a operating temperature determined by the design, but the heat generated by the flow of this charging current will consume some of the cable's thermal capability and hence reduce the capacity for real power flow through the cable. A length of around 30 km for submarine cables and 80 km for buried cables [7] is typically considered the maximum economical distance for AC transmission through a cable for a power above a few hundred megawatts.

In DC transmission systems the voltage polarity remains constant for normal steady-state operation. Consequently, when a DC voltage is applied to a cable, there will be an initial charging current as the stray capacitance of the cable charges to the applied DC voltage. With fixed polarity and magnitude, however, no further charging current will flow. The capability of the cable to carry thermal current is therefore available for real power transmission. In principle, DC transmission via cables is not bound by distance. In practice, however, the transmission distance is usually limited by the cost of the cable and the electrical losses.

Another potential cost advantage of DC transmission is that with one conductor operating at high voltage, the other conductor can be nominally operating at zero voltage (Figure 3). Therefore, the sea or the ground itself can be used as a current path [8], subject to geological conditions and local regulatory restrictions. This solution has been implemented in several projects; both land and sea crossings and can realize significant cost benefits because it reduces the number of cables to one per DC connection.

4. HVDC for bulk power transmission

Similar to power transmission via cables, transmission via overhead wires can be more cost-effective if transmitted by a DC as opposed to an AC. DC permits a number of cost benefits to be realized.

First, the equipment needed for conversion from an AC to a DC for transmission and then a DC to an AC has a cost implication attributable to the cost of the AC/DC converter station. The AC/DC converter station will consist of similar equipment to that found in a conventional AC substation but with the addition of an AC/DC converter and its associated equipment. The cost of a DC converter station is, therefore, higher than that of an AC substation and, as a consequence, the cost of a DC transmission system must be lower than the equivalent AC transmission system to justify the use of a DC.

Figure 4 shows a simple comparative representation of two HVDC bipole circuits on a common tower and a double-circuit, three-phase AC line. The capital cost of the DC transmission line is

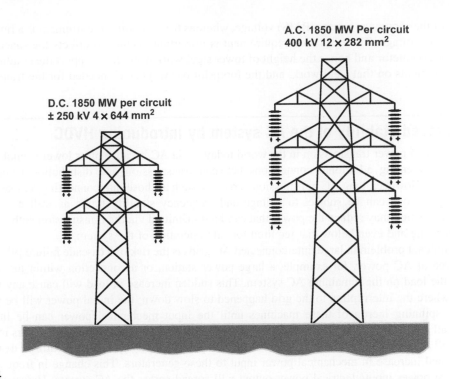

D.C. 1850 MW per circuit
± 250 kV 4 × 644 mm²

A.C. 1850 MW Per circuit
400 kV 12 × 282 mm²

FIGURE 4

Comparison of a direct current (D.C.) and an alternating current (A.C.) transmission tower.

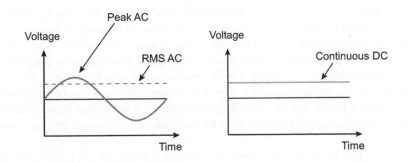

Peak AC

Voltage

RMS AC

Voltage

Continuous DC

Time

Time

FIGURE 5

Alternating current (AC) (left) and direct current (DC) voltages (right). RMS, root mean square.

lower because both the number of conductors and the insulation requirement are reduced. The change in insulation requirement is significant, as can be visualized by considering Figure 5. In a DC transmission scheme both the power transfer capability and the insulation requirement of the line are a function of the DC voltage. However, in an AC transmission circuit the power transfer is a function of the root mean square (RMS) voltage, whereas the insulation requirement is a function of the peak AC voltage. The insulation requirement is important because it effects the space needed around the conductor and hence the height of tower steel work needed to support the conductor, the bending moments on that steel work, and the footprint or "wayleave" needed for the transmission corridor.

5. Improved stability of the AC system by introducing HVDC

The majority of power transmission in the world today is via AC because of its lower capital cost and the ease of creating additional connections between transmission and distribution using power transformers. However, long AC transmission corridors, such as those interconnecting remote loads or generating stations can be subject to voltage and frequency stability limits as well as inter-area oscillations, which may reduce the power that can be transmitted via the transmission path to a value below its rating and even below that required for safe operation of the system.

A significant problem in large, interconnected AC grids is the risk of a cascade failure [9]. The loss of a source of AC power, for example, a large power station, or a connection within the grid will increase the load on the remaining AC system. This sudden increase in load will cause any machine close to where the interruption to the grid happened to slow down; the initial power will be supplied from the spinning inertia of these machines until the input mechanical power can be increased. Because all machines are synchronized in an AC system, all of the synchronous generators in the AC system will automatically follow the reduction in grid system frequency, locally creating a demand for an additional increase in mechanical power input to these generators. This change in frequency and mechanical power input/electrical power output will spread across the AC system. However, as this change ripples across the AC system, parts of it may become overloaded, causing these elements to trip. This puts further strain on the AC system and can potentially lead to a cascade failure across the network, leading to blackouts.

If DC transmission is used to break up the AC grid into "islands", then the DC transmission will provide an asynchronous power interconnection between these AC islands. A change in frequency in one AC island will not automatically lead to a change in frequency in another AC island. Instead, the DC transmission can be programmed to automatically respond to a change in frequency in one AC island by either increasing or decreasing the DC transmission into that island. Hence, the power transmitted by the DC can be increased, if required, to compensate for a loss of generation in one AC island; this increase in power can be limited dynamically to a value that is not detrimental to the stability of the AC island supplying the power. In this way the risk of cascade failures across a network can be avoided.

DC transmission can be used in the form of a so-called back-to-back connection when used to break up the AC system into islands. In a back-to-back scheme, both the sending and receiving AC/DC converter equipment are located in the same place, typically in the same building, and the DC transmission circuit can be just a few meters long. This separation of the AC system on either side of the DC back-to-back link will result in well-damped voltage, frequency, and inter-area oscillations [10], as indicated in Figure 6. Figure 6 shows two scenarios. The first is power through a weak AC tie line connecting to strong AC systems when one of those AC systems is subjected to a severe fault. In this case it can be seen that the fault leads to an under-damped voltage oscillation and eventually voltage collapse. Conversely, when a back-to-back DC transmission link is added to the tie line to

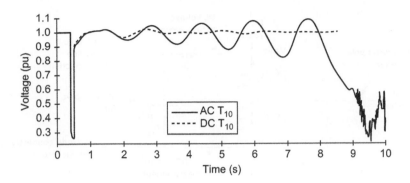

FIGURE 6

Transient recovery improvement with a direct current (DC) transmission link used to separate two alternating current (AC) systems.

provide an asynchronous link between the two AC systems, Figure 6 shows that the voltage oscillation is well damped and voltage stability is restored following the fault.

Many modern forms of renewable generation are connected to AC grids via an electronic- based power converter; hence the actual renewable generation is not directly connected to the AC power grid. This allows power transfer to be controlled, as with HVDC transmission, providing more flexibility and reducing the susceptibility of the renewable source to tripping during power system disturbances.

6. Voltage source converter versus line commutated converter

Until recent years, all installed static DC transmission schemes have been based on so-called line commutated converter (LCC) technology; originally these used mercury-arc valves but since the 1970s have used power thyristors. A simplified LCC HVDC converter is shown in Figure 7 [11]. An LCC HVDC converter typically consists of the LCC valves that convert AC to DC, the power transformer to interface the converter to the AC system, and both AC and DC harmonic filters to improve the power quality of the waveforms produced by the LCC converters. The LCC conversion process inherently causes a lagging power factor (absorbs reactive power) at the converter AC terminals; hence it is necessary to provide capacitive reactive support locally at the converter to minimize the reactive power load on the AC system. This capacitive reactive power supply typically is combined with AC harmonic filters.

Figure 8 shows a simplified present day voltage source converter (VSC) arrangement. What should be immediately obvious when comparing Figures 7 and 8 is that there are no AC or DC harmonic filters. This was not true with early VSC schemes, which used two-level or three-level topologies. However, the modern modular multilevel converter (MMC) topology shown in Figure 8 overcomes the limitations of earlier technology. The MMC VSC generates a high-quality waveform with very little harmonic content, and consequently does not need the additional filtering necessary in LCC. Moreover, the VSC is able to control the reactive power it exchanges with the AC system independent of

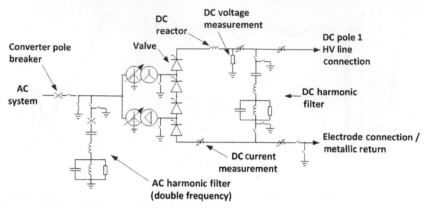

FIGURE 7 Simplified line commutated converter arrangement

AC, alternating current; DC, direct current; HV, high voltage.

the real power; hence, unlike an LCC converter, there is no requirement for additional capacitive (or inductive) reactive power at the converter AC connection point. Figure 9 shows an MMC VSC installation.

As mentioned above, an LCC uses thyristors to perform the conversion between AC and DC. A thyristor is a device that, once in the conducting state, cannot be turned off until the current flow

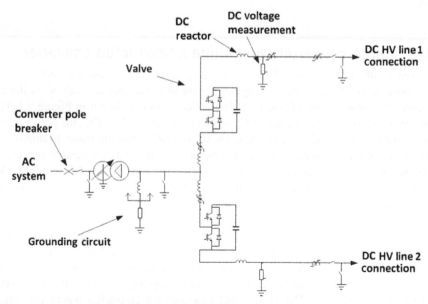

FIGURE 8 Simplified voltage source converter arrangement

AC, alternating current; DC, direct current; HV, high voltage.

FIGURE 9 A voltage source converter installation

(courtesy of ALSTOM Grid)

through it has been brought to zero by the external circuit. Therefore, for DC transmission, an existing AC system is needed to force the current through the converter to commutate between switches—hence the name "line commutated converter": it is the "line" to which the converter is connected that creates the commutation process [12]. This is true even when the converter is exporting power from a DC system to a AC system (an "inverter") and hence the converter is unable to energize the AC power system from its shutdown condition, a process known as a "black start," without the assistance of an rotating AC source such as another energized electrical network or a synchronous compensator [13].

Unlike LCC, VSCs use a power transistor as the main switching device, typically an insulated gate bipolar transistor (IGBT). Such devices can interrupt the current flowing through them independent of the external circuit. Hence a VSC is able to create its own output AC waveform and therefore supply a passive load without the need for an additional rotating plant.

An important difference between a VSC and LCC is that in an LCC circuit the DC current can flow around the DC loop in only one direction. Hence, to reverse the direction of power flow, the DC voltage polarity must be reversed [14]. With a VSC this is not the case: The DC current can change direction; hence the direction of power flow in a converter can be changed without changing the polarity of the DC voltage.

7. Large-scale variable generation integration

HVDC schemes using LCCs have predominantly been point-to-point, with only a few exceptions. A major reason for this is the complication arising from the need to reverse the voltage polarity to reverse the power flow when using an LCC. Hence, a change at one converter has a global impact on the HVDC network. However, because VSCs can change the direction of power flow at an individual converter by changing the direction of current flow and do not affect the DC voltage polarity, there is now the opportunity to interconnect many HVDC VSCs together to create an HVDC network. This gives the potential advantage of being able to connect and share power resources with power loads using a single DC connection at each connection point without the need for—and hence the cost and transmission losses of–repeatedly converting between AC power and DC power. Figure 10 illustrates the advantage of a VSC HVDC network in terms of the significant reduction in converter equipment that is possible. By taking advantage of this reduction in equipment, along with the inherent savings achieved by using a DC for transmission via cables and long overhead lines, it is now possible to economically build large HVDC networks.

HVDC networks can be used to interconnect several remote or offshore renewable energy sources with one or multiple load centers. The inherent controllability of the HVDC converters means that the power available to the HVDC network can be dispatched as determined by the operator, irrespective of the phase or frequency of the connected AC systems. In addition, the HVDC network can be used in support, or as the "backbone," of the existing AC system(s), providing automatic re-dispatch of power where needed to compensate for events in the AC system and provide decoupling between parts of the AC system(s) as necessary.

Such an HVDC "backbone" may, in the future, cross large geographical areas and national boundaries, making a new overarching power network, or "supergrid". A simple analogy for this would be the road transport network, where the AC is represented by the streets around towns and going to homes, whereas the supergrid is the freeway, providing bulk transfer capability over long distances and with geographically defined "on and off ramps" to allow the traffic to flow between the two systems. The junctions between these two road systems are designed to meet the possible levels of traffic, that is, the energy transmission capacity. Further, by considering the freeway as a road requiring payment of a toll to travel on it, a supergrid can be understood as a means of capacity-based revenue.

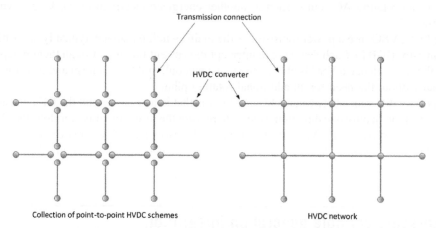

Collection of point-to-point HVDC schemes   HVDC network

FIGURE 10 The advantage of a high-voltage direct current (HVDC) network over a conventional arrangement using multiple point-to-point HVDC schemes

An important aspect of HVDC network design is the control of individual converters to ensure robust operation during dynamic and unpredictable events without the need to rely on fast telecommunications. To solve this, lessons have been learned during the more than 100 years of experience with operating AC networks where generators operate in parallel to share the load. The converters can be operated using a "droop" characteristic. In AC this is frequency versus power, but for a DC, as there is no frequency DC voltage can be used instead. Unlike frequency, however, the DC voltage is not a constant across the network because of voltage drops across the circuits; hence it is more akin to a "phase angle" within an AC network. Therefore, when initially setting the desired load flow of the HVDC network, these voltage drops have to be considered; however, they are not complex—they are simply a consequence of the power transferred across the conductor and the conductor's resistance (Ohm's law: voltage (V) = current (I) × resistance (R)). Moreover, by using an HVDC droop, which is defined by DC voltage and DC current, the slope of the droop represents an artificial resistance, whilst the conductors represent real resistance, making the overall analysis and control of an HVDC network's power flow relatively simple.

An important aspect of HVDC networks is how to clear a fault within the network with the minimum effect on those AC systems connected to the HVDC network. With LCC technology the converter itself is able to block in the presence of a fault on the DC side, removing the source of energy from the fault and allowing the fault to be cleared or removed. However, with early applications of VSCs in HVDC networks, the converters used were unable to block the fault current in-feed from the AC side of the converter to the DC side of the converter; hence the only means of isolating the fault was to open the AC circuit breakers associated with each converter connected to the HVDC network [15]. However, developments in HVDC VSC technology are leading to a new tool box of equipment that can be used in HVDC networks.

Under developed are new generations of VSCs that have the ability to block a DC side fault within less than 500 μs without having to resort to opening the converter AC breaker. Moreover, while blocking the fault current on the DC side, during the fault, these converters can actively take part in reactive power control on the AC side of the converter, helping to minimize any voltage disturbance as a consequence of a fault on the DC side [16] (Figure 11).

Another major tool needed for large HVDC networks is the HVDC circuit breaker. Unlike in an AC system, where the AC breaker interrupts the current by opening at a naturally occurring current of zero, imposing the change in energy flow onto the AC system, there are no current zeros in a DC current. As a consequence, an HVDC breaker has to operate on a totally different principle to that of an AC breaker. A DC breaker has to create a voltage across its own terminals that will oppose the voltage driving the fault current. The breaker must also provide a continuous current path with minimal losses during most of its in-service life. To achieve this, manufacturers have invested heavily in the development of "hybrid breakers," comprising a mechanical switch to provide a normal low-loss current path in combination with a power electronic element to provide an initial current interruption as the mechanical switch opens and surge arresters to provide the subsequent dissipation of energy. Such HVDC breakers are being developed today with current ratings of 7.5 kA and opening times of 2.5 ms [17], thereby minimizing the overall length of disturbance in the network.

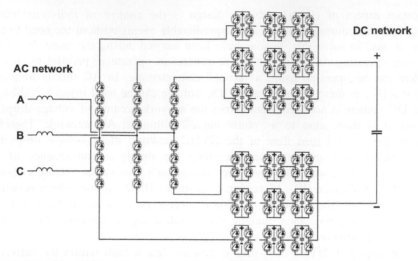

FIGURE 11 Advanced high-voltage direct current (HVDC) voltage source converter topology with independent fault blocking and reactive power control

AC, alternating current; DC, direct current.

(courtesy of ALSTOM Grid)

8. Taking DC to lower power transmission levels

With the introduction of more renewable energy sources at different transmission levels within an AC system, the traditional model for power generation, transmission and distribution is being blurred. Instead of power flow from the "top down," power flow is reversing in some parts of the network, increasing the complexity of distribution- and transmission-level control of the network. This, in turn, leads to a problem with voltage control and reliability of the connection, which lead to a need for more meshing on the interconnections. However, increasing the meshing on the existing distribution grid will increase the short circuit level, in some cases leading to a situation where the short circuit level is too high for conventional AC equipment.

A further problem with new sources of renewable generation injecting power at the distribution level is that of controlling the AC frequency at the transmission level with changing distribution load and newly introduced low-inertia, variable generation.

By introducing a "new" medium voltage class of HVDC transmission into an expanding distribution network, the benefits already being realized at the transmission level can be used, in particular these include increasing cable transmission distances and providing an asynchronous barrier to segment the AC network so that the frequency of one part of the network does not affect another.

How low a DC voltage can we reach? Many domestic products today use a DC as their power source, and most domestic homes contain a plethora of small, individual power connections for computers, phones, televisions, as well as emerging low-loss lighting using light-emitting diodes. Research is already under way into houses in which DC is supplied to sockets.

In many ways, the technology is returning to its origins. Today, however, we have the benefit of power electronics to facilitate smooth manipulation of power flow. We can therefore see a push down from the transmission network to penetrate further toward consumers and a push up from the domestic customer to use more DC power without the need for AC/DC conversion.

9. Conclusion

The power electronics industry has made massive technological advances, allowing new systems to be developed, and current research will inevitably lead to further breakthroughs allowing for more flexible, lower loss, and lower cost DC networks both at high voltage (HVDCs) and at lower distribution levels within the power grid. This, in turn, will open up the access of renewable energy to electric networks.

References

[1] http://www.ieeeghn.org/wiki/index.php/Edison%27s_Electric_Light_and_Power_System.

[2] Lomas R. The man who invented the twentieth century. Headline Book Publishing; 1999.

[3] Kimbark EW. Direct current transmission, vol I. Wiley-Interscience; 1971 [Chapter 1].

[4] Rey MA. Transport d' énergie Moutiers-Lyon par courant continu à 50 000 volts. Cinquième Année No. 56, Communications techniques; December 1908.

[5] Hill EP. Rotary converters, their principles, construction & operation. Chapman & Hall; 1927. 319.

[6] Arrillaga J. High voltage direct current transmission [Chapter 1]. 2nd ed. IEE power and energy series 29; 1998.

[7] BICC Cables, Electric cables handbook, 3rd ed., Blackwell Science, [Chapter 38].

[8] General guidelines for the design of ground electrodes for HVDC Links, CIGRÉ WG 14-21 – TF2 Guide.

[9] http://en.wikipedia.org/wiki/List_of_major_power_outages.

[10] Barker CD, Kirby NM, MacLeod NM, Whitehouse RS. Widening the bottleneck: increasing the utilisation of long distance AC transmission corridors. IEEE 987-1-4244-6547-7; 2010.

[11] HVDC for beginners and beyond, ALSTOM grid, systems-L4-HVDC Basics -2165-V2-EN, www.alstom.com/contactcentre.

[12] Lander CW. Power electronics. 2nd ed. McGraw-Hill; 1987 [Chapter 3].

[13] Kim SI, Lee SJ, Haddock JL, Baker MH. System design characteristics for the 300 MW submarine link to Cheju. Cigré 1993 Regional Meeting, Queensland, SE Asia + Western Pacific; 4 October 1993.

[14] Barker CD, Whitehouse RS, Adamczyk AG, Boden M, Kirby NM. Building a HVDC bipole scheme from one LCC pole and one VSC pole. Cigré Canada; 2013.

[15] Barker CD, Wheeler JD, Mukhedkar RA, Ingemansson D, Danielsson M, Moberg U, et al. Multi-terminal operation of the South-West link HVDC scheme in Sweden. Cigré Canada; 2012.

[16] Merlin MMC, Green TC, Mitcheson PD, Trainer DR, Critchley DR, Crookes RW. A new hybrid multi-level voltage-source converter with DC fault blocking capability. In: ACDC conference. IET; 2010.

[17] http://www.twenties-project.eu/system/files/Feasibility-test-DEMO3.pdf.

Integration of Renewable Energy—The Indian Experience

16

Sushil Kumar Soonee[1], Vinod Kumar Agrawal[2]

[1] *Chief Executive Officer Power System Operation Corporation Ltd. Qutab Institutional Area New Delhi,*
[2] *Executive Director National Load Despatch Center Power System Operation Corporation Ltd. Qutab Institutional Area New Delhi*

1. Introduction

India is among the fastest growing countries in the world, and achieving energy security is an area of prime concern. The generation capacity of currently installed energy sources in India is about 228 GW, of which about 29 GW is from renewable energy sources (RESs). The distribution of energy sources is uneven, with fossil fuel located in central-eastern India, hydro source in northern and northeastern India, and RESs mostly in southern and western India. *In a total estimated installed generation capacity of around 778 GW, the projected renewable capacity by the year 2032 has been estimated to be about 183 GW.*

Figures 1 and 2 show the geographical distribution of solar and wind energy sources, respectively. As can be seen, the wind-based RESs are mainly located in the southern and western part of the Indian peninsula, and solar sources predominate in the western part of the country. Although the present solar capacity is small, it is expected to grow in a big way. The uneven distribution of RESs presents a challenge in terms of promoting renewable energy and its widespread utilization across the country.

The preamble of the Electricity Act of 2003 [3] defines the "promotion of efficient and environmentally benign policies" as an objective; the Act also provides legal mandates toward this goal. India has a federal structure of governance and "electricity" is a concurrent subject, with both the central and state governments having jurisdiction. At the central level, the Ministry of Power [4] and the Ministry of New and Renewable Energy (MNRE) [5] provide the policy framework for promotion of RESs.

Further, the Central Electricity Regulatory Commission (CERC) at the central level and state electricity regulatory commissions (SERCs) at the state level are mandated to provide the regulatory framework. The operation of power systems in the country is hierarchical in nature, with the national, regional and state load despatch centers (LDCs) discharging responsibilities in their respective areas.

2. Policy initiatives

The *National Electricity Policy, 2005* [6] provides direction to the policy initiatives for promoting RESs. The Tariff Policy of 2006 [7] elaborates the role of electricity regulatory commissions and mechanisms for promoting RESs. For the first time in India, Section 3 of the Rural Electrification

Renewable Energy Integration. http://dx.doi.org/10.1016/B978-0-12-407910-6.00016-8

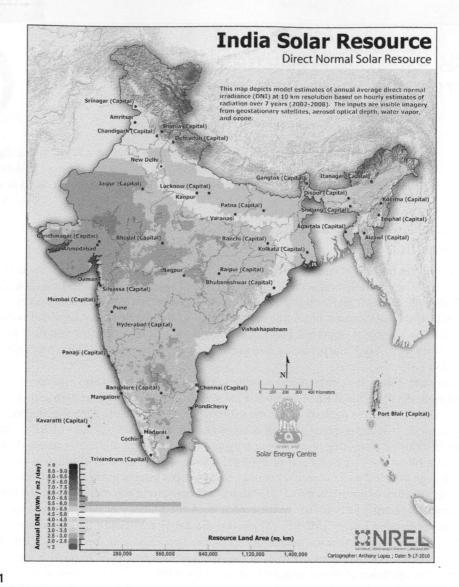

FIGURE 1

Map of solar energy sources in India.

Policy of 2006 [8] provided policy framework for decentralized distributed generation of electricity and thereby provided the relevant regulatory direction for off-grid, stand-alone, small-scale generation of renewable energy. The Integrated Energy Policy of 2006 discusses various policy initiatives for promoting RESs and emphasizes the need to move from capital incentives to performance-based incentives for promoting RESs.

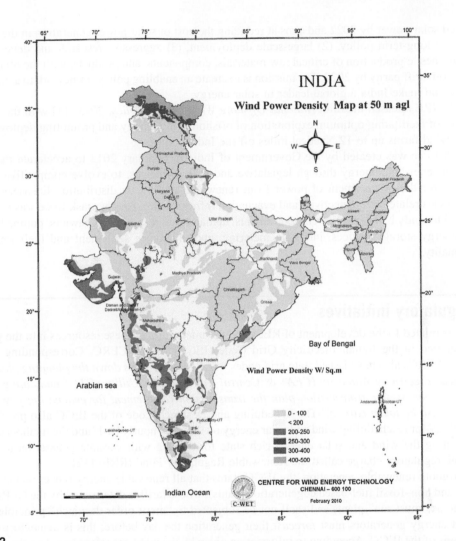

FIGURE 2

Map of wind energy sources in India.

The *National Action Plan on Climate Change* [9] set a target of 5% renewable energy purchase for the fiscal year 2009–10, which is proposed to increase by 1% every year for the next 10 years. SERCs specify a percentage of the total consumption of electricity, in the form of a renewable purchase obligation (RPO), which the distribution licensee must procure from RESs. Various initiatives in the form of non-market-based instruments for the promotion of renewable energy include tax waivers, accelerated depreciation (80% in the first year), preferential tariffs, and generation-based incentives.

The *Jawaharlal Nehru National Solar Mission* [10] was launched on January 11, 2010, by the Prime Minister of India. The Mission has set the ambitious target of deploying 20,000 MW of grid-connected solar power by 2022 and aims at reducing the cost of solar power generation in the country through (1) long-term policy, (2) large-scale deployment, (3) aggressive research and development, and (4) domestic production of critical raw materials, components, and products, with the objective to achieve grid tariff parity by 2022. The mission is to create an enabling policy framework to achieve this objective and make India a global leader in solar energy.

The MNRE released the draft *National Offshore Wind Energy Policy, 2013* [11] with the primary objectives of facilitating optimum exploitation of offshore wind energy and promoting deployment of offshore wind farms up to 12 Nautical miles off the Indian coast.

A task force was created by the Government of India in February 2012 to accelerate the development of renewable energy through legislative and policy changes, to evolve competitive bidding guidelines for the procurement of power from renewable sources by distribution licensees, and to address issues related to connectivity and evacuation infrastructure. Another task force was created in June 2013 to study balancing power requirements, demand side response / negawatts, pumped storage plants, energy storage devices, intrastate metering, and imbalance settlement and their impact on power quality.

3. Regulatory initiatives

Provisions related to the development of RESs and the integration of these resources into the grid has been specified in the Indian Electricity Grid Code (IEGC) by the CERC. Corresponding to this, respective SERCs also have issued state grid codes. "*IEGC clearly lays down the planning philosophy for Central Electricity Authority (CEA) & Central Transmission Utility (CTU) mandating that in formulating perspective transmission plan the transmission requirement for evacuating power from RES shall also be taken care of.*" The Scheduling and Despatch Code of the IEGC also provides the methodology for rescheduling wind and solar energy on three (3) hourly basis and the methodology for compensating the wind and solar energy–rich state for dealing with variable generation through a renewable regulatory charge called the Renewable Regulatory Fund (RRF) [12].

To promote renewable generation, the IEGC states that all renewable energy power plants (except biomass and non–fossil fuel-based cogeneration plants), whose tariff is determined by the CERC, shall be treated as "must run" plants and shall not be subjected to "merit order despatch" principles.

Wind energy generators must *forecast* their generation the day before; this is mandatory per the stipulations of the IEGC. According to information given by the wind generator based on the forecast, the concerned LDC prepares the dispatch schedules. The accuracy of forecast is important, and the wind generators are responsible for forecasting their generation with up to 70% accuracy. For actual generation within ±30% of the schedule, no charges for deviation would be payable/receivable by the generator, and the host state bears the charges for this variation within ±30%. If the actual generation is beyond ±30% of that scheduled, the wind generator bears the charges for the deviation. The charges for deviation (caused by the variability of wind generation) that are borne by the host state are to be shared among all the states in the country according to the ratio of their peak demand in the previous month, based on the data published by the CEA; these shared costs take the form of a regulatory charge, the renewable regulatory charge, operated through the RRF.

These provisions are applicable to wind farms with a collective capacity of ≥ 10 MW and those connected to the grid at ≥ 33 kV. Furthermore, from the point of view of grid security, maximum wind energy generation of only 150% of the schedule is allowed in a time block. Solar generation ≥ 5 MW, with connectivity to the grid at ≥ 33 kV, is exempted from any charges because of deviations from the schedule; these charges are to be paid by the host state and socialized through the RRF.

The *CERC's Sharing of Interstate Transmission Charges and Losses Regulations, 2010* specified that no transmission charges or transmission losses shall be levied on solar generation projects commissioned up to June 15, 2013.

CERC Grant of Connectivity, Long Term Access (LTA) and Medium Term Open Access (MTOA) in inter-State Transmission Regulations allows any renewable energy generating station of 5 MW capacity & above but less than 50 MW capacity developed in an existing generating station to connect to the inter-state transmission system, provided the existing generating station agrees to act as the "Principal Generator" on behalf of the renewable energy generating station.

4. Transmission planning initiatives

4.1 Transmission planning criteria

The CEA revised the Transmission Planning Criteria [13] in January 2013 and envisages large-scale integration of RESs throughout the country. These criteria state that the maximum generation at a wind/solar aggregation level may be calculated using the capacity factors specified in the Transmission Planning Criteria (reproduced in Table 1). To address further the variability of wind/solar projects, the planning and implementation of other aspects such as reactive compensation, forecasting, and the establishment of renewable energy control centers also are envisaged.

4.2 Report on the green energy corridors

Considering the large-scale requirements for integrating renewable energy, the MNRE and the Forum of Regulators [14] commissioned a study to identify transmission infrastructure and other control-related requirements for the integration of RESs, especially in renewable-rich states such as Tamil Nadu, Karnataka, Andhra Pradesh, Gujarat, Maharashtra, Rajasthan, and Himachal Pradesh; CAPEX requirements; and models for funding. The outcome is in the form of the "Report on the Green Energy Corridors" [15], which provides an overview of the renewable energy in the country, operational trends, reactive power requirements, the renewable energy management center, international experiences, and a perspective plan including cost estimates and challenges in the integration of renewable energy (see Figure 3).

Table 1 Capacity Factors for Renewable Energy Sources	
Voltage Level/Aggregation Level	**Capacity Factor (%)**
132 kV/Individual wind/solar farm	80
220 kV	75
400 kV	70
State (as a whole)	60

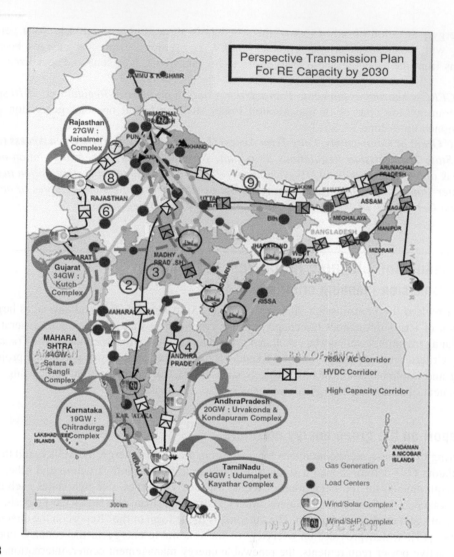

FIGURE 3

Map of green energy corridors in India.

5. Experience with RECs in India

To promote the consumption of energy from RESs, the SERCs specify RPO targets as a percentage of the total energy consumption of electricity that must be procured from RESs.

In light of this, and to promote further investment and development of RESs, the country needed a market-based mechanism. RECs provide such a market-based mechanism for promoting renewable energy in the country. Under this mechanism, a REC is issued against the energy injection by grid-connected RESs; the RECs can be traded, thereby providing a means of fulfilling RPO targets.

The CERC created the REC Regulations 2010 [16] on January 14, 2010, to facilitate the implementation of the REC mechanism in the country and designated national LDCs as the central agency in the implementation of the REC mechanism in India.

A certain category of grid-connected RESs are considered eligible under the REC mechanism; the energy generated from such units is demarcated into two components: the brown component, which is the electricity component, and the green component, which is represented by the REC. One REC corresponds to 1 MWh of energy. Both the brown and the green components can be traded separately in the pan-India Electricity Market on designated Power Exchanges. At the state level, the SERCs are responsible for providing the regulatory framework for implementation of the REC mechanism. The REC process comprises four stages, namely, accreditation, registration, issuance, and redemption. The entire REC process is paperless through an integrated web-based application [17–19].

5.1 Experience gained

Participation in the REC mechanism has increased steadily. As of September 2013, more than 800 projects have been accredited, with a total renewable energy capacity of about 4000 MW. In the short period of about 3 years since the inception of the REC mechanism, nearly 8 million RECs have been issued, of which about four million RECs have been traded, amounting to about Indian Rupee 809 crores (equivalent to approximately US$135 million, assuming Rs 60 = US$1). It has been observed that the seller's market is buoyant, but the response from the buyers has been slightly sluggish, resulting in lower demand for RECs; as a consequence, both solar and nonsolar RECs are being sold at floor price.

6. Challenges

6.1 Integration

The transmission network is a critical infrastructure required for the integration of renewable, and at the bulk transmission level, high-capacity green energy corridors are being planned and implemented. A lot needs to be done in terms of strengthening the subtransmission network and the last mile connectivity to the renewable energy generators. There is a need for forecasting for effective scheduling of renewable energy generation, both at the local level by the renewable energy generator and at the state/regional level by the system operators. The operational frequency band also needs to be tightened further, and the CERC has initiated steps in this direction.

6.2 Addressing variability

In light of the variable nature of renewable energy generation, the balance of power requirements needs to be addressed. This is essential when considering grid security, which must be maintained at all times. The existing deviation settlement (Unscheduled Interchange (UI) mechanism) is being reviewed, and the CERC has floated draft Deviation Settlement Mechanism Regulations for consultation by stakeholders.

6.3 REC mechanism

A major challenge for the REC market in India is stimulating the demand for RECs or, in other words, the "buy" side of the REC market. This is primarily happening because of a lack of effective enforcement of the regulatory provisions and RPO compliance.

6.4 Institutional arrangements

The MNRE created a task force in March 2012 to examine and resolve issues related to the implementation of the RRF mechanism in the country. The task force submitted its report in June 2012 and identified the following key issues:

1. Need for a suitable institutional arrangement for intrastate metering and settling imbalances.
2. Need for a registry/database for renewable generation in India.
3. *Need for a recognized legal entity to act as nodal agency on behalf of the developers & individual owners of the wind generators.*

7. Concluding remarks

Renewable energy and its integration is a key focus area at all levels, be it policy, regulation, or implementation. Transmission planning is being done, keeping in mind large-scale integration of renewable energy, and green corridors are being implemented. A number of financial incentives also have been provided. Forecasting and scheduling are being emphasized for integrated grid operation.

India is among the few countries in the world to have implemented the market-based REC mechanism with a pan-Indian market. Discovery of REC prices in the Power Exchanges has provided price signals for investment as well as investor confidence and comfort. The REC mechanism has facilitated an alternate route for fulfilling RPO compliance requirements, especially by the states that are less endowed in terms of RESs.

Various concerns being faced with respect to the REC mechanism include comprehensive monitoring and enforcement of RPO compliance; creating a secondary market for RECs, allowing them to be retrading; as well as inclusion of off-grid RESs under the REC category. Furthermore, institutional arrangements, a central registry/database for renewable sources, and legal arrangements in terms of a nodal agency on behalf of a group of RE generators for scheduling, metering, and settlement are being debated and considered for future implementation.

References

[1] Expert Committee on Integrated Energy Policy, Planning Commission, Government of India. http://planningcommission.gov.in/reports/genrep/rep_intengy.pdf.
[2] Working Group on Power for Twelfth Plan (2012–17), Planning Commission, Government of India. http://planningcommission.nic.in/aboutus/committee/wrkgrp12/wg_power1904.pdf.
[3] Indian Electricity Act, 2003, Ministry of Power, Government of India. http://www.powermin.nic.in/acts_notification/electricity_act2003/pdf/The%20Electricity%20Act_2003.pdf.
[4] Website of Ministry of Power, Government of India. www.powermin.nic.in.

[5] Website of Ministry of New and Renewable Energy, Government of India. www.mnre.gov.in.

[6] National Electricity Policy, Ministry of Power, Government of India. http://www.powermin.nic.in/whats_new/national_electricity_policy.htm.

[7] Tariff Policy, Ministry of Power, Government of India. http://www.powermin.nic.in/whats_new/pdf/Tariff_Policy.pdf.

[8] Rural Electrification Policy, Ministry of Power, Government of India. http://www.aegcl.co.in/RE%20Policy%2023_08_2006.pdf.

[9] National Action Plan on Climate Change, Government of India. http://pmindia.nic.in/climate_change.htm.

[10] Jawaharlal Nehru National Solar Mission (JNNSM). http://www.mnre.gov.in/solar-mission/jnnsm/introduction-2/.

[11] National Offshore Wind Policy, 2013 (draft). http://mnre.gov.in/file-manager/UserFiles/draft-national-policy-for-offshore-wind.pdf.

[12] Indian Electricity Grid Code Regulations, 2010, Central Electricity Regulatory Commission. http://www.cercind.gov.in/Regulations/Signed-IEGC.pdf.

[13] Manual on Transmission Planning Criteria, Central Electricity Authority. http://www.cea.nic.in/reports/articles/ps/tr_plg_criteria_manual_jan13.pdf.

[14] Website of Forum of Regulators. www.forumofregulators.gov.in.

[15] Green Energy Corridors Report, Power Grid Corporation of India Ltd. www.powergridindia.com/_layouts/PowerGrid/WriteReadData/file/Green_Energy_Corridors.zip.

[16] Central Electricity Regulatory Commission (terms and conditions for recognition and issuance of renewable energy certificate for renewable energy generation) Regulations. http://cercind.gov.in/Regulations/CERC_Regulation_on_Renewable_Energy_Certificates_REC.pdf; 2010.

[17] REC Registry of India. www.recregistryindia.nic.in.

[18] Soonee SK, Minaxi Garg, Satya Prakash. Renewable energy certificate mechanism in India, 16th national power systems conference, December 2010. http://nldc.in/papers/Renewable%20Energy%20Certification%20Mechanism%20in%20India_2010.pdf.

[19] Soonee SK, Minaxi Garg, Saxena SC, Satya Prakash. Implementation of renewable energy certificate (REC) mechanism in India, C5–PS1, CIGRE – 2012. https://www.recregistryindia.nic.in/pdf/Others/C5_113_2012.pdf.

System Flexibility

Long-Term Energy Systems Planning: Accounting for Short-Term Variability and Flexibility

Manuel Welsch, Dimitris Mentis, Mark Howells

Division of Energy Systems Analysis, KTH Royal Institute of Technology, Stockholm

1. Introduction

According to the International Energy Agency (IEA), the world continues to diverge from a pathway toward meeting internationally agreed climate change targets. Global average temperatures are expected to increase by 2.8–4.5 °C by 2100 if no countermeasures are taken [1]. This calls for a major transformation of our energy systems, in which renewable energy technologies play an important role to mitigate climate change.

The electricity production of renewable energy sources, such as wind and solar power, is variable as a function of the availability of the renewable resource at hand. Variable renewables certainly provide secure quantities of energy when considered over longer time periods, but they do not guarantee the secure delivery of power as and when needed [2]. The variability they introduce adds to the overall fluctuations in power systems. For example, on the supply side, these may be due to outages in conventional power plants, and on the demand side, they may be due to the time dependency of loads. As the shares of renewable electricity generation rise, future power systems need to be increasingly flexibility to cope with such fluctuations to balance supply and demand. Energy policies and strategies are required that facilitate the transformation to such increasingly flexible power systems.

Energy models have successfully proven their use in informing the development of energy policies and strategies from the early 1980s on [3–6]. They commonly serve as test-beds to investigate developments or system configurations that would be impractical, too expensive or impossible to test in real-world conditions [7]. Numerous types of energy models have been developed. These tools are specialized and adjusted to create the best model for a particular situation within a given context, scale, and time frame. This chapter focuses on long-term energy system models and their application to assess energy and investment strategies. They may guide decisions makers in designing flexible and robust energy systems, which facilitate future adaptations and perform well under various potential developments [8].

Section 2 describes the provision of flexibility in power systems through supply- and demand-side operating reserves and storage options. Section 3 provides a concise outline of several modeling approaches and presents limitations of long-term models regarding their temporal resolution and the metrics applied to ensure the power system's reliability. Section 4 describes selected modeling approaches that address the gap between the short time scales required to assess operational issues and the long time horizons of the investment decisions calculated by these models. This chapter concludes in Section 5.

2. Flexibility in power systems

Flexibility in power systems mainly is required to balance unpredictable variations in demand and generation. Such variations may be due to, for example, load forecast errors, forced power plant and transmission outages, and wind forecast deviations. Variations in supply and demand are compensated by drawing on short-term balancing services provided by so-called operating reserves. These usually consist of power plants that can adjust their generation at fast ramp rates or at loads that can be increased or decreased at short notice. Numerous denominations for these balancing services exist, yet the same term might entail different technical requirements in different countries. Most commonly, three types of operating reserves are distinguished based on the time scales within which they can be activated [9].

Primary reserve is used to limit the change in the power system's frequency due to an imbalance between supply and demand. It is locally automated and reacts to deviations in the nominal system frequency. *Secondary reserve* is used to restore the frequency to its design value. It is automated centrally and serves to release the primary reserve for future operation. Finally, the *tertiary reserve* is used to restore the secondary reserve. The time frames associated with these reserves vary. Commonly, primary reserve can be activated within 30 s, while secondary and tertiary reserves have to be fully available within 5–15 min [10]. If any one of these reserves is drawn on to increase the system's frequency (e.g. by increasing generation), this is referred to as upward reserve. If the system's frequency needs to be reduced, downward reserve is activated.

Conventionally, operating reserves are activated by adjusting generation. Yet, demand response measures may offer a largely untapped potential, which usually are available at faster rates than supply-side options [11]. The ability to realize this potential draws to a large extent on the increased communication, automation, and control facilitated by Smart Grids [12].

Examples of such demand-response measures include the automated control of industrial loads, such as interruptible heating or cooling demands. Furthermore, about half of private household demand is estimated to provide some flexibility regarding the timing of when to meet this demand [13]. An example includes the control of the electricity supply of refrigerators that hold enough thermal storage to withstand interruptions. At the municipal level, pumps linked to water supply reservoirs may offer operating reserve. In addition to consumers, flexible appliances also may be operated by power plant owners. For example, several Danish combined heat and power plants are equipped with additional electrical boilers to increase their flexibility (see Chapter 19).

Apart from adjusting generation and demand-response measures, storage technologies provide another option to balance the variability in demand and generation. As renewable electricity portfolios expand in many countries, it is essential to gain a better understanding of the available storage options [14]. System planners need to optimize not only the overall storage capacity but also the mix of different technologies, from bulk to distributed storage, and from electricity to fuel and heat storage [15].

In the near term, according to the Electric Power Research Institute (EPRI), the most cost-competitive energy storage options may include pumped storage hydropower, compressed air energy storage (CAES), and vehicle-to-grid (V2G) storage [16]. CAES uses off-peak electricity to compress air and store it in a reservoir, an underground cavern, surface pipes, or vessels. When electricity is needed, the compressed air is heated, expanded, and directed through a conventional gas turbine to generate electricity. V2G refers to a system that enables a controllable, bidirectional electricity flow between a vehicle and the grid [17]. This allows electric vehicles to provide peaking power or contribute to operating reserves.

With increasing shares of renewable power generation, all sources of flexibility within the power system gain in attractiveness, from supply-side to demand-side and storage options. Too little flexibility in the power system might result in the best case in the need, for example, to curtail wind power generation. In the worst-case scenario, it increasingly will cause outages due to an unreliable power system that is unable to curb deviations from the systems design frequency.

3. Modeling approaches and their limitations

Energy and power system tools are applied to model the impacts of increasing shares of variable generation at various levels of detail. Long-term energy system models analyze the evolution of the energy system over several decades and include nonelectricity sectors, such as heat or transportation. The investment decisions and policy recommendations derived from such models may serve as input for a more detailed analysis of electricity markets based on power system models (Figure 1).

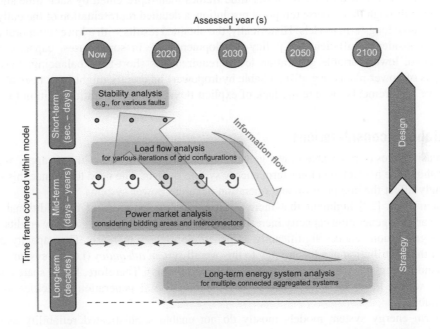

FIGURE 1

How energy system models may inform power systems analysis [18]. Note the schematic nature of this diagram, which represents a general tendency rather than a strict categorization.

Commonly, these models focus exclusively on electricity and model shorter time horizons up until several years. On the basis of their calculations, electrical engineering types of models may analyze the implications of increasing shares of renewables on the grid (e.g. by assessing the resulting load flows or potential faults).

3.1 Temporal resolution

Usually, long-term energy system models do not attempt to consecutively model all days or hours within a year, but rather analyze a limited number of temporally independent so-called time slices [19]. A time slice combines a fraction of the year with specific load and supply characteristics. The coarse temporal resolution applied in such models is to a large extent due to their long-term time horizon and the computational power required for enhancing their temporal resolution. Assessing the daily dispatch in detail might create a false precision compared with the overall uncertainties associated with long-term projections.

Long-term models may be based on as little as 6–12 time slices [20,21]. An example for the latter is the model used to inform the IEA's Energy Technology Perspectives report [22]. Several models, however, apply a more detailed temporal resolution [23,24]. The models developed by Pina et al. [25] and Kannan and Turton [26] are even based on 288 time slices per year. This increased number of time slices can facilitate a more accurate depiction of variable renewable energy resources, storage constraints, and demand-side options. Balancing the variability of supply and demand, however, requires a power system response within much shorter time frames than represented by such time slices.

As implied through their coarse temporal resolution, a detailed representation of the daily dispatch is not the focus of long-term models. Historically, the obtained results with a lower temporal resolution often were considered sufficient for policy development and, in some cases, capacity expansion planning. Also, lower variable generation levels required less short-term balancing. Systems with certain types of power plants (e.g. dispatchable hydropower) inherently might have been able to meet most reserve requirements, despite the lack of explicit flexibility considerations within the models.

3.2 Reliability considerations

With increasing rates of renewable power generation, the short-term variability of supply and demand no longer should be neglected in long-term energy system models. The need to balance this variability may strongly affect the composition and operation of future power systems.

Holttinen et al. [27] highlight that there is an increasing requirement for additional operating reserve as variable generation capacity increases. Furthermore, the range of capacity credits attributed to variable generation reduces significantly as penetration levels increase. The capacity credit is a measure of the contribution of a technology to the overall *system adequacy* (i.e. the system's ability to meet demand throughout the year under steady-state conditions). Therefore, larger shares of variable renewables with a lower capacity credit require increased total generation capacities to maintain system reliability.

Long-term energy system models mostly do not enable sophisticated reliability assessments. Rather, a simple metric is applied by entering a system reserve margin as an input parameter. This margin ensures that the capacity credit of all power plants within the system always exceeds the load by a certain percentage. The capacity credit usually is defined as an input value and is not calculated within the model as a function of renewable generation penetration rates. Subsequently, such long-term energy models may misrepresent the investment implications of maintaining system adequacy.

Furthermore, despite the fact that unpredictable variations are a key driver for system costs associated with balancing, they rarely are considered in long-term energy system models. Therefore, no statement about the level of *system security* achieved with such models can be made. System security refers to the ability of a power system to dynamically respond to disturbances from within the system. Assessing system security usually requires the use of separate power system models. These models commonly provide limited insights for capacity expansion planning and exclude nonelectricity sectors. Therefore, it is apparent that a gap exists between the two modeling families [25]. This gap should be bridged to reach an appropriate approach for energy planning that considers those short-term dynamics that influence long-term investments.

4. Addressing the gap between short-term and long-term models

Several successful attempts focused on bridging this gap by interlinking long-term energy system models with short-term power system models [8,28–32]. Although linking models may enable a more accurate analysis of an energy system, the level of effort to set up two independent models might be a deterrent to a more frequent application. Furthermore, although long-term capacity investments easily can be fed into short-term models, the information flow back to the long-term model appears to be more challenging and sometimes is omitted.

An approach therefore may be useful that allows drawing on the advantages of long-term models with a lower temporal resolution, without ignoring the main dynamics introduced by variable generation. The next sections provide selected examples of how the short-term variability introduced by renewable energy sources can be considered in long-term models. It includes both interlinked energy and power system models, as well as long-term models that were enhanced to capture the implications of short-term variability.

4.1 Renewable electricity futures study

The National Renewable Energy Laboratory's (NREL's) Renewable Electricity Futures Study analyzes the U.S. electricity system's ability to meet customer demand until 2050, drawing on high levels of renewable electricity generation [30]. The Regional Energy Deployment System (ReEDS) was used for the underlying analysis. ReEDS is a multiregional long-term capacity-expansion model for the deployment of electric power generation technologies and transmission infrastructure throughout the United States [33].

ReEDS uses statistical methods to quantify the impacts of variability on the power system for each 2-year optimization period. The capacity credit of each technology is calculated for every time period before calculating the linear program optimization. This draws on exogenously defined Pearson correlations between the expected generation at predefined pairs of potential plant sites. In ReEDS, designated power plants are allowed to operate in between their minimum operating level and their maximum seasonal output to meet reserve requirements. Although the model captures the increasing need for operating reserves as greater levels of variable generation are integrated, it does not explicitly deal with subhourly or subminute events, such as frequency regulation. Furthermore, downward reserves were not assessed in the ReEDS model used for the NREL study. Also, ramping characteristics of various generation technologies were not explicitly considered. The ReEDS model therefore was interlinked with a separate hourly electricity dispatch model to analyze one year in more detail.

The key result of this study is that renewable electricity generation from technologies that are commercially available today would be able to supply 80% of the total electricity production in 2050. To realize this potential, a more flexible electricity system is needed. A broad portfolio of supply- and demand-side options are stated in Ref. [30] to support such a system, including flexible conventional generation, grid storage, new transmission infrastructure, more responsive loads, and improved power system operations.

4.2 Interlinking TIMES-PLEXOS

A recent study [29] interlinked an operational power system (PLEXOS) with a long-term energy system model (TIMES) for Ireland. The Irish TIMES model was used to assess the expansion of the Irish energy system. Input assumptions are described in Ref. [20]. The power system model PLEXOS was used for a detailed operational analysis of the unit commitment and dispatch in the year 2020. Interlinking these models assessed how the dispatch and emissions would change with more temporal and technical detail.

The Integrated MARKAL-EFOM System (TIMES) is a bottom-up technoeconomic linear optimization model [34]. It usually is applied to analyze the entire energy sector of a country or a region, subject to environmental, policy, and technical constraints. PLEXOS is a deterministic mixed-integer linear programing model [35]. The model considers various constraints on unit operations, including limits on the generation, reserve provision, up and down times, and ramp rates [36].

Both TIMES and PLEXOS were set up to model Ireland's 40% renewable generation target. In line with a publication by the Irish Transmission System Operators (2010) [40], the technically acceptable instantaneous maximum wind share in the generation mix was limited to 70% of the load. The Irish TIMES model draws on 12 yearly time slices. PLEXOS instead was set up as a chronological, hourly model. A soft-linking model was implemented by feeding the power plant capacity mixes for 2020 as derived from the TIMES model into PLEXOS. PLEXOS then was used to evaluate the overall operational reliability of this particular capacity mix. The soft-linked models are referred to as TIMES-PLEXOS in Table 1 and in Figure 3.

Table 1 Model Set-ups and Their Main Characteristics

OSeMOSYS Simple	OSeMOSYS 70% Wind	TIMES-PLEXOS Simple	OSeMOSYS Enhanced	TIMES-PLEXOS Enhanced
• Uses the core code of OSeMOSYS, as available at osemosys.org • Comparable to stand-alone TIMES model	• External hourly wind data analysis to ensure a max. wind penetration of 70%	• Comparable input data to OSeMOSYS 70% wind • Hourly, chronological simulation	• External wind data analysis • Min. stable generation • Reserve contribution • Operating reserve	• Hourly (wind power) simulation • Start-up costs • Min. stable generation • Ramp rates • Operating reserve

The work by Deane et al. [29] demonstrated that results from the Irish TIMES model represented a reliable and adequate power system configuration. Yet, linking TIMES with PLEXOS showed that the need for flexibility in energy systems was significantly underestimated in the stand-alone TIMES model: Too little wind curtailment occurred and the roles of storage and older combined cycle gas turbines were undervalued.

4.3 OSeMOSYS

OSeMOSYS is an open-source energy systems model with a medium- to long-term time horizon. It is available for download at www.osemosys.org. In a recent effort, it was enhanced to consider the short-term variability of supply and demand without the need to interlink it with a more detailed power system model [9].

The enhanced OSeMOSYS model considers both upward and downward primary and secondary operating reserves. As in conventional long-term models, a reserve margin can be specified to additionally ensure the availability of sufficient reserves that do not have any specific requirements regarding the speed of their activation. Any technology, including supply- and demand-side technologies as well as storage options, can be assigned to meet reserve requirements. In addition to cost parameters, emission profiles, thermodynamic efficiencies, investment limits, and generation constraints, each technology is characterized by exogenously defined maximum reserve contributions and minimum stable generation levels.

On the basis of these characteristics, the model decides which technologies need to be online (spinning) to provide operating reserve and which ones can be started up from zero output if needed. A minimum share of upward reserve requirements can be specified that has to be met by online technologies. Startup costs were not considered explicitly to avoid mixed-integer programing and minimize calculation times. Instead, cycling characteristics can be defined that may limit changes in the online capacity and generation of a technology. The modeling of storage and demand-side management options within OSeMOSYS has been explained in detail by Welsch et al. [19].

Figure 2 demonstrates some of the changes in results when adding secondary and primary reserves to a conventional long-term model. Results are presented for a test-case application. The most significant changes are the increased investments in open and combined cycle gas turbines (OCGT and CCGT). A more flexible operation of nuclear power may compensate for some of the investments in gas turbines (see graph on the bottom right). Further background on this test case is provided by Welsch et al. [9].

To quantify the dispatch improvements, the enhanced OSeMOSYS model was further applied and compared with the TIMES-PLEXOS model mentioned in Section 4.2 [37] in a separate model run. Several set-ups were compared with each other. A succinct summary of these set-ups is provided in Table 1.

The various OSeMOSYS models were compared with several TIMES-PLEXOS models to enable a differentiation of the accuracy gains due to a more detailed external wind data analysis (as considered in OSeMOSYS 70% Wind) from those due to an hourly resolution (as considered in TIMES-PLEXOS Simple) and those due to additional operational detail (as considered in OSeMOSYS Enhanced and TIMES-PLEXOS Enhanced). Special attention was paid to ensure that the assumptions and data used in OSeMOSYS matched those of the TIMES-PLEXOS models to ensure the comparability of the results between the different models.

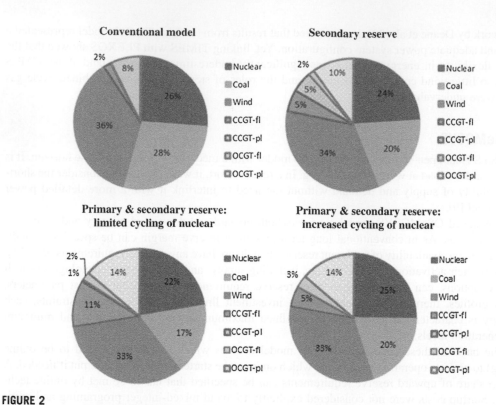

FIGURE 2

Capacity mix in 2040 for selected model set-ups [9].

Results of the simple OSeMOSYS model were first compared with those of the stand-alone TIMES model. The annual electricity generation by fuel type for 2020 proved to be similar in the two models. This demonstrated a consistent representation of Ireland in the two modeling tools. Results of the stand-alone TIMES model are not shown in Figure 3. Compared to the most accurate enhanced TIMES-PLEXOS model, however, >20% of the yearly generation as calculated by the simple OSeMOSYS model was not attributed to the correct power plant types.

This discrepancy decreased when considering an external wind data analysis to ensure that a maximum wind penetration of 70% is not exceeded (OSeMOSYS 70% wind). Applying the enhanced OSeMOSYS model that considers operating requirements further improved the results, which are now close to the more complex enhanced TIMES-PLEXOS model: 95.0% of the dispatch results of the enhanced OSeMOSYS model matched those of an interlinked model with a 700 times higher temporal resolution and more operational detail.

Extending the analysis until 2050 indicated that ignoring short-term variability of supply and demand may result in suboptimal investments in individual generation technologies. In the simple OSeMOSYS model without operational constraints, up to 23.5% of the total capacity was assigned to different power plant types as compared with the enhanced model.

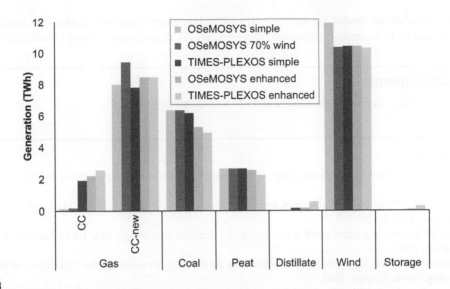

FIGURE 3

Annual generation of the modeled power plant types in 2020 [37]. OSeMOSYS results in shades of green, TIMES-PLEXOS results in shades of blue.

5. Conclusion

Clearly, long-term energy models may underestimate the importance of flexibility within the power system if the implications of variability in demand and generation are not considered adequately. If policies were to be derived from such long-term models, they might promote energy systems that do not ensure that expected reliability standards are met.

The underestimation of flexibility may as well be observed in energy-only electricity markets, where dispatchable technologies face increasingly unstable and on average lower electricity prices and capacity factors because of the growing shares of renewable electricity generation [38]. The profitability of investments in such dispatchable technologies may drop, and retired capacities may not be replaced. On the other hand, as indicated by Figures 2 and 3, dispatchable power plants play a key role in the technology mix. They are required to ensure the system's adequacy and security by balancing the variability introduced by, inter alia, renewable electricity generation. Markets might have to deal with a gap between the need for flexibility and the incentives to invest in flexible technologies [39]. Modeling approaches may help evaluate this gap in economic and technical terms [37].

Soft-linking models will ensure the most accurate results regarding the generation mix, but this requires setting up two separate models: a long-term model, which focuses mainly on optimizing capacity expansions; and a power system model, which assesses the dispatch. Because there is no overall optimization across the two models, the identified capacity investments may not present the most economically efficient technology mixes. This limitation may be addressed by enhancing a single long-term optimization model, for example, by combining a more detailed external data analysis with simplified metrics to represent operational requirements, as demonstrated using OSeMOSYS.

Acknowledgments

Abhishek Shivakumar (KTH) for his valuable input to writing this chapter.

References

[1] IEA. Redrawing the energy–climate map – world energy outlook special report. Paris: International Energy Agency; 2013.

[2] Milborrow D. Wind power on the grid, in: renewable electricity and the grid: the challenge of variability. Earthscan; 2008:31–54.

[3] Häfele W. Energy in a finite world: a global systems analysis. Cambridge, MA, USA: Ballinger Publishing Company; 1981.

[4] Huntington HG, Weyant JP, Sweeney JL. Modeling for insights, not numbers: the experiences of the energy modeling forum. Omega; 1982.

[5] Meier P. Energy modelling in practice: an application of spatial programming. Omega 1982;10:483–91.

[6] Rath-Nagel S, Stocks K. Energy modelling for technology assessment: the MARKAL approach. Omega 1982;10:493–505.

[7] Howells MI. Analyzing sustainable energy in developing countries: selected South African case studies; 2008.

[8] Heinrich G. A comprehensive approach to electricity investment planning for multiple objectives and uncertainty [Ph.D. thesis]. Cape Town, South Africa: University of Cape Town; 2008.

[9] Welsch M, Howells M, Hesamzadeh M, Ó Gallachóir B, Deane JP, Strachan N, et al. Supporting security and adequacy in future energy systems – the need to enhance long-term energy system models to better treat issues related to variability, under review.

[10] Rebours Y, Kirschen D. What is spinning reserve? The University of Manchester; 2005.

[11] Ela E, Milligan M, Kirby B. Operating reserves and variable generation – a comprehensive review of current strategies, studies, and fundamental research on the impact that increased penetration of variable renewable generation has on power system operating reserves (NREL/TP-5500-51978). Golden, CO, USA: National Renewable Energy Laboratory (NREL); 2011.

[12] Welsch M, Bazilian M, Howells M, Divan D, Elzinga D, Strbac G, et al. Smart and just grids for sub-Saharan Africa: exploring options. Renew. Sustain Energy Rev 2013;20:336–52.

[13] Block C, Neumann D, Weinhardt C. A market mechanism for energy allocation in micro-chp grids; 2008. Presented at the 41st Hawaii International Conference on System Sciences, pp. 172–180.

[14] Roberts BP, Sandberg C. The role of energy storage in development of smart grids. Proc. IEEE 2011;99:1139–44.

[15] Østergaard PA. Comparing electricity, heat and biogas storages' impacts on renewable energy integration. Energy 2012;37:255–62.

[16] Rastler D. Electricity energy storage technology options; a white paper primer on applications, costs and benefits. Palo Alto, California: EPRI; 2010.

[17] Briones A, Francfort J, Heitmann P, Schey M, Schey S, Smart J. Vehicle-to-grid (V2G) power flow regulations and building codes. Idaho National Laboratory; 2012.

[18] Welsch M. Enhancing the treatment of systems integration in long-term energy models [Ph.D. thesis]. Stockholm: KTH Royal Institute of Technology; 2013.

[19] Welsch M, Howells M, Bazilian M, DeCarolis J, Hermann S, Rogner HH. Modelling elements of smart grids – enhancing the OSeMOSYS (Open Source Energy Modelling System) code. Energy 2012;46:337–50.

[20] Chiodi A, Gargiulo M, Rogan F, Deane JP, Lavigne D, Rout UK, et al. Modelling the impacts of challenging 2050 European climate mitigation targets on Ireland's energy system. Energy Policy 2013;53:169–89.

[21] Kannan R. The development and application of a temporal MARKAL energy system model using flexible time slicing. Appl Energy 2011;88:2261–72.

[22] IEA. Energy technology perspectives 2012: pathways to a clean energy system. Paris: International Energy Agency; 2012.

[23] Howells M, Alfstad T, Victor DG, Goldstein G, Remme U. A model of household energy services in a low-income rural African village. Energy Policy 2005;33:1833–51.

[24] Nelson J, Johnston J, Mileva A, Fripp M, Hoffman I, Petros-Good A, et al. High-resolution modeling of the western North American power system demonstrates low-cost and low-carbon futures. Energy Policy 2012; 43:436–47.

[25] Pina A, Silva C, Ferrão P. Modeling hourly electricity dynamics for policy making in long-term scenarios. Energy Policy 2011;39:4692–702.

[26] Kannan R, Turton H. A long-term electricity dispatch model with the TIMES framework. Environ Modell Assess; 2012.

[27] Holttinen H, Meibom P, Orths A, Van Hulle F, Lange B, O'Malley M, et al. Design and operation of power systems with large amounts of wind power, IEA WIND Task 25. International Energy Agency; 2009.

[28] Chaudry M, Ekins P, Ramachandran K, Shakoor A, Skea J, Strbac G, et al. Building a resilient UK energy system (REF UKERC/RR/HQ/2011/001). UK Energy Research Centre (UKERC); 2011.

[29] Deane JP, Chiodi A, Gargiulo M, Ó Gallachóir BP. Soft-linking of a power systems model to an energy systems model. Energy 2012;42:303–12.

[30] Hand MM, Baldwin S, DeMeo E, Reilly JM, Mai T, Arent D, et al. (NREL/TP-6A20–52409). Renewable electricity futures study, vol. 1–4. Golden, CO, USA: National Renewable Energy Laboratory (NREL); 2012.

[31] Möst D, Fichtner W. Renewable energy sources in European energy supply and interactions with emission trading. Energy Policy 2010;38:2898–910.

[32] Rosen J, Tietze-Stöckinger I, Rentz O. Model-based analysis of effects from large-scale wind power production. Energy 2007;32:575–83.

[33] Short W, Sullivan P, Mai T, Mowers M, Uriarte C, Blair N, et al. Regional energy deployment system (ReEDS) (NREL/TP-6A20-46534). Golden, CO, USA: National Renewable Energy Laboratory (NREL); 2011.

[34] Loulou R, Labriet M. ETSAP-TIAM: the TIMES integrated assessment model part I: model structure. Computational Management Science 2007;5:7–40.

[35] Foley AM, Ó Gallachóir BP, Hur J, Baldick R, McKeogh EJ. A strategic review of electricity systems models. Energy 2010;35:4522–30.

[36] Energy Exemplar. Plexos for power systems – leading the field in power market modelling; 2013.

[37] Welsch M, Deane JP, Howells M, Rogan F, Ó Gallachóir B, Rogner HH, et al. Incorporating flexibility requirements into long-term models – a case study on high levels of renewable electricity penetration in Ireland, submitted for publication.

[38] NEA. Nuclear energy and renewables: system effects in low-carbon electricity systems. Paris: OECD Nuclear Energy Agency; 2012.

[39] Traber T, Kemfert C. Gone with the wind? – electricity market prices and incentives to invest in thermal power plants under increasing wind energy supply. Energy Econ 2011;33:249–56.

[40] EirGrid, SONI. All island TSO facilitation of renwables studies; 2010.

Role of Power System Flexibility 18

Andreas Ulbig, Göran Andersson

Power Systems Laboratory, ETH Zurich, Zürich, Switzerland

1. Introduction

In recent years, power system dispatch optimization and real-time operation are becoming more and more driven by several major trends, which notably include the following:

1. The widespread deployment of variable renewable energy sources (RESs; i.e. wind turbines and photovoltaic units), in many countries worldwide, has led to significant relative and absolute shares of power generation, which is significantly fluctuating and not perfectly predictable nor fully controllable. Well-known operational issues caused by variable RES power in-feed are nondeterministic power imbalances and power flow changes on all grid levels.

2. The growing power market activity on the increasingly integrated national and transnational power markets has led to operational concerns of its own (i.e. deterministic frequency deviations caused by transient power imbalances due to more frequent changes in the now market-driven operating set-point schedules of power plants and more volatile (cross-border) power flow patterns).

3. The emergence of a smart grid notion or vision as a driver for change in power system operation: by using the reference frame of control theory, the term smart grid can be understood as the sum of all efforts that improve observability and controllability over individual power system processes (i.e. power in-feed to the grid and power out-feed from the grid, as well as power flows on the demand/supply side, happening on all grid levels).

Altogether, these developments constitute a major paradigm shift in the management of power generation and load demand portfolios. Operating power systems optimally in this more complex environment requires more detailed assessment of the available operational flexibility at every point in time.

Operation flexibility in power system operation and dispatch planning is of importance and has a significant commercial value. Ancillary service markets enable system operators the cost-effective procurement of needed control reserve products. In the case of frequency control schemes that are in essence a set of differently structured flexibility services provided to system operators for achieving active power regulation on different time scales [1], the overall remuneration for providing control power and energy on ancillary service markets is usually significantly higher than for bulk energy from spot markets [2].

Renewable Energy Integration. http://dx.doi.org/10.1016/B978-0-12-407910-6.00018-1

The value of operational flexibility can also be shown indirectly by looking at the *inflexibility costs* incurred by conventional generation units in the form of ramping costs and plant start/stop costs.

In some power markets, the real or merely perceived inflexibility of generator units to reduce their power output from planned set-points appears in the form of negative bids in the supply-side curve of the merit order [3]. The negative bids may then either reflect the inflexibility costs that would be incurred in case a plant's power output is lowered (e.g. lower efficiency and wear and tear) or simply the more general goal to keep a certain generation unit online (e.g. must-run generation units that provide ancillary services or RES units that have in-feed priority).

Several sources of power system flexibility exist, as is illustrated in Figure 1. Operational flexibility can be obtained on the generation side in the form of dynamically fast-responding conventional power plants (e.g. gas- or oil-fueled turbines or rather flexible modern coal-fired power plants) and on the demand side by means of adapting the load demand curve to partially absorb fluctuating RES power in-feed. In addition to this, RES power in-feed can also be curtailed or, in more general terms, modulated below its given time-variant maximum output level. Furthermore, stationary storage capacities, e.g. hydrostorage, Compressed Air Energy Storage (CAES), stationary battery, or fly-wheel systems, and time-variant storage capacities, e.g. Plug-In Hybrid Electric Vehicle (PHEV) and Electric Vehicle (EV) fleets, are well suited are well suited for providing operational flexibility.

Additional flexibility can be obtained from other grid zones via the electricity grid's tie-lines in case that the available operational flexibility in one's own grid zone is not sufficient or more expensive than elsewhere. Matter of fact, power import and export, nowadays facilitated by more and more integrated transnational power markets, is used in daily power system operation to a certain degree as a "slack bus" for fulfilling the active power balance and mitigating power flow problems of individual grid zones by tapping into the flexibility potential of other zones.

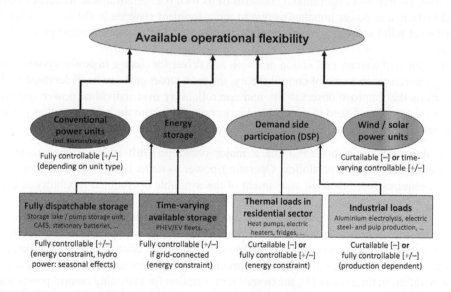

FIGURE 1

Sources of power system flexibility.

For power system operation, importing needed power in certain situations and exporting unde-sirable power in-feed in other situations to neighboring grid zones is, for the time being, probably the most convenient and cheapest measure for increasing operational flexibility. However, power import/export can only be performed within the limits given by the agreed line transfer capacities between the grid zones. In the European context, this corresponds to the so-called net transfer capacity values [4], which are a rather conservative measure of available grid transfer capacity.

2. Metrics for operational flexibility

The term operational flexibility in power systems, or simply flexibility, is often not properly defined and may refer to different things, ranging from the quick response times of certain generation units (e.g. gas turbines), to the degree of efficiency and robustness of a given power market setup.

In the following, the focus is on the basic technical capability of individual power system units to modulate power and energy in-feed into the grid and then power out-feed out of the grid.

For analysis and assessment purposes, this technical capability needs to be characterized and categorized by appropriate flexibility metrics. A valuable method for assessing the needed operational flexibility of power systems (e.g. for accommodating high shares of wind power in-feed) has been proposed by Y. Makarov et al. [5]. There, the following metrics have been characterized:

- *Power capacity π* (MW) for up/down power regulation
- *Power ramp rate ρ* (MW/min)
- *Storage energy ε* (MWh)
- *Ramp duration δ* (min)

Their respective role in modulating the operation point of a power plant and with it the relative power flow into the grid (>0) and out of the grid (<0) with respect to the nominally planned operation point is shown in Figure 2. Herein, the deliberate deviation between the nominal power plant output trajectory and the actual power output trajectory is bounded by the maximum flexibility capability (i.e. the three metrics ρ, π, and ε) of the power plant in question. In the following, we will stick to the same notation as was originally proposed by Makarov et al. in [5] for the sake of simplicity and clarity.

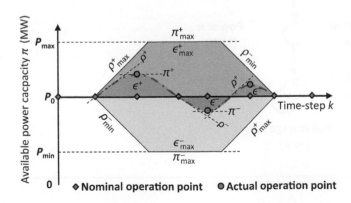

FIGURE 2

Flexibility metrics in power systems oper-ation: ramp-rate ρ, power π, and energy ε. The colored areas indicate the technically feasible region for the trajectory of the operation point for going above the nom-inal plant output (blue region) or below the nominal plant output (orange region).

Having a closer look on the proposed flexibility metrics, the following two things can be observed:

- The ramp duration δ is actually dependent on the power ramp rate ρ and power capacity π (and vice versa) as $\delta = \pi/\rho$. Three of the four previously described metrics are entirely sufficient for describing operational flexibility. We will thus only consider the power-related metrics ρ, π, and ε.
- An intriguing feature is that the metric terms ρ, π, and ε are closely linked via integration and differentiation operations in the time domain, as shown by Eqn (1). The interaction of the individual metrics clearly exhibits so-called double integrator dynamics: energy is the integral of power, which, in turn, is the integral of power ramping. These three flexibility metrics constitute a *flexibility trinity* in power systems, because they cannot be thought of independently in power system operation because of the intertemporal linking.

$$
\underset{\rho}{\text{Ramp-rate (MW/min)}} \quad \underset{\substack{\frac{d}{dt} \\ \rightleftarrows \\ \int dt}}{\text{Power (MW)}} \quad \underset{\substack{\frac{d}{dt} \\ \rightleftarrows \\ \int dt}}{\pi} \quad \underset{\varepsilon}{\text{Energy (MWh)}} \tag{1}
$$

Using these three flexibility metrics instead of, say, only one (i.e. power-ramping capability ρ) allows a more accurate representation of power system flexibility over a time interval. The power ramping for absorbing a given disturbance event, measured in MW/min, in a power system may be abundant at a certain time instant. But, for a persistent disturbance over time, the maximum regulation power that can be provided by a generator is limited, as is the maximum regulation energy that can be provided in the case of storage units, which are inherently energy constrained. Because the share of storage units in power systems and their importance for the grid integration of RES in-feed is increasing, the intertemporal links between providing ramping capability and eventually reaching power/energy limits cannot be neglected when assessing the overall available operational flexibility of a power system.

Having defined these flexibility metrics and the causal interlinking between them, as illustrated in Figure 3, allows the assessment of the available operational flexibility of an individual power system

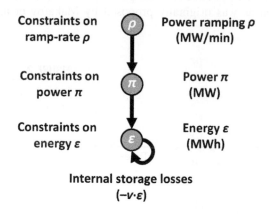

FIGURE 3

Intertemporal linking of flexibility metrics, including internal storage losses (dissipation).

unit and for whole power systems. The operational constraints (i.e. minimum/maximum (min/max) ramping) and power and energy constraints of individual power system units have to be considered when assessing their available operational flexibility.

3. Modeling power system flexibility via the power nodes modeling framework

The analysis and assessment of operational flexibility necessitates a modeling framework that allows to explicitly include information on the degree of freedom for shifting operation set points so as to modulate the power in-feed and out-feed patterns of individual power system units. This includes information on whether a unit has storage and is thus energy constrained, whether a unit provides fluctuating power in-feed, and what type of controllability and observability, including predictability (i.e. full, partial, or none), does a system operator have over fluctuating generation and demand processes. All these properties combined define the operational flexibility of individual units.

For our modeling and assessment purposes, we make use of the Power Nodes modeling framework, a unified framework for the detailed functional modeling of power system units, such as the following:

- diverse storage units (e.g. batteries, pumped hydro, etc.),
- diverse generation units (e.g. fully dispatchable conventional generators and variably in-feeding power units), and
- diverse load units (e.g. conventional (noncontrollable), interruptible, or thermal loads (controllable within their constraint sets), etc.).

It also includes their operational constraints and relevant information of their underlying power supply-and-demand processes. Operational constraints, such as min/max ramp rates, min/max power set points, and energy storage operation ranges, information of the underlying power system processes (i.e. fully controllable, curtailable/sheddable, or noncontrollable), and information on observability and predictability of underlying power system processes (i.e. state measures and/or state estimation and prediction of fully or only partially observable/predictable system and control input states), can also be included.

We illustrate the workings of the Power Node notation by looking at the Power Node model representation of an energy storage unit (Figure 4). The provided and demanded energies are lumped into an external process termed ξ, with $\xi < 0$ denoting energy use and $\xi > 0$ denoting energy supply. The term u_{gen} describes a conversion corresponding to a power generation with an efficiency η_{gen}, whereas u_{load} describes a conversion corresponding to consumption with an efficiency η_{load}. The introduction of generic energy storages in the Power Nodes framework adds a modeling layer to the classical power system modeling. Its energy storage level, the state of charge, is normalized to $0 \leq x \leq 1$ with an energy storage capacity $C \geq 0$. We can see that the illustrated storage unit serves as a buffer between the external process ξ and the two grid-related power exchanges $u_{\text{gen}} \geq 0$ and $u_{\text{load}} \geq 0$. Internal energy losses associated with energy storage (e.g. physical, state-dependent dissipation losses) are modeled by the power dissipation term $v \geq 0$, whereas enforced energy losses (e.g. curtailment/shedding of a power supply or demand process) are denoted by the waste

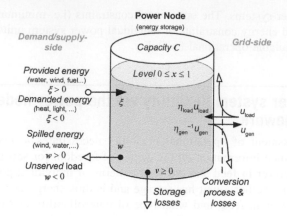

FIGURE 4

Power Node model of an energy storage unit with power in-feed (u_{gen}) and out-feed (u_{load}).

power term w, where $w > 0$ denotes a loss of provided energy and $w < 0$ denotes an unserved load demand process. The dynamics of an arbitrary power node i which can be nonlinear, are given as follows:

$$C_i \dot{x}_i = \eta_{\text{load},i} u_{\text{load},i} - \eta_{\text{gen},i}^{-1} u_{\text{gen},i} + \xi_i - w_i - v_i,$$

$$
\begin{aligned}
\text{s.t.} \quad &\text{(a)} \quad 0 \leq x_i \leq 1, \\
&\text{(b)} \quad 0 \leq u_{\text{gen},i}^{\min} \leq u_{\text{gen},i} \leq u_{\text{gen},i}^{\max}, \\
&\text{(c)} \quad 0 \leq u_{\text{load},i}^{\min} \leq u_{\text{load},i} \leq u_{\text{load},i}^{\max}, \\
&\text{(d)} \quad \dot{u}_{\text{gen},i}^{\min} \leq \dot{u}_{\text{gen},i} \leq \dot{u}_{\text{gen},i}^{\max}, \\
&\text{(e)} \quad \dot{u}_{\text{load},i}^{\min} \leq \dot{u}_{\text{load},i} \leq \dot{u}_{\text{load},i}^{\max}, \\
&\text{(f)} \quad 0 \leq \xi_i \cdot w_i, \\
&\text{(g)} \quad 0 \leq |w_i| \leq |\xi_i|, \\
&\text{(h)} \quad 0 \leq v_i.
\end{aligned}
\tag{2}
$$

Depending on the specific process represented by a Power Node, each term in the Power Node equation may be controllable, observable, and driven by an external process. Internal dependencies, such as a state-dependent loss term $v_i(x_i)$, are possible. Charge and discharge efficiencies may be nonconstant and possibly also state dependent: $\eta_{\text{load},i} = \eta_{\text{load},i}(x_i)$, $\eta_{\text{gen},i} = \eta_{\text{gen},i}(x_i)$. Nonlinear conversion efficiencies can be arbitrarily well approximated by a set of piece-wise affine linear equations. The constraints (a)–(h) denote a generic set of requirements on the variables. They are to express that (a) the state of charge is normalized, (b)–(e) the grid power in-feeds and out-feeds and their time derivatives (ramp rates) are nonnegative and constrained, (f) the supply or demand and the curtailment need to have the same sign, (g) the supply/demand curtailment cannot exceed the supply/demand itself, and (h) the storage losses are nonnegative.

The explicit mathematical form of a power node equation depends on the particular modeling case. The notation provides technology-independent categories that can be linked to evaluation functions for energy and power balances. Power nodes can also represent processes independent of energy storage, such as fluctuating RES generation. A process without energy storage implies an algebraic coupling between the instantaneous quantities ξ_i, w_i, $u_{\text{gen},i}$, and $u_{\text{load},i}$. Storage-dependent losses do not exist in this case ($v_i = 0$). Equation (2) thus degenerates to the algebraic constraint

$$\xi_i - w_i = \eta_{\text{gen},i}^{-1} u_{\text{gen},i} - \eta_{\text{load},i} u_{\text{load},i} \tag{3}$$

More detailed information on the Power Node modeling framework can be obtained from [6–9].

4. Assessment and visualization of operational flexibility

The functional representation of complex power system interactions using the Power Nodes notation allows a straightforward assessment of the three metrics of operational flexibility (i.e. power ramping capability ρ, power capability π, and energy storage capability ε). Taking, as an example, the operational flexibility of a generation unit i that also has an inherent storage function and the possibility for curtailment (e.g. a hydrostorage lake), given by following Power Node modeling equation

$$C_i \dot{x}_i = -\eta_{\text{gen},i}^{-1} u_{\text{gen},i} + \xi_i - w_i - v_i, \tag{4}$$

for providing power regulation is accomplished by calculating the set of all feasible power regulation points $\{\pi_i^{\pm}(k)\}$ of this unit i at time-step k, where up/down power regulation is denoted by "$\pm$" respectively, based on equation

$$
\begin{aligned}
\{\pi_i^{\pm}(k)\} &= \left\{ u_{\text{gen},i}^{\text{feasible}}(k) \right\} - u_{\text{gen},i}^0(k) \\
&= \left\{ \eta_{\text{gen}} \cdot (\xi - w - v_x - C\dot{x}) \right\}_{k,i} - u_{\text{gen},i}^0(k) \tag{5}
\end{aligned}
$$

$$\text{s.t.} \quad 0 \leq u_{\text{gen},i}^{\min}(k) \leq \left\{ u_{\text{gen}}^{\text{feasible}}(k) \right\} \leq u_{\text{gen},i}^{\max}(k).$$

Herein, $u_{\text{gen},i}^0(k)$ denotes the nominal set point of the generation unit and the term $u_{\text{gen},i}^{\text{feasible}}(k)$ represents any operation point from the set of all feasible operating points $\{\cdot\}$. Both terms can be chosen to be time variant (i.e. changing from one time-step k to the next). The set of all feasible operation points thus depends on the internal status of the generation unit, as defined by the terms $\xi_i(k)$, $w_i(k)$, $v_i(x_i(k))$, and $C_i(x_i(k))$, and is bounded by the unit's power-ramping and power-rating constraints (Eqn (2) (b)–(d)).

The maximum for up/down power regulation for this generation unit type is given analytically as

$$
\begin{aligned}
\pi_{\max,i}^+(k) &= \min\left[\eta_{\text{gen}} \cdot (\xi^{\max} - w^{\min} - v_x - C\dot{x}), \quad u_{\text{gen}}^{\max} \right]_{k,i} - u_{\text{gen},i}^0(k) \\
\pi_{\min,i}^-(k) &= \max\left[\eta_{\text{gen}} \cdot (\xi^{\min} - w^{\max} - v_x - C\dot{x}), \quad u_{\text{gen}}^{\min} \right]_{k,i} - u_{\text{gen},i}^0(k),
\end{aligned} \tag{6}
$$

in which the terms $w_i^{\min}(k)$ and $w_i^{\max}(k)$ define the minimum/maximum allowable curtailment for this generation unit. In case the primary fuel supply is controllable, the terms $\xi_i^{\min}(k)$ and $\xi_i^{\max}(k)$ define the minimum/maximum allowable primary power provision. Sign of the storage power term $C\dot{x}$ is negative when providing positive flexibility (i.e. discharging) and positive when providing negative flexibility (i.e. charging $(C\dot{x} > 0)$).

This flexibility assessment for the power metric π (Eqns (4–6)) can be extended to the other two metrics, ρ and ε, via time differentiation and integration, respectively. The flexibility assessment for all other power system unit types can be accomplished in a similar manner. In this way, the maximum available flexibility is calculated without any consideration of how long a certain power system unit would need to reach a new operation point that allows the provision of this flexibility.

The three thus calculated flexibility metrics span a so-called *flexibility volume*, which can be represented in its simplified form as a flexibility cube for a generic power system unit i, with the terms $\{\pi_i^+, \pi_i^-, \rho_i^+, \rho_i^-, \varepsilon_i^+, \varepsilon_i^-\}$ as its vertices or extreme points. A qualitative illustration of this is shown in Figure 5, where the flexibility volume is cut into eight separate sectors.

The evolution over time of the (maximum) available operational flexibility from a generic storage unit with both load and generation terms, $u_{\text{load}}(k)$ and $u_{\text{gen}}(k)$, is illustrated in Figure 6. The plots show that the available operational flexibility is highly time variant because of the actual storage use over time.

However, when taking into account the internal double-integrator dynamics, the flexibility volume becomes a significantly more complex polytope object. An illustration of the more realistic, but significantly more complex, polytope flexibility volume is given in Figure 7. Herein, the information of how long it takes to *reach* a certain new operation point providing a required set $\{\rho, \pi, \varepsilon\}$ of operational flexibility is explicitly given. The set of reachable operation points providing additional flexibility (green) becomes larger when the available time span is longer. The flexibility set remains, however, always smaller or becomes at most equal to a set of maximal flexibility (red), as defined by the underlying technical constraints of a given power system unit. Calculating the available set of operational flexibility that is achievable after a given number of time-steps k is equivalent to reach a set calculation, which is in general more computationally expensive than the analytic approach sketched out by Eqn (6).

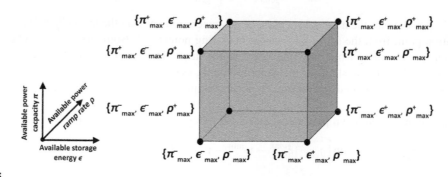

FIGURE 5

Flexibility cube of maximum available operational flexibility of a generic power system unit.

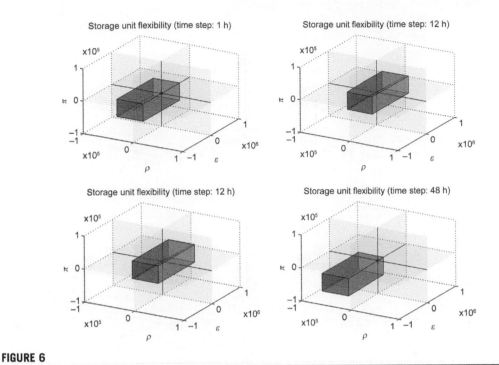

FIGURE 6

Time evolution of maximum available operational flexibility ($k = 1$, 12, 24, and 48 h).

For the sake of simplicity in representing flexibility volumes, we will stick to the simplified flexibility cubes for the remainder of this chapter.

5. **Aggregation of operational flexibility**

An important question in power system analysis is how a group or pool of power system units act together in achieving a given objective (i.e. delivering a scheduled power trajectory or providing ancillary services by tracking a control signal). Pooling together different power system units to provide a service that they cannot provide individually is an active research field. A prime example is to combine a dynamically slow power plant with a dynamically fast, but energy-constrained, storage unit to provide fast frequency regulation that neither of the units could provide individually [10] because of the lack of one flexibility metric (i.e. the missing fast ramping capability ρ of the power plant) or another (i.e. the small energy capability ε of the storage unit).

Obtaining the aggregated operational flexibility that a pool of different power system units provides is equivalent to aggregating the flexibility volumes of the individual units. Because these are given by more or less complex polytope sets, depending on the chosen calculation approach presented in the previous section, a well-known polytope operation, the Minkowski sum, can be used for calculating the aggregated flexibility of the pool. In the following, we illustrate the aggregation of a

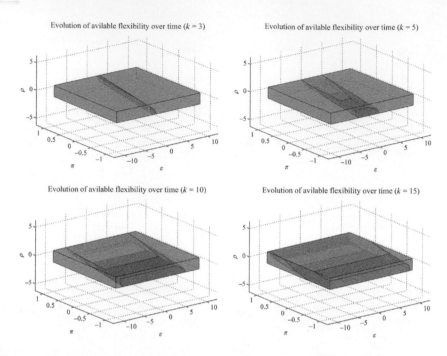

FIGURE 7

Time evolution for reaching available operational flexibility from a generic storage unit at the nominal operation point ($k=0$). (Green: evolution of available flexibility after $k=3$, 5, 10, and 15 h; red: maximum available flexibility ($k \rightarrow \infty$)).

slow-ramping power plant, together with a fast-ramping but energy-constrained storage unit, in Figure 8. The aggregation of two or more power system units leads to the addition of individual flexibility metrics:

$$\{\rho, \pi, \varepsilon\}_{\text{agg}} = \{\rho, \pi, \varepsilon\}_{\text{slow}} + \{\rho, \pi, \varepsilon\}_{\text{fast}}. \tag{7}$$

The aggregation of the operational flexibility of both units, given individually by polytope objects, is accomplished via the Minkowski sum approach

$$
\begin{aligned}
\rho_{\text{agg}}^{+} &= \sum_i \rho_i^{+}, \quad \rho_{\text{agg}}^{-} = \sum_i p_i^{-} \\
\pi_{\text{agg}}^{+} &= \sum_i \pi_i^{+}, \quad \pi_{\text{agg}}^{-} = \sum_i \pi_i^{-} \\
\varepsilon_{\text{agg}}^{+} &= \sum_i \varepsilon_i^{+}, \quad \varepsilon_{\text{agg}}^{-} = \sum_i \varepsilon_i^{-}.
\end{aligned}
\tag{8}
$$

The slow-ramping unit (e.g. a thermal power plant, with $\{\rho, \pi, \varepsilon\}_{\text{slow}}$) is assumed to have an unlimited fuel supply, which implies that no energy constraints exist and that the energy storage capability is infinite ($\varepsilon_{\text{slow}} \rightarrow \infty$). Also, the potential power output ρ is large. Dynamically slow means in this

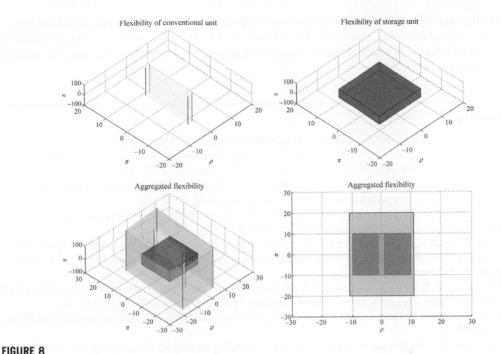

FIGURE 8

Aggregation of maximum operational flexibility of individual power system units (yellow: flexibility of conventional unit with no energy constraint; blue: flexibility of energy-constrained storage unit; green: aggregated flexibility of both units).

context that the power ramping ρ is small. The fast-ramping storage unit (e.g. a fly-wheel or battery system, with $\{\rho, \pi, \varepsilon\}_{\text{fast}}$) has a limited run time bounded by energy constraints of the storage unit and thus only a limited energy storage capability exists ($0 < \varepsilon_{\text{fast}} \ll \varepsilon_{\text{slow}}$). As is often the case for storage units, ramping ρ is large, whereas power capability π is comparatively small. Depending on the storage technology in use, time-dependent storage losses, $v(x)$, can be significant. This is notably the case of fly-wheel energy storage systems, where storage losses become large when going beyond a storage cycle duration of a few minutes as the result of bearing friction.

6. Conclusion

Herein, the presented techniques allow, in a first phase, the modeling and definition of operational flexibility of individual power system units by building up on our previous work on the Power Nodes modeling framework [6,7] and combining it with the valuable work of others, notably in [5].

In a second phase, the analysis and visualization of the operational flexibility of individual power system units is presented for some illustrative examples. The approaches are, however, also applicable for more complex, larger-scale power system setups.

In a third phase, the aggregation of operational flexibility from several, different, individual power system units is explained and illustrated. This allows notably the analysis of the overall flexibility properties of unit pools, in which different power system units are aggregated and work together to achieve a common control objective.

References

[1] Kundur P. Power system stability and control. New York: McGraw-Hill Inc.; 1994.
[2] German Ancillary Service Transparency Platform, www.regelleistung.net [last accessed 01.10.13.].
[3] European Power Exchange (EPEX), www.epexspot.com [last accessed 01.10.13.].
[4] European Network of Transmission System Operators for Electricity (ENTSO-E). Operation handbook www.entsoe.eu/resources/publications/entso-e/operation-handbook/; 2009.
[5] Makarov Y, Loutan C, Ma J, de Mello P. Operational impacts of wind generation on California power systems. Power Syst IEEE Trans May 2009;24(2):1039–50.
[6] Heussen K, Koch S, Ulbig A, Andersson G. Energy storage in power system operation: the power nodes modeling framework. In: IEEE PES conference on innovative smart grid technologies (ISGT) Europe, Gothenburg; October 2010.
[7] Heussen K, Koch S, Ulbig A, Andersson G. Unified system-level modeling of intermittent renewable energy sources and energy storage for power system operation. Syst J IEEE March 2012;6(1):140–51.
[8] Ulbig A, Andersson G. On operational flexibility in power systems. IEEE PES general meeting, San Diego, USA; July 2012.
[9] Ulbig A, Koch S, Andersson G. The power nodes modeling framework – modeling and assessing the operational flexibility of hydro power units. XII SEPOPE, Rio de Janeiro, Brazil; May 2012. 20–23.
[10] Chunlian Jin, Ning Lu, Shuai Lu, Makarov YV, Dougal RA. A Coordinating Algorithm for Dispatching Regulation Services Between Slow and Fast Power Regulating Resources. Smart Grid, IEEE Transactions on March 2014;5(2):1043–50.

The Danish Case: Taking Advantage of Flexible Power in an Energy System with High Wind Penetration

19

Sune Strøm[1], Anders N. Andersen[2]

[1] *The Danish Wind Industry Association, Denmark,* [2] *EMD International A/S, Denmark*

1. Introduction

The Danish electricity system is tailored both to ensure the energy for 5.4 million people and to be a part of the North European electricity system. As of 2012, the annual gross consumption of electricity in Denmark is approximately 34.000 GWh. The country is situated between the central European thermal power system and the hydro-based power system in Norway and Sweden. The western part of Denmark is synchronized with the central European area and the eastern part of Denmark is synchronized with the Nordic system. The different systems in the western and eastern parts are a consequence of the nonsynchronized systems divided into two separate price areas. Since 2010 the two areas have been connected by a small direct current interconnector. More than 75% of wind power generation capacity is mainly situated in the western part of Denmark.

Denmark's power generation is characterized by both centralized and decentralized thermal power plants and a patchwork of wind turbines. Figure 1 shows the distribution of capacity between thermal and wind power.

Besides its own generation capacity, Denmark is connected to its neighboring countries—Norway, Sweden, and Germany. In 2012 Denmark had approximately 11.5 GW of generation capacity, of which thermal power plants accounted for 64% and wind power for 36%. In terms of electricity production, wind covered 30% in 2012; it is estimated that by 2020 wind will provide 50% of electricity.

Denmark's consumption level is between 2 and 6.5 GW, and the interconnectors to Norway, Sweden, and Germany have a total export capacity of 5.7 GW (from Denmark to its neighbors) and 5.2 GW of import capacity. It is, then, technically possible for Denmark to cover approximately 80% of its peak demand with energy from its neighbors; on the other hand, Denmark can export large amounts of electricity when the market prices are better in neighboring countries than at home.

2. Distribution of generator capacity in Denmark

During the period from 2001 to 2012, Denmark imported and exported electricity equal to an amount of approximately 30% of its electricity consumption.

Renewable Energy Integration. http://dx.doi.org/10.1016/B978-0-12-407910-6.00019-3

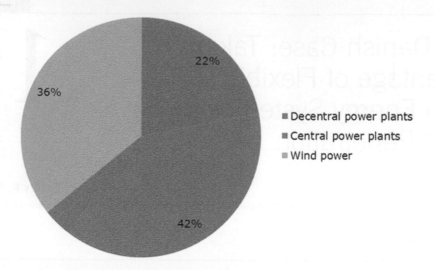

FIGURE 1

Distribution of generator capacity in Denmark.

In short, the Danish electricity system is well connected with that of its neighboring countries, and electricity is continuously traded across the borders. There is a high proportion of fluctuating energy, with 36% of the generator capacity stemming from both onshore and offshore wind turbines. Thermal power plants provide 64% of the capacity using coal, gas, waste, biomass, and biogas, which to different degrees can be flexible in their production.

It is within this framework this chapter describes the potential of using wind farms and decentralized combined heat and power plants (D-CHPs) to contribute to an efficient balancing of the power system. Other areas around the world will have other generator and consumption mix than the Danish; however, the Danish experience can be used as inspiration to increase wind penetration within the existing power system by challenging the barriers and conventional thinking that block the realization of the power systems' flexibility reserves.

In Denmark the Transmission System Operator (TSO) Energinet.dk has set up three balancing markets that, together with two wholesale markets, sustain the balance of the power system. The five markets are shown in Figure 2.

3. The Danish markets for balancing the electricity system

The fastest responding market is the primary reserve market, followed by the secondary reserve market. Independent wind farms are not allowed to participate in these two markets because of the time lag between when the market is settled and the hour of activation. Technically—and if the wind is blowing—wind turbines can deliver primary reserve services. Larger utilities with a widespread portfolio of generators may win activation based on generators other than wind

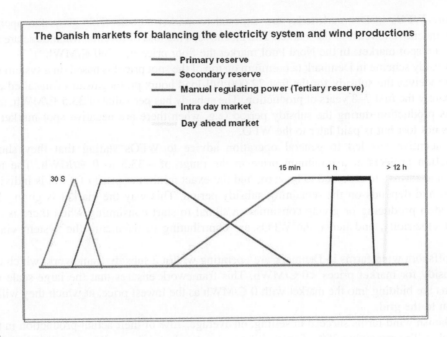

FIGURE 2

This illustration of the Danish electricity markets is made by EMD International A/S.

farms and then use the wind farms to deliver the activation if this is the best economic solution. The manual regulating power market or the tertiary reserve market is the balancing market where independent wind farms can offer activation in Denmark. Wind farms participating in this market are described in the following section. The fourth market is the intraday market, where expected deviations can be bought and sold to adjust for errors in the production forecast. Here, wind farm owners (WFOs) participate on equal terms with other generator owners. Most independent generators use a balancing production responsible party (PRP) to handle their participation in the different markets.

4. Wind is a part of the balancing solution—not the problem

Historically, wind power generation delivered electricity only when the wind was blowing, without taking much notice of market signals. In general this works fine when the penetration level is low, but with more wind penetration comes a growing need for more market responsiveness from not only the other actors on the power market but also from WFOs.

A modern wind farm is technically able to regulate and control the production up- or downward within seconds, depending on the availability of wind. In most cases it is relevant to offer downward regulation, as upward regulation requires the wind farm to be producing below maximum production at the given wind speed.

Onshore WFOs in Denmark today are exposed to market signals. This gives the operators an economic incentive to react to the market. The incentive is the loss of revenue when there are negative prices on the spot market. In the Nord Pool market the floor price is −200 €/MWh.

The subsidy scheme in Denmark (a premium on top of the spot price) is based on a system whereby the WFO receives the subsidy for the first 22.000 full-load hours pr. megawatt of installed capacity, approximately the first 7–8 years of production. The subsidy has per value of 33.5 €/MWh, and if the WFO cuts production during the subsidy period (e.g. when there are negative spot market prices), subsidy is not lost but is paid later to the WFO.

This incentive has led to general operation advice to WFOs stating that they should bid on production projects at a minimum price in the range of −33.5 to 0 €/MWh. The marginal cost for wind farms are assumed to be zero, and the exact minimum price of a bid is individual for the WFOs and depends on the remaining subsidy period. This way the market is giving WFOs a signal to stop producing or giving consumers a signal to start consuming when there is an over-supply of electricity, and hence the WFOs are contributing to balancing the system via market signals.

New offshore wind farms in Denmark are operating within a subsidy framework, which does not pay a subsidy for market prices <0 €/MWh. This framework ensures that the large-scale offshore wind farms are bidding into the market with 0 €/MWh as the lowest price, at which they will deliver production to the grid.

The Danish wind farms succeed in selling, on average, 75% of their actual production in the day-ahead market. The remaining 25% of production in a certain hour of operation deviating from the bids. To handle these deviations, WFOs need to either use the intraday market or await "punishment" in the balancing markets, depending on the economic evaluation of the two options. WFOs with deviations are after the settlement of the hour of operation, receiving a bill for their deviations based on the amount of the deviations and the price for balancing services.

Until 2011 it was possible for only independent WFOs to handle their deviations in the intraday market, during which WFOs can buy or sell the deviations as the hour of operation gets closer and hence the forecast would be more precise.

Since 2011 the Danish TSO Energinet.dk has made it possible for independent wind farms to participate in the tertiary reserve market by offering activation of typically downward regulation in the hour of operation, hence allowing them be a part of solving the balancing challenges within the actual hour of operation.

The main challenge of the participation of WFOs in the reserve market is the time lag from trading to the hour of operation; the TSO needs to be sure that the ancillary services will be delivered as planned. The 45-min time span from trading to the hour of operation is, in this sense, enough: the wind forecast for the next up to 1 h 45 min is precise enough to estimate the wind farm's potential for regulating its power delivery.

Therefore WFOs now have three possibilities to maximize profit after bidding into the spot market and until the hours of operation are over:

- Let turbines run at maximum capacity
- Handle possible deviations in the intraday market
- Offer activation in the tertiary reserve market (Box 1)

BOX 1 SEQUENCE FOR OFFERING ACTIVATION IN THE TERTIARY RESERVE MARKET

- Bids by the PRP into the tertiary reserve market are made no later than 45 min before the hour of operation. The PRP delivers an aggregated bidding of the potential downward regulation, measured as capacity in megawatts, which can be shut down.
- The TSO pays to the generators who delivered the activation that covered the imbalances through the hour of production. After this the TSO finds the generators with imbalances and send them the bill covering the TSO's activation cost.
- For small generators the bill is send to their PRP who pays the TSO first and then passes on the bill to owner of the generator.
- In short the PRP is the middleman between the TSO and the small generator owner.

---

5. Case example: an hour with negative prices for downward regulation

Figure 3 shows the prices in the electricity markets and the consumption and production in West Denmark during the 24 h of February 14, 2012. During the first hours of the day there was low consumption, and as the consumption gradually increased (red line) the spot market price (green line) increased and the central and decentral power plants increased their production because there was hardly any wind that morning. In hour 19 there is a steep drop in the production from the central power plants, reflecting a good price of downward regulation of the production (yellow line).

In the last hour of the day the price for downward regulation or electricity consumption is at a very attractive level—approximately −500 DKK/MWh or −66.8 €/MWh—and here a 21-MW offshore wind farm won activation of downward regulation of the entire wind farm to 0 MW. During this last hour of the day the wind farm produced only 6.7 MWh of a possible 21 MWh (full production). Figure 4 shows the actual production of this wind farm on February 14, 2012.

The red line shows the wind velocity, and from midnight until 6 p.m. there is only low wind. Then the wind begins to increase, and during the evening the wind farm reaches maximum production at 21 MWh. The WFO bids on and wins activation of downward regulation during the last hour of the day. This is shown by the blue line, which has a steep drop from maximum production to zero production and back to maximum production, when the activation period ends. The case of the participation of this wind farm in the tertiary reserve market also shows the general ability of wind farms to ramp production up- and downward to balance the electricity system and be a part of the generators that deliver these kinds of services.

The wind farm's participation in the tertiary reserve market is based on an economic calculation of when it is profitable to offer activation in the tertiary reserve market instead of producing.

The profit of being proactive in hour 24 on February 14, 2012, is shown in Table 1 by comparing the proactive participation with the case of no reaction to the market.

The standard action of the WFO is to do nothing, just produce and pay the balancing cost for the deviation from the original production plan cleared by the spot market. Based on Table 1, this would have generated an income of 1093 € during this hour of operation. In the spot market the expected production of 17.8 MWh resulted in an income of 603 €. If the WFO did not react to the surplus production, they had to pay 214 € in imbalance cost for the delivery of an extra 3.2 MWh to the grid. The income from the subsidy was 704 € and the total income was 1093 €.

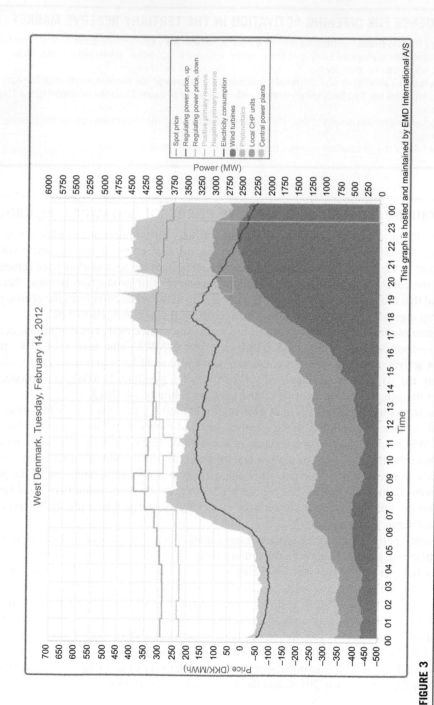

FIGURE 3

The electricity market on February 14, 2012, from www.emd.dk/el.

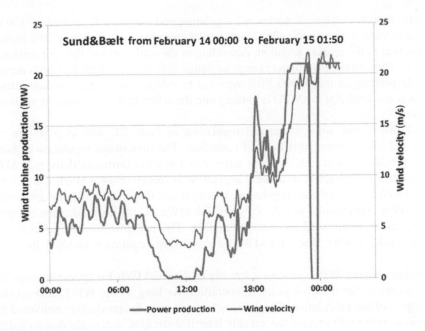

FIGURE 4

A 21-MW offshore wind farm won activation of downward regulation on February 14, 2012.

Table 1 Income Without and With Activation in Hour 24 of February 14, 2012

	MWh	€/MWh	€
Income without activation			
Sold at the spot market	17.8	33.9	603
Production if no downward regulation	21		
Surplus imbalance	3.2	−66.8	−214
Subsidy at potential production	21	33.5	704
Income without activation			**1.093**
Income with activation			
Sold at the spot market	17.8	33.9	603
Production if no downward regulation	21		
Surplus imbalance	3.2		
Production with downward regulation	6.7		
Downward regulated power			
Gross	−14.3		
Net	−11.1	−66.8	741
Subsidy for actual production	6.7	33.5	225
NPV of delayed subsidy (postponed production)	14.3	20.1	288
Income with activation of downward regulation			**1856**

The income with activation of downward regulation ends at 1856 € instead of 1093 €, which means that the income with activation in this particular case was approximately 70% higher than the case of no reaction to the market situation. According to the settlement of the spot market, the WFO received 603 €. This settlement will not change no matter how the actual production is during the hour of operation. Depending on the WFO's PRP, there can be differences in exact income calculations. In the example it is assumed that the WFO is bidding into the different markets directly without a PRP as market operator.

The WFO offers and wins downward regulation in hour 24, and in total the wind farm delivers only 6.7 MWh during this hour of operation. The downward regulation is based on the last available production forecast, and this states that the wind farm can deliver 21 MWh during the hour of operation. The gross downward regulation is therefore 14.3 MWh (21–6.7 MWh), but to deliver 14.3 MWh of downward regulation, it is assumed that the wind farm delivered 21 MWh and not 17.8 MWh. This results in a deviation of 3.2 MWh in surplus production compared to the amount of megawatt hours sold on the spot market. The net downward regulation is therefore 11.1 MWh (14.3–3.2 MWh). The 11.1 MWh of downward regulation resulted in an income of 741 €.

In the Danish context, WFOs receive a subsidy of 33.5 €/MWh for approximately 22,000 full-load hours, similar to the first 7–8 years of operation. As long as the WFO receives the subsidy, this will be a part of the calculation. The subsidy is given only for production delivered to the grid, and therefore the WFO will receive the subsidy later if it decides to activate downward regulation. The reduced net present value (NPV) of the delayed subsidy, therefore, must also be taken into account when calculating the benefit of activation. In this case the NPV of the 33.5 €/MWh subsidy is set to 20.1 €/MWh. The NPV will, in general, be less than 33.5 €/MWh, depending on the years left with the subsidy and the interest rate for the owner. The more years with the subsidy and the higher the interest rate, the closer the NPV will be to 0, and vice versa.

It is, however, not very often that either the spot prices are negative or there are negative prices for downward regulation. In Denmark's western price zone, from January 1, 2011, until August 31, 2013 (a total of 23,376 h) there have been:

- 35 h with spot market prices between 0 and −33.5 €/MWh;
- 19 h with spot market prices lower than −33.5 €/MWh.

This may seem like only a few hours. However, had an onshore wind farm delivered 20 MWh to the spot market during the 19 h with prices lower than −33.5 €/MWh, the expenses due to negative spot market prices would have been more than 38,000 € for the WFO. A wind farm with a full subsidy would have regained only a little less than 13,000 € in the same period. This indicates a clear economic incentive for WFOs to adjust their production based on the needs of the spot market with negative prices.

In the same period there were:

- 131 h with downward regulation prices between 0 and −33.5 €/MWh;
- 77 h with downward regulation prices lower than −33.5 €/MWh.

As shown in the case example above, income increased by approximately 70% during the hour of operation with participation in the tertiary reserve market. Especially during the 77 h with less than −33.5 €/MWh, there was extra profit to be earned by being proactive rather than inactive.

5.1 Challenges to participation in the tertiary reserve market

The main challenges for the participation of wind farms in the tertiary reserve market are:

- The subsidy system's influence on the marginal prices for wind farms.
- The acceptance by WFOs and society of cutting wind energy production when the wind is blowing.
- The TSO's willingness to set up a sufficiently flexible tertiary reserve market and allowing independent wind farms to participate in it.
- The technical ability of turbines to be fast and precisely controlled through up- or downward regulation of production. For modern turbines this is not a problem, but for turbines installed before 2004 reacting to market signals can be a challenge because of less technologically advanced control systems.
- The subsidy scheme has a direct impact on the incentives to adjust production based on the market price signals. Both the amount of the subsidy and the system itself influences owners' production decisions. A feed-in tariff protects the generator owner entirely against cutting production because of economic valuation. On the other hand, all systems where the owner is exposed to spot market price signals give the owner a more or less strong incentive to react to price signals. Negative spot market prices and lower subsides increase the incentive to respond to market price signals.

6. Decentralized combined heat and power plants are a part of the balancing solution

With a comprehensive penetration of wind energy in the power system, a large amount of energy storages is needed, but technologies such as batteries and pumped hydroelectricity storage (PHES) are rather expensive. In energy systems with central heating and cooling solutions, it is more economically efficient to promote thermal stores and cold stores since in most cases these are much cheaper than electricity storages. For example, it may cost around 1/10 to store energy as heat compared to storing it as electricity in a PHES.

In Denmark there are many D-CHPs that are equipped with large thermal stores. This gives them a high operational flexibility on the wholesale and balancing markets. Most Danish D-CHPs participate in both the day-ahead spot market and the tertiary reserve market.

In recent years many of the D-CHPs have been equipped with electric boilers and heat pumps, allowing them to consume electricity when the market prices are low because of a relative high supply of compared to demand for electricity. Together with large thermal stores, the heat pumps and electric boilers are basic instruments for the flexible and efficient participation of D-CHPs in the Danish energy system.

During hours with relatively low market prices, electric boilers and heat pumps produce heat based on electricity instead of traditional fuels such as gas, coal, or biomass. These situations can occur when there is a high wind penetration and/or low consumption. The use of electricity as a source for heating production in such cases is a shortcut toward a low-carbon energy system because the wind power substitutes fossil fuels as energy source.

We show online the operation of some of these D-CHPs. One typical example of a D-CHP is the Hvide Sande CHP, which can be seen at www.emd.dk/plants/hvidesande. It has two identical 3.7-MW

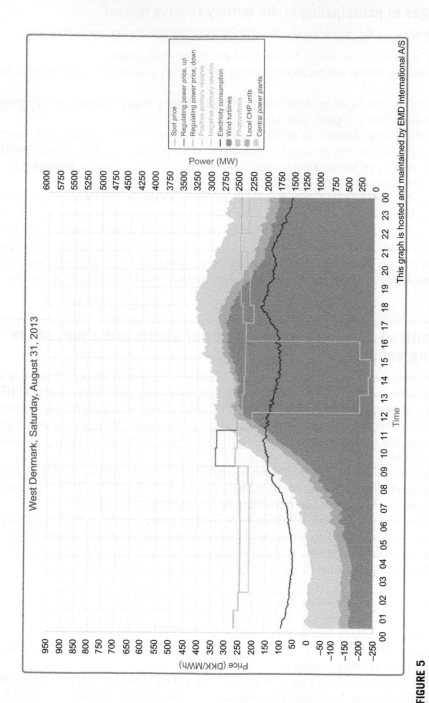

FIGURE 5

West Danish electricity consumption and production (available online at http://www.emd.dk/el/).

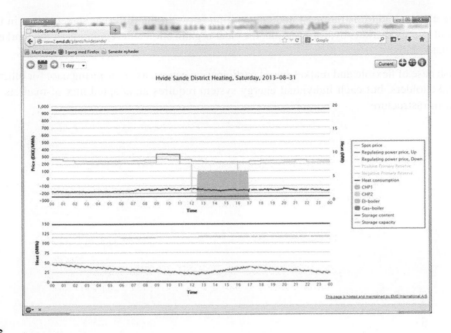

FIGURE 6

The online operation of the Hvide Sande CHP on August 31, 2013 (available online at www.emd.dk/plants/hvidesande).

Caterpillar engines, a 6-MW electric boiler, boilers that use only natural gas and produce heat, and a thermal store with a volume of 2000 m³, equal to a heat capacity of 130 MWh.

The challenge of integrating major amounts of wind is well demonstrated online at our homepage www.emd.dk/el/. On August 31, 2013 (see Figure 5), the prices in the Scandinavian regulating power market (the tertiary reserve market) was negative during more hours because of a fast unscheduled rise in wind production. Figure 6 shows that the electric boiler at the Hvide Sande CHP was active during these hours.

7. Conclusions and recommendations based on the Danish experience

Listed here are the three most important lessons from the Danish case of integrating high amounts of wind power into the electricity system.

1. Create markets that ensure economic incentives for both generators and consumers to be flexible to the price signals of the market. In this article, flexible participation is shown by the examples of an active wind farm and an active D-CHP participating in the ancillary service markets.
2. Integrate electricity markets with those of neighboring countries/areas and take advantage of higher socioeconomic benefits along the road of fewer physical and economic barriers to the cross-border trade of electricity.

3. Ensure efficient administration of and cooperation between the different stakeholders in the different energy markets by, for example, the clear organization and regulation of the markets and high-class skills.

The Danish case of flexible and market-based participation can act as an inspiring case for other energy system stakeholders, but each individual energy system requires an adapted mix of markets, regulations, and infrastructure.

Demand Response and Distributed Energy Resources

DR for Integrating Variable Renewable Energy: A Northwest Perspective

20

Diane Broad, Ken Dragoon
Ecofys, US

1. Role of demand-response in integrating variable energy resources

The increasing mandated levels of renewable energy resulting from state-adopted renewable energy standards in the Northwest United States are expected to result in an approximately two-thirds increase in qualifying renewable resources by 2020, and wind generation is likely to fulfill well more than 50% of the new demand. The 2007 Northwest Wind Integration Action Plan recognized that accommodating the variability of wind generation on the Northwest power grid would require increasing levels of flexible resources capable of responding to variations in wind generation. Demand-response (DR) programs can be implemented that take advantage of new capabilities in communication and control, enabling two-way communication with loads as small as 5 kW. The loads can be controlled on short time scales, and verification of their performance is nearly instantaneous. These new DR programs are dubbed herein "Smart DR." By providing flexible demand resources in various categories of ancillary services, Smart DR can alleviate the balancing authority's need to rely on conventional balancing resources from generation, potentially preventing the need to expand system reserve capacity under increased penetration of variable renewable energy sources (Figure 1).

Constructing flexible natural gas combustion turbines as a balancing resource is an option. However, the fuel costs of operation are volatile. Furthermore, this configuration will result in wind power intrinsically relying on fossil fuel—the thing that is to be displaced by wind integration. It has been suggested [1], and this work supports the assertion, that Smart DR can be a more cost-effective approach than other possible solutions, such as battery energy storage, at least for now, while also supporting environmental goals.

Balancing reserves are deployed in response to imbalances in generation and load that occur in distinct time periods. Load-following reserves, which are of most interest to the Bonneville Power Administration (BPA) in the near future, are deployed over time periods ranging from 10 to 60 min (sometimes up to 120 min). BPA wants a Smart DR resource to behave like a put-and-call option for capacity with 10-min notice, or like a balancing reserve resource with ability to increase load and shed load.

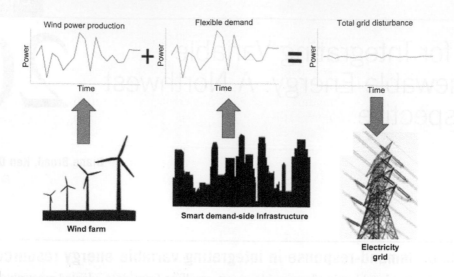

FIGURE 1

Smoothing wind power variability with Smart DR enabled demand-side infrastructure.

2. DR in the Northwest today

The BPA is the primary producer of electricity in the Pacific Northwest region of the United States, providing power to approximately 2.5 million customers (87% residential, 12% commercial, and 1% industrial). The area is blessed with access to low carbon energy. Beyond the 22 GW of seasonal hydrocapacity, BPA currently has more than 5 GW of wind resources within its balancing area. Significant expansion of wind power is further predicted over the next decade, as shown in Figure 2.

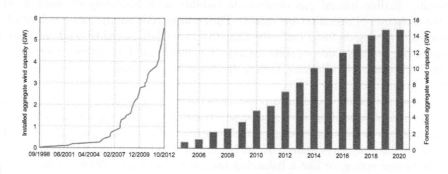

FIGURE 2

Current (left) and forecasted (right) wind power capacity in the Bonneville Power Administration.

BPA studies suggest that the extensive Columbia River hydrosystem that it oversees may not be able to provide the balancing reserves future regional wind capacity will require. Capacity and ancillary services based on available capacity are beginning to get the attention of planners in the search for resources to balance the variability of renewables, and provide capacity for system demand peaks, hydroconstraints, or transmission congestion.

A further issue is that of springtime oversupply. During the spring run-off, hydrocapacity is high and, in the event that wind power production is also high, there are times when too much clean energy is available. Fish and wildlife regulations prevent BPA from reducing hydro output lower than certain levels; instead, wind energy has been curtailed. Having a place to store this excess energy would avoid low carbon resource waste. However, conventional energy storage systems are expensive and can be inefficient. For this reason, current energy storage technology is not considered a viable option at this time.

2.1 Utility-sponsored DR programs

Since the 1980s, the City of Milton-Freewater, OR, has been a pioneer in DR in the Northwest. The city implemented conservation voltage reduction, a robust and cost-effective program that has limited peak demand. From 2009 through 2013, BPA, together with many load-serving utilities, has been conducting field testing, pilot projects, modeling, and analysis to investigate the potential of DR in the Northwest. An overview of the programs is provided in Figure 3.

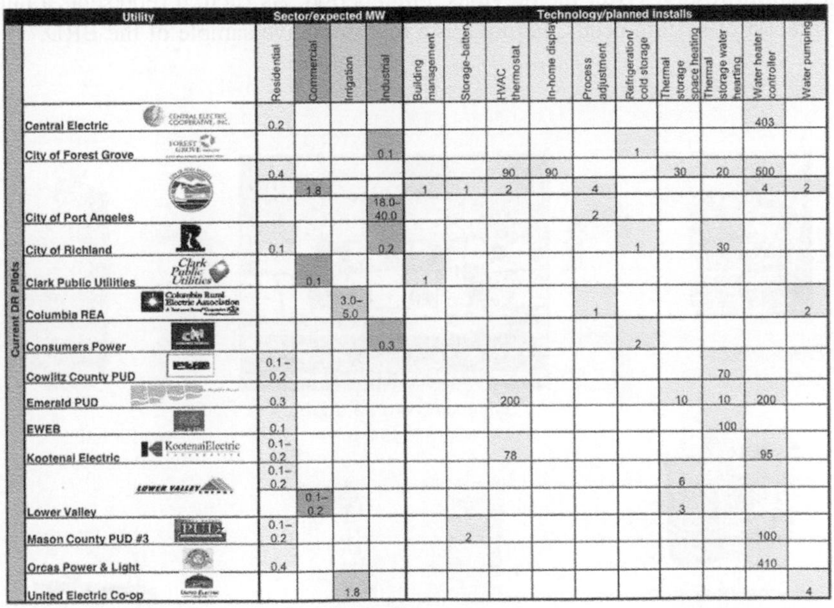

Utility	Residential	Commercial	Irrigation	Industrial	Building management	Storage-battery	HVAC thermostat	In-home display	Process adjustment	Refrigeration/cold storage	Thermal storage space heating	Thermal storage water heating	Water heater controller	Water pumping
Central Electric	0.2												403	
City of Forest Grove				0.1						1				
City of Port Angeles	0.4	1.8		18.0–40.0	1	1	90 / 2	90		4 / 2	30	20	500 / 4	2
City of Richland	0.1			0.2						1		30		
Clark Public Utilities		0.1			1									
Columbia REA			3.0–5.0							1				2
Consumers Power				0.3						2				
Cowlitz County PUD	0.1–0.2												70	
Emerald PUD	0.3						200				10	10	200	
EWEB	0.1												100	
Kootenai Electric	0.1–0.2						78						95	
Lower Valley	0.1–0.2	0.1–0.2										6		
Mason County PUD #3	0.1–0.2				2							3		
Orcas Power & Light	0.4												100	
United Electric Co-op			1.8										410	4

FIGURE 3

Northwest demand-response pilot projects sponsored by Bonneville Power Administration.

Among the investor-owned utilities, PacifiCorp and Idaho Power have had active DR programs of ~300 MW (each) with irrigation loads. For reducing peak demand, Portland General Electric (PGE) has embarked, in 2013, on a DR program with commercial and industrial customers, expected to build out to a resource of 100+ MW.

2.2 DR and integration of renewable energy: demonstration projects

The BPA Technology Innovation program has sponsored several important DR projects from 2009 through 2013. Among those was a project led by Ecofys, "Smart End-Use Energy Storage and Integration of Renewable Energy." Many end-use technologies have an element of energy storage, and their electricity use can be managed with no reduction in service quality. Electric water heaters, space heating, and cold storage warehouses are great examples of end uses with thermal energy storage capacity or operational flexibility.

A variety of end-use loads in the commercial and residential sectors were enabled and tested for their potential to contribute to cost-effective integration of variable energy resources, such as wind. The project's 1.2-MW portfolio covered the service territory of seven Northwest utilities and included control of more than 100 electric water heaters (residential), seven electric thermal storage furnaces (residential/commercial), heating, ventilation, and air conditioning systems in two public school buildings, and five cold storage warehouse facilities [2] (Figure 4).

The need for reserves in the BPA balancing area is quantified in the Balancing Reserves Deployed (BRD) data set. The BRD, made available on BPA's public Website [3] and updated every 5 min, is a record of the call for reserves over the previous 5-min period. As such, it represents a fairly accurate persistence forecast of system needs. Figure 5 is a representative sample of the BRD.

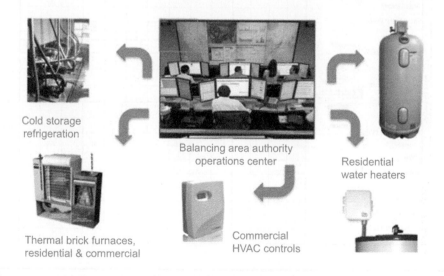

Cold storage refrigeration

Balancing area authority operations center

Residential water heaters

Thermal brick furnaces, residential & commercial

Commercial HVAC controls

FIGURE 4

Integrating control of flexible end-use loads with grid operations.

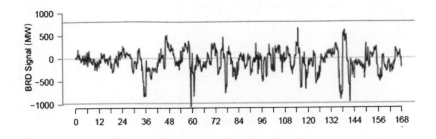

FIGURE 5

Sample balancing reserves deployed (in MW) per hour over 1 week. BRD, balancing reserves deployed.

Negative BRD values correspond to decreased generation (DEC) requests and, conversely, positive values correspond to increased generation (INC) requests. The red and blue horizontal lines demark the magnitude of the DEC and INC values contained in the BRD signal at or above which 100% of the reserves will be dispatched. DEC requests can be met with an increase in load, and INC requests can be met with a decrease in load.

The general goals of controlling end-use loads were three-fold:

1. Maintain operation such that the end-user services remain within acceptable levels. It is extremely important for any DR project that the customers are not inconvenienced by the project.
2. Provide benefits to the load-serving utilities. Top priorities are peak shaving and load shaping, but benefits may also be found in the added efficiencies of transmission and distribution at night.
3. Provide balancing services, or at least show proof of concept. The most innovative aspect of the project is demonstrating the abilities of these resources to provide balancing services to the balancing area authority.

2.2.1 Residential grid-interactive water heater control results

The pilot program included Steffes Corp grid-interactive water heater controllers and Carina WISE water heater controllers installed on a mix of new and existing tanks. Tanks included 105- and 50-gallon units. Steffes controllers keep the tank temperature between 90 and 170 °F, with a thermostatic mixing valve on the output of the water heater to provide hot water to the customer at a safe temperature (not exceeding 130 °F). Communications were achieved through the broadband Internet connection in the home.

The Carina controller provided a reliable resource for peak load reduction and load shaping, following an hourly control schedule provided on a day-ahead basis. The Steffes controller proved to be the most adept at adjusting to a fast-moving control signal, and by the end of the pilot, the water heaters with this controller were following the BRD with impressive precision (Figure 6). From each water heater, the capacity to provide INCs was approximately 0.6 kW, whereas the capacity to provide DECs was approximately 3.0 kW.

Along with field demonstrations, the water heaters with controls were modeled in a high-penetration network of "smart charging" electric water heaters (EWHs) and simulation studies were

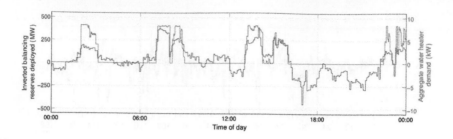

FIGURE 6

Balancing reserves deployed (inverted) and aggregate response from two Steffes thermal storage water heaters over a 24-h period.

conducted. Studies focused on three perspectives: the transmission services level, the distribution utility's level, and the consumer level. The key results are summarized as follows:

- Overall load profiles are "flattened" (i.e., the extreme loads (both the peaks and troughs) are smaller in magnitude relative to the mean load).
- Peak load management is critical to guarantee protection of the distribution system.
- The available INCs and DECs are strongly dependent on the charging scheme; striking a useful balance will depend on the economics and transmission constraints on any given system.

One surprising and encouraging finding of the water heater program is that with larger water heaters and Steffes control technology, virtually all of the load can be served in providing DEC services. In other words, water heater loads can be served in the process of providing an important balancing service—potentially removing water heater loads from peak demand and load forecasts. The prospect of simultaneously reducing load while providing balancing services is especially enticing.

2.2.2 Cold storage DR and energy storage results

The cold storage demonstration included five facilities, each receiving individualized calls for INC and DEC events. At each facility, project partner EnerNOC installed a two-way communications solution that (1) captures near real-time electricity consumption data on 1-min intervals and (2) relays the event signals to the centralized refrigeration control system at the participating facilities (Figure 7 and 8).

During Phase I of the project, EnerNOC's Network Operations Center was used to dispatch event signals to all the facilities. During Phase II, one of the facilities received signals dispatched from a DR automation server (DRAS) operated by Utility Integration Solutions. This single facility was enabled with an OpenADR (Open Automated Demand Response) compliant gateway. The events were called based on a preset testing schedule created by Ecofys, initially shared with the facility operators but later used in the pilot study without their previous knowledge. The facility aimed to respond within 10 min of a call, to perform the requested increase or decrease in load for a period of 30–90 min, and to respond to a maximum of 10 events per week.

Overall, the sites delivered 77% of goal in Phase II. The average curtailment performance was 101% of goal, whereas the average increase was 47% of goal. The Site C responses were the least

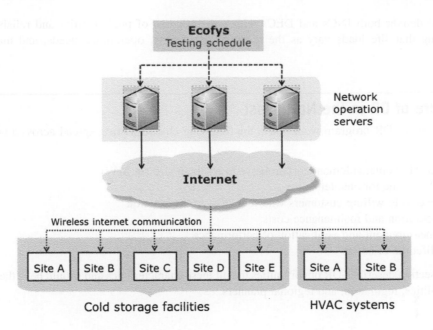

FIGURE 7

Testing configuration for the cold storage facilities.

Site	Controlled equipment	Average demand	Target increase	Target decrease
A	Compressors	400–700 kW	200 kW	200 kW
B	Compressors, evaporators	400–700 kW	100 kW	200 kW
C	Compressors, evaporators	100–450 kW	50 kW	200 kW
D	Compressors, evaporators	700–1100 kW	100 kW	200 kW
E	Compressors, evaporators	700–900 kW	200 kW	200 kW
Total	–	2300–3850 kW	650 kW	1000 kW

FIGURE 8

Characteristics of the five cold storage pilot sites.

consistent. If the response of Site C during increased events is disregarded, then the average performance during increased events jumps from 47% to 67%.

Although the performance at Site C was particularly volatile, overall the portfolio consistently delivered both INC and DEC performance. It is anticipated that a 10-MW (or larger) portfolio would be able to deliver both INCs and DECs with a high degree of predictability and reliability, even considering that site loads vary as the result of seasonality, operational needs, and maintenance events.

3. Future of DR in the Northwest

The ideal Smart DR program would have the following characteristics, spread across a portfolio of assets:

- Low installation/enablement costs relative to resource size (in MW)
- Short lead time for enablement
- Easy access to willing customers who feel "part of the solution"
- Low operation and maintenance costs
- Easy measurement and verification of performance
- Long lifetime of assets

The next section describes DR opportunities that fit these characteristics and should be at the top of the list for utilities, regulators, and regional planners.

3.1 District energy

District energy systems provide heating and cooling generated in a centralized location and distributed (usually through underground piping) to residential and commercial consumers. The heat is often obtained from a cogeneration plant burning fossil fuels. Increasingly, the central plant energy source is renewable energy: biomass, geothermal heating, and central solar heating. With the addition of electric boilers or heat pumps, a district energy system can use renewable energy supplied from the electric grid. District energy plants can provide higher efficiencies and better pollution control than localized boilers. The Northwest has many legacy district energy systems, notably at multiple university campuses and in the downtown Seattle, WA, core. Several of these systems will be upgraded or abandoned in the coming decade. With emphasis on the value of DR, the district energy systems could be expanded to emulate advanced district energy systems in operation in Northern Europe: multiple combined heat and power (CHP) plants are dispatched, electric and gas boilers turn on and off, industrial waste heat is used, and energy is stored.

Princeton University has a district energy system for heating and cooling, including CHP and thermal energy storage. In 2005, the campus peak demand for electricity from the grid was 27 MW. By 2006, the campus peak demand was reduced to 2 MW. The district energy system "freed up" 25 MW of local grid capacity. The system both reduces peak load on the power grid and enhances reliability of service for the campus. During Superstorm Sandy, Princeton used its district energy CHP system to provide uninterrupted power, heating, and cooling.

3.2 Industrial programs

A standout cost-effective Smart DR potential resource exists in cold storage warehouses. There are more than 300 frozen food processing facilities in the Pacific Northwest (PNW) and more than 100 could be good candidates to supply the types of DR services BPA has prioritized. Cold Storage is attractive because of the following factors:

1. Favorable economics—$100–$500/kW upfront cost, which compares favorably to the $610/kW cost for a frame combustion turbine [4]. Cold Storage should be one of the first Smart DR resources to be targeted regionally. The DR programs seem to be a good fit with energy-efficiency programs, and synergies between the two objectives should be actively sought out.
2. Mature controls technology—Control vendors are sophisticated in achieving energy efficiency and have implemented DR in other parts of the country.
3. Multiple benefits—Cold storage promises to provide BPA with a load-following resource, and to provide the distribution utility with demand reduction and load shifting. Multiple benefits will depend on careful program implementation, including DRAS.
4. Ease of enablement—Industrial loads are easier to recruit and manage because the commercial and operational arrangements are done with only a few parties.

3.3 Residential and commercial programs

With the onset of Advanced Meter Infrastructure (AMI), utilities will have much better load data to identify customers and loads that are suitable for Smart DR technologies.

Utilities with the most success will enlist multiple DR services: peak shaving, load shaping, and INC and DEC balancing services. Figure 9 demonstrates the distribution of project revenues across the multiple DR services, which can be obtained from a Smart DR resource, such as a grid-interactive electric water heater, given a modest indicative pricing of balancing services between $2/kW-month and $7/kW-month.

3.4 Role of market prices and structures

In October 2011, with BPA's rate changes, many utilities in the Northwest began to see a clear price signal that encourages them to reduce their peak demand. Through the collaboration with BPA customer utilities in the development of a business case tool (part of the Ecofys TI project [2]), it became clear that if Smart DR can be enabled for prices close to the $200/kW-year range, there is a strong incentive for implementation. In comparison, variable and fixed operational costs for a gas

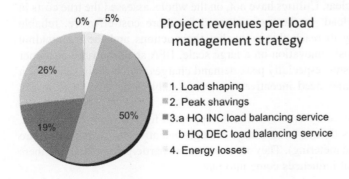

Project revenues per load management strategy

0% ⌐5%
26%
19%
50%

1. Load shaping
2. Peak shavings
3.a HQ INC load balancing service
   b HQ DEC load balancing service
4. Energy losses

FIGURE 9

Sample composition of total revenue to the distribution utility from a Smart demand-response resource.

Financial results per device (incl. costs for comm. & data kit)	Additional investment	Add. annual maintenance	Annual revenue	Payback time (years)	Annual return on investment	Costs	benefits
Steffes IWHC with 50 gallon tank	$ 1058	11	$ 231	4.8	20%	$ −1270	$ 4611
Steffes IWHC with 105 gallon tank	$ 1705	$ 17	$ 270	6.7	14%	$ −2046	$ 5404
Steffes ETS furnace (forced air) (incl. air source heat pump)	$ 4382	$ 74	$ 728	6.7	14%	$ −5858	$ 14,564
Steffes ETS furnace (hydronic) (incl. air source heat pump)	$ 6671	$ 97	$ 795	9.5	8%	$ −8605	$ 15,910
Carina WISE controller 50 gallon retrofit	$ 630	$ 6	$ 208	3.1	32%	$ −756	$ 4160
One-way WH controller switch	$ 300	$ 3	$ 90	3.5	29%	$ −360	$ 1795

FIGURE 10

Representative costs and benefits to the distribution utility of various DR technologies.

turbine are estimated to be $267/kW-year [5]. Utilities or other program designers must have confidence there will be stable peak demand charges (from BPA).

Currently, the Northwest lacks a real-time market for balancing services or any visible price. Utilities need to see a market for capacity and balancing services, including clear price signals. With the arrival of a promising market, manufacturers will invest to integrate the controls into the end-use loads. This represents a step function in reduction of program costs, especially in the residential end uses. As seen in Figure 10, a decrease is needed in the up-front investment to improve KPI's (e.g., payback time and annual return on investment).

Another structural roadblock is customer demand charges for C&I customers. These customers typically pay their electric utility based on energy consumed (kWh) and peak demand during the month (kW). If a C&I customer is participating in an innovative Smart DR program and accepting load increases when that benefits the grid, the rate structure for that customer segment must be changed and the AMI data used so the participant is not penalized for participation.

3.5 Policy opportunities

There is still no single universally accepted definition of "smart grid." Utilities fear making a large investment in an AMI system or particular DR communication technology while the risk is high that it could become obsolete in the near future. Convergence on a common Smart DR resource management strategy is needed, as well as a scalable and secure automation infrastructure for control, communication, and verification.

Costs and benefits to utilities are unclear. Utilities have not, on the whole, assessed the true costs in addressing peak demand, shaping of load, or the regional need for more cost-effective, reliable balancing resources. BPA is beginning to reach this point, but their actions and the surrounding messages are not focused enough to push innovation on a large scale. BPA should ensure customer utilities that DR cost recovery mechanisms, especially peak demand charges, will remain stable for a period of more than a few years. Utilities need incentives to remain stable long enough to justify investment in Smart DR assets.

Both Northwest load-serving utilities and BPA need a better approach for customer engagement. Consumers, individuals, and businesses do not have a large enough incentive to participate or are lacking the infrastructure (e.g., advanced metering). They are not likely to participate in large numbers unless financial, convenience, and social influences come into play.

4. Thoughts on the way forward

The Northwest has a long history of managing a variable energy renewable resource, the hydro-system, which comes with huge amounts of storage capability and flexibility compared with most US systems. That inherent flexibility and storage capability diminished the value of storage and of peaking capability compared with other systems. As a result, the Northwest has been a little late to the party with respect to demand management resources. The economics have not been there. On the other hand, the good news is that demand management represents a relatively vast and untapped resource.

The problem of getting to a low-carbon grid is no longer "how," but "how best." Demand management is one of a growing list of measures to help integrate variable generation into the grid. It will compete with dedicated energy storage, such as pumped storage, batteries, flywheels, and the myriad of other technologies that show promise—not all of it yet realized. The Danish district energy systems spur the imagination—massive flexibility to produce or consume power, all while serving its customers' demand for heat. In a future world that is largely, or entirely, powered by variable renewable resources, there are likely to be niches for many of the approaches we envision today. Our ability to figure out how best to get to the low-carbon grid may indeed be limited only by our imagination—certainly not technology at this point. As tempting as it is to think this is the problem of energy storage, it is at least equally a problem of demand profile. To what extent can we shape demand to the whims of variable generation?

Years ago, when a large fraction of Northwest demand was composed of aluminum smelters, we considered stockpiling aluminum in high hydroyears, and shutting down the smelters during droughts. Today, we might stockpile high-temperature heat in molten salts, as is done at solar plants, or at lower temperatures to serve our space and water-heating loads. Wherever there have been high penetrations of variable renewable energy (including our beloved hydro), there have been times of wild excess and low prices—down to zero and even negative prices. That will not go away. How can we make best use of that resource? This is another way of looking at managing demand.

In terms of more traditional demand resources, the Northwest Power and Conservation Council targeted 1500–1700 MW of DR resources in its regional Sixth Power Plan (2010). That estimate is relatively conservative compared with demand resource developed in other regions, and only approximately one tenth of the targets are currently being realized [6] in the Northwest. The plan notes that electric water heaters alone account for a load that varies from a low of 400 MW to a peak of approximately 5300 MW. Controlling the operation of those water heaters in a range of 120–135 °F, the Council estimates an energy storage capability of more than 6000 MWh—the equivalent of approximately 240,000 all-electric vehicle batteries.

Commercial and industrial customers in the region used to have both electric and natural gas boilers to rely on the least expensive source of heat, but the electric boilers were largely decommissioned in the 1980s. Let us consider paralleling today's gas boilers with electric ones, giving utilities an optional electric load. In effect, we would displace gas turbines, and as prices decline further, we can displace gas boilers. This would multiply the energy storage capability count by a factor of two or so.

Although technology and cost are not insurmountable obstacles, institutional issues and inertia are holding us back. Markets in the Northwest do not reflect the value of storage—certainly not the future value in a low carbon grid. We need policies to get us there. California took a brave step in overcoming institutional inertia in addressing the flexibility issue, mandating utilities to build some 1300 MW of energy storage, largely in anticipation of the targeted 33% of power coming from renewable energy. The Federal Energy Regulatory Commission stepped in, too, requiring utility tariffs to pay for flexibility offered by system resources. Closer to home, the Oregon Public Utility Commission now requires utility planning processes to determine a supply and demand of "flexibility resources," including plans for meeting that demand.

Although more needs to be done, these are promising signs.

References

[1] Meeting renewable energy targets in the west at least cost: the integration challenge. Western Governors' Association; June 2012. p. 76.
[2] Smart end-use energy storage and integration of renewable energy. Bonneville Power Administration and Ecofys; September 2012. TI 220 Project Evaluation Report, http://www.bpa.gov/Energy/N/pdf/TI_220_Project_Ecofys_Evaluation_Report.pdf.
[3] Bonneville power administration transmission services website: http://transmission.bpa.gov/Business/Operations/Wind/default.aspx
[4] Sixth Northwest conservation and electric power plan. Northwest Power and Conservation Council; February 2010. Table 6-3, pp. 6–45.
[5] Cost and performance data for power generation technologies. National Renewable Energy Laboratory; February 2012.
[6] Sixth Northwest conservation and electric power plan. Northwest Power and Conservation Council; February 2010 [Chapter 5].

Case Study: Demand-Response and Alternative Technologies in Electricity Markets

21

Andrew Ott

Executive Vice President, Markets, PJM, USA

1. Overview of PJM wholesale market

PJM Interconnection is a regional transmission organization that coordinates the movement of wholesale electricity in all or parts of 13 US states and the District of Columbia. Acting as a neutral, independent party, PJM operates the largest competitive wholesale electricity market in the world and manages the operation of the high-voltage electricity grid to ensure reliability for more than 60 million people. PJM operates wholesale markets for capacity, Day-Ahead electricity, Real-Time electricity, Day-Ahead Scheduling Reserve, Synchronized Reserve, and Frequency Regulation.

The capacity market is a 3-year physical forward market with annual incremental auctions to allow market participants to adjust their forward positions. Capacity is a product designed to ensure resource adequacy and reliability. Each load-serving entity is required to purchase forward capacity contracts to satisfy its load obligation on a 3-year forward basis either through the market or bilaterally. The capacity product is essentially a call contract on energy production or curtailment during times when the PJM grid is experiencing capacity emergency shortage conditions.

The Day-Ahead electricity market is a Day-Ahead hourly market for electricity that allows market participants to schedule energy sales, purchases, and deliveries at binding Day-Ahead prices for each hour.

The Real-Time electricity market is a balancing market that allows participants to coordinate the continuous buying, selling, and delivery of electricity. The Real-Time electricity market is based on the 5 min security-constrained economic dispatch of resources to balance supply and demand. Locational electricity prices are posted on the Internet every 5 min.

The Day-Ahead Scheduling Reserve market is an ancillary services market operated and cleared simultaneously with the Day-Ahead electricity market. The Day-Ahead Scheduling Reserve product is a commitment to provide 30-min response resources to ensure reliable scheduling of power grid operations.

The Synchronized Reserve market is an ancillary service market operated and cleared simultaneously with the Real-Time electricity market and the Frequency Regulation market. The Synchronized Reserve product is a 10-min response product, and resources are committed on an hourly basis and prices are posted every 5 min. The resources scheduled to provide Synchronized Reserve commit

Renewable Energy Integration. http://dx.doi.org/10.1016/B978-0-12-407910-6.00021-1

to provide a specified amount of energy production from generation or reduction from demand response as the result of a call from PJM operators within 10 min. This product is deployed by grid operators to maintain reliable grid operations during system loss of generation events or during periods of high ramp.

The Frequency Regulation market is an ancillary service market operated and cleared simultaneously with the Real-Time electricity market and the Synchronized Reserve market. The Frequency Regulation product is a 4-s automated response product, resources are committed on an hourly basis, and prices are posted every 5 min. The resources scheduled to provide Frequency Regulation are put on automatic control to regulate frequency within established criteria. This product is deployed by grid operators to maintain reliable grid operations during all operating periods.

PJM's wholesale markets are generally focused on entities that produce, buy, or sell electricity, but do not actually consume the electricity. In contrast, the retail market is focused on entities that buy electricity from the wholesale market and then sell the electricity to a customer that physically consumes the electricity. For example, a local utility may purchase wholesale electricity in the PJM market and then sell it to customers for their retail electricity consumption needs. PJM generally does not interact directly with electricity consumers, although some large consumers can participate directly in the wholesale markets. The PJM market has more than 800 member companies that are certified to participate in the wholesale markets.

2. Opportunities for demand-response in the wholesale market

Demand-response is consumers' ability to reduce electricity consumption at their location when wholesale prices are high or when the reliability of the electric grid is threatened. Common examples of demand-response include the following: increasing the temperature of the thermostat so the air conditioner does not run as frequently, slowing down or stopping production at an industrial operation, or dimming/shutting off lights. Demand-response is any explicit action taken to reduce load in response to short-term high prices or a control signal from PJM, but it does not include the reduction of electricity consumption based on normal operating practice or behavior. For example, if a company's normal schedule is to close for a holiday, the reduction of electricity as the result of this closure or scaled-back operation is not considered a demand-response activity in most situations.

Demand-response is important because it is another competitive resource that can be used to maintain demand and supply in balance for grid operations and the associated wholesale markets. Retail electricity consumers tend to be unresponsive to wholesale prices under traditional average rate tariffs. Therefore, as demand increases, more expensive generators are called on to serve this demand. By reducing demand during these periods and encouraging price-responsive demand, the wholesale market power potentially can avoid using more expensive generation resources to meet high demand.

Demand-response opportunities in the wholesale market enable electricity consumers to earn a revenue stream for reducing electricity consumption during periods of high wholesale prices or when the reliability of the electric grid is threatened. Technology enabled, automated demand-response can also earn revenue by providing grid services. Demand-response participation is divided into three classifications (economic, emergency, and ancillary services). An electricity consumer may participate in each of these, depending on the circumstances.

2.1 **Emergency demand-response: capacity market**

This product is generally committed on a 3-year forward basis in the capacity market or in the subsequent incremental auctions. The revenue stream derived from participation is largely driven by the Capacity market, and the revenue earned is a function of the relevant capacity market price and the load reduction commitment. The resource is paid to be available for curtailment by system operators during expected emergency conditions on a monthly basis for a commitment that is made for 1 year, which starts on June 1 and ends on May 31 of the following year. Demand-response participation in the Capacity market is voluntary, but once the demand-response is committed as a capacity resource, then the commitment represents a mandatory commitment to reduce load or only consume electricity up to a committed certain level when PJM needs assistance to maintain reliability under supply shortage or expected emergency operations conditions. Penalties will be applied for noncompliance. The demand-response resources must be available to respond to PJM's request to reduce load when the availability depends on the product committed in the capacity market, as follows:

- Limited demand-response: resource is available for up to 10 weekdays from June through September, where each request may be up to 6 h in duration.
- Extended Summer demand-response: resource is available for all days from May through October, where each request may be up to 10 h in duration.
- Annual demand-response: resource is available for all days from June through May of the following year, where each request may be up to 10 h in duration.

From a capacity perspective, PJM considers demand resources similar to a generator and expects them to perform at the time when the grid most needs it to avoid brownouts and/or rolling blackouts within the PJM service territory.

2.2 **Economic demand-response: real-time market and day-ahead market**

This product primarily represents a voluntary commitment to reduce electricity demand when the wholesale price is high. PJM publishes a net benefit price that represents the price at which the benefits incurred by a reduction in wholesale prices from the economic demand-response will exceed the cost to pay for the economic demand-response. Because the economic demand-response displaces a generation resource, PJM expects the demand-response resource to perform if committed and will assess deviation charges if the amount of load reductions realized is significantly different than the amount of load reductions dispatched by PJM.

2.3 **Ancillary service markets: day-ahead scheduling reserves, synchronized reserves, and frequency regulation**

These products are another potential revenue stream for demand-response resources. Demand-Response resources may provide ancillary services in the wholesale market with the appropriate automated infrastructure and qualification by PJM. There are three Ancillary Services markets in which economic demand-response resources may participate: Synchronized Reserves (the ability to reduce electricity consumption within 10 min of PJM dispatch), Day-Ahead Scheduling Reserves (the ability to reduce electricity consumption within 30 min of PJM dispatch), and Frequency Regulation (the ability to follow PJM's Frequency Regulation signal). Participation in the ancillary service markets is voluntary; however, if a resource clears and is assigned as an ancillary service resource, performance is mandatory. Penalties for nonperformance will apply, and a resource that fails to perform may be disqualified from future participation.

Demand-response resources can be offered and cleared as Day-Ahead Scheduling Reserve if they have the ability to curtail in less than 30 min notice in the Real-Time market. Such resources are paid the Day-Ahead hourly clearing price for each hour they are committed, and they are expected to be available for curtailment in the Real-Time energy market if needed.

Demand-response resources can be offered and cleared in the Synchronized Reserve market if they have the capability to curtail consumption within 10 min of an operator call. Such resources are offered into the Synchronized Reserve market each hour of the operating day, and the clearing prices for Synchronized Reserve are calculated and posted every 5 min. Generally, Synchronized Reserve resources are scheduled hourly and, on average, PJM operators call on Synchronized Reserve resources once per week. Therefore, the Synchronized Reserve market has been attractive to demand-response resources because they are paid the clearing price every hour for standing by as a reserve resource and are called to curtail relatively infrequently.

PJM has recently enhanced the Frequency Regulation market to provide more compensation for better-performing Frequency Regulation resources. This pay-for-performance compensation approach provides enhanced economic incentive for fast-responding Frequency Regulation resources. In this market, resources are assigned as regulating resources each hour and they are paid the clearing price for being on automatic frequency control. In addition, regulating resources receive an additional payment based on how much they are requested to move during their regulating assignment.

3. PJM experience with demand-response

3.1 Emergency demand-response

Demand-response participation as an emergency resource has increased significantly since PJM implemented a forward capacity market. Before the implementation of a forward capacity market, approximately 2000 MW of emergency demand-response resources was committed as capacity resources in the PJM region. After implementation of a forward capacity market in 2007, demand-response participation as a capacity resource has demonstrated a remarkable increase. The amount of demand-response committed as capacity resource in the PJM region for the 2013/2014 delivery year is more than 10,000 MW. For the year 2015/2016, more than 14,000 MW of demand-response resources is committed, which represents more than 8% of total capacity supply. These trends, illustrated in Figure 1, indicate the forward capacity market has provided a strong investment incentive for development of demand-response as an emergency resource. These market results indicate that emergency load curtailment can be effective as a competitive alternative to installation of peaking generation that would operate only a few days per year.

Because capacity is a physical, reliability-based, product, PJM requires all capacity resources to perform when called and adheres to established standards to ensure compliance. For demand-response resources, PJM has enhanced measurement and verification procedures and has implemented testing requirements, comparable to generation resources, to demonstrate the ability to achieve the curtailment quantity that is committed. PJM's experience is that demand-response resources have performed well during compliance periods and in aggregate have met or exceeded performance requirements

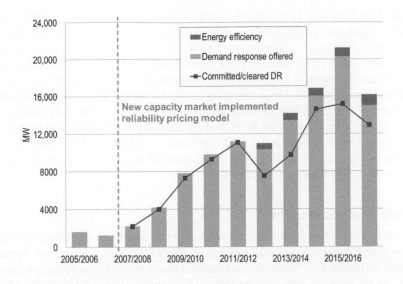

FIGURE 1

Demand-resource participation in capacity market.

consistently. A summary of both emergency event performance and testing performance for demand resources for the period 2009–2012 is shown in Table 1.

For example, PJM dispatched emergency demand-response two times during 2012: July 17th (Tuesday) and 18th (Wednesday). A summary of the events, in Table 2, illustrates the aggregate

Table 1 Testing Performance Summary

Performance Summary		
Year	Event Performance	Test Performance
2009	No events	118%
2010	100%	111%
2011	91%	107%
2012	104%	116%

Table 2 Summer 2013 Demand-Response Events

Event Date and Zones	Committed MW	Reduction MW	Performance
7/17, AEP, DOM	1670	1736	104%
7/18, AECO, BGE, DPL, JCPL, METED, PECO, PENELEC, PEPCO, PPL, PSEG	2135	2203	103%

performance on July 17th was 103% and the performance on July 18th was 104%. Summer 2012 performance was significantly higher than performance for the single event in July 2011 (91%).

The two summer 2012 events varied in size and length. The July 17th event was a long lead time event (resources have up to 2 h to reduce) called in two zones (AEP and DOM), lasting for almost 4 h, for 1670 MW of demand-response resources. In comparison, the July 18th event was a combination of long and short lead times (short lead time resources respond in up to 1 h) across 10 mid-Atlantic zones, lasting less than 2 h, for 2135 MW of demand-response resources. The July 18th event had the potential to be a longer event, but storms developed and the associated decrease in load shortened what would have otherwise been a longer event. Although some curtailment service providers did not meet their curtailment obligations, most did and, in fact, demand resources in aggregate performed at a higher level than obligated based on commitments.

3.2 Economic demand-response

In the fall of 2008, PJM wholesale electricity market prices decreased significantly because of economic slowdown and decreasing natural gas prices. Therefore, recent economic demand-response activity has been relatively low. However, economic demand-response participation has experienced some growth since the implementation of the Federal Energy Regulatory Commission Order 745 rules that compensate economic demand-response resources at the full wholesale electricity price for their location. The market rules before implementation of Federal Energy Regulatory Commission Order 745 compensated economic demand-response resources based on the difference between the wholesale price (locational marginal pricing) and the retail price for Generation and Transmission services (G&T). Economic demand-response for the 7-month period from April through October of 2012, for which compensation under Order 745 was in place, received $8.7 million of revenue from 133,466 MWh of response. This is an average of $1.24 million and 19,067 MWh per month. In contrast, for the 41-month period from November 2008 through March 2012, for which economic demand-response was compensated at LMP-G&T, averaged only $0.173 million of revenue and 4056 MWh per month. In addition, the amount of economic demand-response activity in the Day-Ahead energy market compared with the Real-Time energy market has increased during the Order 745 period. Currently, 44% of all load reductions were based on commitments made in the Day-Ahead market, whereas only 11% were based on the Day-Ahead market before Order 745. This indicates an improvement because scheduling demand-response on a Day-Ahead basis is more efficient. These trends are illustrated in Figure 2.

Although economic demand-response activity has increased relative to recent years, the amount of participation is low relative to the overall economic demand-response capability that has been registered and approved to participate in the wholesale market. The total amount of demand-response registered to participate in the economic market in July 2012 was 2300 MW, but, on average, the amount of load reduction was only 70 MW. This amounts to only 3% of the Economic demand-response resource capability used during the month of July. For the July nonemergency peak day, 305 MW was delivered, which is 13% of the total Economic demand-response capability. This trend may indicate that economic demand-response resources could not reduce load in the middle of a hot summer day or did not think the market prices justified economic curtailment.

Although overall participation is low, aggregate economic demand-response performance has significantly increased since the implementation of Order 745 market rules. For April through October,

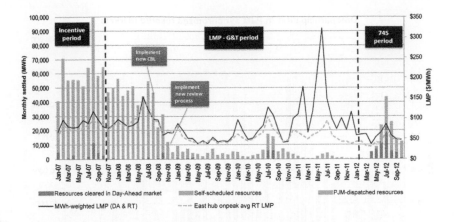

FIGURE 2

Economic demand-resource participation in the PJM market.

the monthly average performance in Day-Ahead and Real-Time energy markets has been between 80% and 110%, where performance is measured as actual load reduction divided by the cleared Day-Ahead offer or Real-Time dispatch amount. Curtailment Service Providers are doing a better job of determining when their customers can provide a load reduction and the quantity of load reductions they expect to deliver to the wholesale market.

In summary, the implementation of Order 745 rules has resulted in the following: an increase in energy market participation from demand-response (most load reductions are from large customers located in higher-priced regions); an increase in the amount of demand-response activity in the Day-Ahead market; and better demand-response performance (actual delivered load reductions closer to amount dispatched in Real-Time market or cleared in Day-Ahead market). Furthermore, there is the potential for a significant increase in economic demand-response activity because most qualified resources have not submitted offers into the Real-Time or Day-Ahead market.

3.3 Ancillary services markets

The Synchronized Reserve market provides a unique opportunity for competitive development of demand reduction response through investment in demand-response infrastructure. The payments to resources that clear in the Synchronized Reserve market are compensation for the demand reduction resource to be available to respond within 10 min. Therefore, although demand reduction resources must install infrastructure to allow them to curtail their consumption of electricity within 10 min, they will only be requested to curtail when system conditions require the 10-min response. Because the PJM market operators have historically requested a 10-min response, on average, once every 3 days, the demand-response customer may provide the service with limited disruption to its business processes.

In 2006, PJM adapted the reliability criteria and market rules to allow demand-response participation in the Synchronized Reserve market. As illustrated in Figure 3, the demand-response participation in the Synchronized Reserve market grew substantially over time, reaching a peak penetration

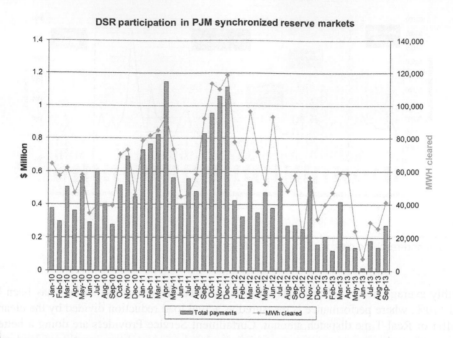

FIGURE 3

Demand-resource participation in the synchronized reserve market.

in 2011, when, on average, 8% of total supply was supplied by demand-response and the highest hourly penetration level was 22% of total supply. Since this period, the increased competition and lower energy prices have caused the Synchronized Reserve prices to decline, which, in turn, has decreased demand resource participation.

Demand resources undergo a compliance verification process similar to the process that generation resources must complete to receive compensation. After each event, the demand resource's power consumption at the start of the Synchronized Reserve event is compared with the power consumption at 10 min after the event was declared. If the measured values indicate that the demand reduction resource fails to respond to the event, the same penalties currently applicable to generation resources would apply to demand reduction resources. Demand reduction resources providing Synchronized Reserve are required to provide metering information at no less than a 1-min scan for the period during and surrounding a Synchronized Reserve event. To reduce barriers to participation, metering information for demand resources is not required to be sent to PJM in Real-Time; daily uploads at the end of the operating day that an event has occurred are acceptable. Because demand reduction resources tend to be small and distributed, PJM does not require Real-Time data feeds because they would create a prohibitive cost burden to participation for demand reduction resources and Real-Time response tracking of individual resources is not required for operational reliability. The overall response of Synchronized Reserve events is tracked in Real-Time by measuring ACE and system frequency; individual response information is only used for market settlements and the assessment of performance penalties. The meter information supplied to PJM is subject to audit by PJM. The reliability of

Synchronized Reserve response is measured based on Real-Time meter data supplied by generation resources and by daily file uploads of 1-min scan meter readings from demand reduction resources. PJM experience indicates that demand-resource response is generally as good as or better than generation response.

The implementation of enhanced Frequency Regulation compensation in October 2012 has begun to attract demand-response resources and Alternative Technologies to provide Frequency Regulation. Although this market is still in the early stages, PJM has seen substantial benefit from the additional fast response resources; in fact, PJM was able to reduce the Frequency Regulation requirement, in large part because of better-performing resources.

PJM expects continued growth in demand-response participation because of technology improvements. PJM's initial experience with demand-response as a frequency-regulating resource is that such resources perform well and have rapid response characteristics.

4. Experience with alternative technologies in the wholesale market

Across the PJM market region, alternative technologies have developed in response to competitive wholesale market opportunities. Most alternative technologies that are developing are based on energy storage or technology-enabled demand-response. The storage-based technologies include stationary batteries, flywheels, thermal, compressed air, water heaters, and electric vehicles. Because most of the storage technologies have relatively limited energy storage capabilities, the most significant market participation opportunities have been in the ancillary services markets, which can take advantage of the quick response capability of such technologies but does not necessarily require sustained energy output.

Alternative resource technologies, such as large-scale battery projects, are beginning to penetrate into the electricity industry and have demonstrated the capability for a nearly instantaneous response to a control signal sent by a grid operator. Current installations have a limited duration with which they can sustain a response, but they are capable of following extremely rapid and volatile control signals. The resource has proved its capability to respond much more quickly and accurately to Frequency Regulation signals than conventional regulating resources. Although the resource has a 15-min limit on delivering at full capacity, PJM experience indicates that such resources can follow specially designed fast regulation signals for an extended period. The battery technology appears to be commercially viable at current Frequency Regulation market prices. PJM has seen recent growth in participation of large-scale battery projects in the market; we currently have 37.3 MW of battery resources participating daily in the Frequency Regulation market, providing almost 90 MW of effective regulation. Given the economies, the PJM region is experiencing growth in this technology and future projections indicate expanded growth rates as technology costs decrease.

5. Potential future evolution for demand-response and alternative technologies

As the US power industry deals with the significant and unprecedented trends that are driving the evolution of supply and demand in the electric power industry in the United States, the development of advanced technologies will play a role in maintaining grid reliability. Transformational changes include increasing penetration of intermittent, renewable resources, rapid expansion of shale gas and gas-fired power generation, significant coal-fired generation retirement, and distributed resource penetration. These transformations are reducing supply-side flexibility in power system operations. The growth of renewable, natural gas–fired, and alternative resources does not appear to be capable of replacing all of the functionality of current resource mix without increasing costs substantially. In addition, grid services, such as black start, frequency response, and voltage control, that, to some extent, are implicitly supplied by the conventional generation resources of today, may need to be secured in a more comprehensive manner in the future.

These increases in cost, combined with smart grid technology and innovation in retail rate design, will drive electric demand to be more flexible and to respond to price and environmental signals to enhance reliable grid operation and potentially replace some of the operational flexibility lost on the supply side. As a result, the demand side of the power balance equation will incur the largest state change. Some amount of the relatively inelastic demand we have today will likely be replaced by Price Responsive Demand. The development of automated price responsive demand will add an additional degree of "load-following" capability in power system operations.

The PJM market has experienced substantial increases in demand resource and alternative technology participation in response to price signals for reliability-based grid services products. These resources have provided enhanced operational flexibility, which can be coordinated through automation and communication technology. PJM Future grid operations will likely require changes in the amounts of grid services, such as Frequency Regulation, 10-min Synchronized Reserve, 30-min operating reserve, black start, frequency response, and voltage control. Such a trend will increase value of these products, creating more opportunity for development in alternative technologies to supply these products. Forward market development may need to be augmented to include commitments and incentives for resources that can provide these services to ensure adequacy.

The Implications of Distributed Energy Resources on Traditional Utility Business Model

22

Fereidoon P. Sioshansi

Menlo Energy Economics, USA

1. The evolution of traditional utility business model

With minor variations, the electric supply industry (ESI) evolved in different parts of the world propelled on the one hand by the growing demand for electricity and the wonderful services it provides and on the other by the remarkable and persistent decline in prices achieved through technological innovations and centralized generation coupled with economies of scale and scope.

From the beginning, the industry's growth was accompanied by pervasive regulations that recognized the ESI's massive need for investment to build, operate, and maintain the infrastructure. During its first century, the ESI succeeded not only in keeping up with the robust growth in demand, but it managed to do so at falling per-unit costs over extended periods of time. In turn, the persistent falling prices encouraged increased demand growth, bringing prosperity, higher standards of living, and comfort to billions of customers around the world.

In this context, the industry's business model gradually evolved around two basic premises:

- First, centralized power generation, delivering the output of large plants through a massive transmission and distribution network designed to deliver power to consumers while adjusting supply to match variable load in real time; and
- Second, a simple mechanism to collect revenues from consumers on the basis of a flat tariff applied to volumetric consumption.[1]

The former encouraged the industry's vertical integration[2] whereas the latter was not only simple in accounting terms but it could be administered with a primitive, inexpensive, and hardy spinning disk meter that measured cumulative consumption and nothing more.[3] Before the advent of smart meters, it

---

[1]For example, refer to J.C. Bonbright, *Principles of public utility rates*. New York: Columbia University Press, 1961.
[2]The pattern was to build a few central power plants away from major load centers while transmitting power via a transmission and distribution network.
[3]Over time, many variations evolved, including a fixed monthly fee based on the capacity, but for the most part, the revenue recovery mechanism was straightforward.

Renewable Energy Integration. http://dx.doi.org/10.1016/B978-0-12-407910-6.00022-3

275

was not easy or feasible to measure other attributes of consumption, say the pattern of electricity use, which turns out to be a better determinant of the costs imposed on the network.[4]

The regulated rate of return (ROR) model that emerged had two noteworthy features:

- First, the regulated monopoly "utility" was allowed, in fact encouraged, to invest large sums to expand the infrastructure necessary to meet its obligation to serve customers' needs, be it power plants, transmission lines, or distribution network[5]; and
- Second, all prudent investments made by the monopoly utility and approved by the regulator were considered legitimate because they were made to serve their needs and they could be recovered in tariffs set by the regulator.[6]

This was, and still is, known as the "regulatory compact." Utilities referred to consumers as rate payers. In rate-setting proceedings, regulators and utilities mutually agreed on rates (not prices) that would allow utilities to recover their revenue requirements. These familiar concepts, still in place in many parts of the United States, were, and remain, alien to competitive industries, where there are customers, prices, and no guaranteed ROR on investment.

2. Gradual transformation of the ESI

Starting in the 1970s, several technical and external developments compelled policy-makers to reexamine, and in some cases question, the fundamentals of how the ESI operated and, more important, recovered sufficient revenues from its customers to stay viable.

For the most part, the stodgy ESI did not welcome regulatory or policy changes or anything that would affect the status quo. For example, the incumbent utilities were hostile to independent power producers, treating them as competitors because the newcomers had a lower cost structure and could generate power at lower costs than they could. Likewise, the ESI has been opposed to the introduction of environmental regulations or proposals to reduce greenhouse gas emissions. The industry has more recently been coerced to engage in energy efficiency programs funded by customers and forced to meet obligatory renewable targets, to name a few.

The ESI has recently been confronted with several other challenges that are affecting the business fundamentals that have historically served the industry, including a few highlighted as follows, not necessarily in order of significance.

- The first major turning point was the gradual end of the industry's golden age and the realization that investing in bigger central plants no longer necessarily resulted in lower average costs. The ESI had essentially run out of the economies of scale. This reality was particularly acute in the case of nuclear power plants. As larger nuclear plants became more sophisticated and safety

---

[4]Proponents of dynamic pricing have been arguing for some time that consumers should be charged not based on total usage but rather their pattern of usage, including consumption during expensive-to-serve peak periods.

[5]Investments made by utilities could be put in the so-called rate-base, on which they could earn an allowed ROR. In fact, this incentive provided a perverse incentive to overinvest, to earn more returns. The scheme also led to incentives for utilities to encourage more, rather than less, consumption.

[6]This provision turns out to be critical, as explained later in the chapter. Under the arrangement, regulated utilities did not risk having stranded assets. All investments approved by the regulator were deemed to be prudent; therefore, they could be recovered under regulated tariffs.

requirements became more stringent, building larger reactors resulted in increased per unit costs, not the other way around.

- The second major change was the introduction of new environmental restrictions on what could and could not be emitted in the power generation process leading to increased costs.
- The third development was the broadly rising costs of fossil fuels and network maintenance leading to higher retail electricity costs.
- Fourth was the gradual—and usually reluctant—acknowledgment on the part of the ESI and their regulators that investing in energy efficiency could be preferable to increased investment in infrastructure—that negawatts were better than megawatts.[7] However, for the most part, the prevailing regulations remain unchanged, which means that selling more kilowatt-hours is more profitable than encouraging consumers to use energy more efficiently. This remains a challenging barrier to further energy efficiency investments in many parts of the world.
- Finally, the increased concerns about climate change and sustainability have given rise to increased interest not only in energy efficiency but also low-carbon and renewable energy technologies.

These powerful forces have brought many time-tested assumptions about the traditional utility business model into question. For example, why should the ESI continue to rely on its outdated revenue collection model, which has remained virtually unchanged since the days of Thomas Edison? This formula, which well served the industry and society when sales were growing and per-unit costs were falling, makes little sense in an age in which consumption in many advanced economies is barely growing, described in Section 3, whereas retail rates are broadly rising.[8]

3. Why the rise of distributed energy resources?

As the preceding discussion explains, the traditional "utility" business model has been under increasing pressure on several fronts for some time. However, the urgency of coming up with an alternative business model did not seem urgent or pressing until the rapid rise of distributed energy resources[9] (DERs) in the last few years.

The rapid rise of DERs can be traced to several inter-related factors:

- Low and falling electricity demand growth rates, most pronounced in mature Organisation for Economic Co-operation and Development economies, means that rising fixed costs must be spread among fewer kilowatt-hours, putting upward pressure on tariffs;
- Generally rising retail rates, mostly as a result of the push toward a low-carbon energy mix, aggressive support of renewables, and environmental regulations;
- Rising popularity of investment in energy efficiency, including appliance efficiency standards, stringent building codes, requirements such as zero net energy (ZNE), or passive buildings;

---

[7]A recent edited volume by the author, *Energy efficiency: toward the end of demand growth*. New York: Elsevier; 2013, covers this topic in more detail.
[8]The United States is an exception to this global trend because of the shale gas boom and low natural gas prices, which have kept retail rates flat in the past few years.
[9]DER includes distributed self-generation and energy efficiency.

- Behavior and demographic changes resulting in flat or possibly lower per-capita electricity consumption rates in mature economies; and
- Expectations for higher levels of power quality and service reliability, which the existing grid can barely meet.

The key driver for DERs is that in several key markets, retail tariffs have been rising while the cost of DERs has been steadily falling. These trends are encouraging an increasing number of consumers to become *prosumers*—cutting down on consumption by investing in energy efficiency while meeting some or all of their needs from on-site generation. Of course, the key question is what caused high and rising tariffs in the first place and what makes DERs attractive?

To be sure, retail tariffs have risen significantly in the recent past only in a few places, and they can usually be traced to specific causes or reasons. Three particular examples are examined below:

- In the case of Germany, residential retail tariffs have historically been high, partly because of high taxes that are applied to all forms of energy. However, these rates have risen in the recent past because of the aggressive push toward high renewable penetration levels and the political decision to phase out the country's nuclear plants by 2022 after the Fukushima disaster in 2011. Making matters worse is the government's policy to shield the country's energy-intensive industries from higher tariffs, which puts an even larger burden on residential rates. The net result is that German residential retail rates are currently approximately 3 times higher than their U.S. counterparts and are projected to rise even higher.
- Similar to Germany, California has an ambitious climate bill, including statewide greenhouse gas emissions reductions to 1990 levels by 2020, mandatory renewable portfolio standards to 33% by 2020, and several other requirements that put upward pressure on tariffs. However, what drives consumers to conserve energy and self-generate is the rising tiered tariffs, which means that residential consumers pay higher per-unit costs as they consume more kilowatt-hours. For many, the marginal price at higher tiers is as high as 30–35 cents/kWh, which makes DERs an attractive bargain.[10]
- Some of the same factors explain what makes DERs attractive in Australia, namely renewable targets and rapidly rising network costs. Overall, retail tariffs in many parts of Australia have more than doubled over the past 5 years. This and overgenerous subsidies for solar rooftop photovoltaics (PVs) have resulted in over 1 million solar roofs in record time.

As the previous examples illustrate, DERs have become attractive in areas where retail tariffs are high and prevailing regulations and/or subsidies encourage energy efficiency and/or self-generation. Looking at average tariffs in the United States (Figure 1), it is easy to guess where energy efficiency and/or self-generation would be attractive.

What typically happens in such areas is lower consumption coupled with increased self-generation. Well-insulated buildings making use of natural lighting, ventilation, and state-of-the art appliances use very little energy to begin with. Adding a solar rooftop PV, a solar water heater, a ground-source heat pump, or other options can make the building virtually ZNE, eliminating utility bills.

---

[10]Tiered rates vary among the state's three large investor-owned utilities, but the top rates for all currently exceed 30 cents/kWh.

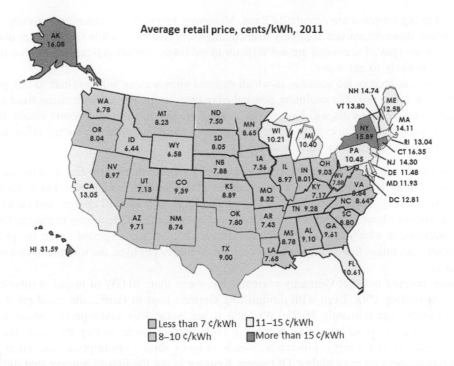

Average retail price, cents/kWh, 2011

Less than 7 ¢/kWh | 11–15 ¢/kWh
8–10 ¢/kWh | More than 15 ¢/kWh

FIGURE 1

Average retail electricity tariffs in the United States (in cents/kWh): 2011 data.

Although this is a blessing for the consumers who can take advantage of DERs, it is a nightmare for the local utility that loses the revenues that used to come from these customers. What makes this scenario worse is not just the loss of revenues, but the fact that prosumers continue to impose high costs on the distribution network while contributing little to its upkeep. According to critics, the prevailing laws allow prosumers to become *free-riders* that are subsidized by the remaining customers.

4. Rethinking the fundamentals

These developments are beginning to undermine the basic tenants of the "traditional" utility business—namely, the regulated ROR in which investments in infrastructure were recovered through a flat tariff on consumption. This basic model is becoming less tenable with the passage of time for at least two basic reasons, and possibly others:

- First, the model no longer applies to certain segments of the business in jurisdictions where utilities have already been vertically disaggregated. For example, in places where distribution companies operate as "stand-alone" poles and wires enterprises (e.g. in Australia or Texas), charging consumers on a volumetric basis makes little or no sense because virtually all of the costs of doing business are essentially fixed. Moreover, given the developments already mentioned, these enterprises are likely to encounter increased costs while facing flat or declining sales. These types of scenarios are not difficult to envisage, and all indications are that the situation is likely to get worse.

- Second, if one accepts the scenario in which demand growth rates will continue to fall, perhaps resulting in flat or possibly declining demand over time, accompanied with rising fixed and variable costs, retail tariffs will have to rise to keep the incumbent stakeholders viable. Rising rates are likely to result in even lower demand growth, more investment in energy efficiency, and more distributed generation—a vicious cycle.[11]

This topic was recently debated at a summit organized by Bloomberg New Energy Finance in New York City.[12] Urban Keussen, Senior Vice President of Technology and Innovation at E.ON, Germany's largest utility, said he wonders for how much longer utilities will charge, and customer will pay, by the kilowatt-hour for electricity. He envisions the evolution of electricity pricing to follow the path of telecoms, in which mobile service providers have gradually migrated to flat fee plans for a bundle of services rather than charging customers per minute or per byte, the traditional cost-recovery mechanism.

Keussen pointed out that Germany currently has more than 30 GW of installed solar capacity, mostly solar rooftop PVs. Even with diminishing German feed-in tariffs, the retail price of electricity, currently approximately \$0.35/kWhr and rising, solar PVs make perfect sense for many customers. However, as an increasing number of customers invest in rooftop PVs and other means of self-generation and energy efficiency, which reduces their consumption, the costs for the remaining customers go even higher. Of course, Keussen is not the first to observe that the costs of maintaining and operating the grid—mostly fixed—will have to be spread among a shrinking number of customers, those without self-generation and who are unwilling or unable to invest in energy efficiency.

5. New definition of service

What is needed in this context is a new definition of *service* for prosumers, for whom the most valued "service" is likely to be *connectivity* to the grid and the ability to feed into it or take energy out of it. In an extreme case, such a customer may use few, if any, net kilowatt-hours while potentially putting enormous stress on the distribution network for having to constantly balance the variable supply and demand.

This suggests the need for a redefinition of service for the ESI. Just as mobile network providers have adjusted their business model to set significant fixed charges on the basis of bandwidth, speed, signal strength, and service ubiquity—moving away from per-minute charges—utilities need to make a similar transition to significant fixed charges and for the network's ability to balance customers' load and on-site generation. It is not surprising that this is the favorite solution of many regulated utilities in the United

---

[11]A report published by the Edison Electric Institute concluded that the growth of small-scale solar systems represents the "largest near-term threat" to the ESI. Refer to *Disruptive challenges: financial implications and strategic response to a changing retail electric business*, January 2013.
[12]Have flat tariffs outlived their usefulness? *EEnergy Informer*, June 2013.

States, who are pressuring regulators to allow them to raise fixed charges to self-generating customers and/ or reduce the credit given for kilowatt-hours fed into the grid when the meter spins backward, metaphorically speaking, under prevailing Net Energy Metering (NEM) laws, currently in effect in 43 states.

6. Responding to disruptive technologies

Opinions vary on what may be the best approach to respond to the rapid rise of DERs, a few of these are as follows:

- Fight NEM laws tooth and nail: This option would keep the current utility business model—flat tariffs applied to volumetric consumption—while pleading with state-level regulators to adjust/ modify current NEM laws and/or DER subsidies to make it less attractive for consumers to abandon grid-provided electricity. If successful, then this strategy will merely temporarily reduce the ESI's revenue hemorrhage.[13] It will not address the fundamental issue, which is that alternatives to grid-provided electricity are becoming more affordable whereas the ESI's costs are rising because of lower or flat-demand growth. For example, a report by Navigant Research concludes that at the current rate of falling prices, solar PVs will be at grid parity, without subsidies "in all but the least expensive retail electricity markets" by 2020.[14]
- Reconsider rate design to accommodate growing DER: This option would help reduce the ESI's current revenue hemorrhage, especially in places such as California, where the rising residential tiered rates and no flat fees make DER attractive. In this context, rate design that better reflects the actual costs of delivering electricity at different times of day to consumers with different load profiles—cost causality—would be an improvement, reducing uneconomic bypassing. However, the fundamental problem is the rise of disruptive technologies that appears to render centrally generated power delivered through an extensive infrastructure expensive. For example, retail prices in Germany, Italy, Denmark, Japan, and Australia are too expensive to compete with the falling price of DERs. The same would be true in many countries where electricity prices are heavily subsidized, such as most Middle-East countries. In Hawaii, DERs will beat utility-generated power today with a significant margin, no subsidies required.
- Introduce dynamic, cost-reflective pricing: This option, a subcategory of improved rate design (previous bulletpoint), has been proposed by several experts for a long time, but it has not been widely adopted to date partly because smart meters were not in place until recently and partly because the regulators are equivocal on whether consumers would or would not like dynamic prices. Some observers believe that many of the ills of the ESI, including noneconomic[15] solar PV bypass, are self-inflicted because prevailing tariffs are not cost-reflective. However, dynamic pricing may turn out to be a two-edged sword. For example, for summer peaking utilities, the value of PV-generated power during peak demand hours may end up being even higher than what

---

[13]California's three investor-owned utilities estimate that collectively they could lose $1.4 billion/year under the current NEM law because of the solar bypass. To put the scale of this loss in perspective, to make up the loss, the 7.6 million nonsolar customers would have to pay an extra $185/year assuming it is evenly spread.

[14]Solar PV market forecast, 3rd Qtr. 2013, Navigant Research.

[15]Noneconomic bypass in this context refers to consumers choosing DER because the prevailing rates do not reflect actual costs of service.

consumers are currently paid under existing NEM laws.[16] Economic theory suggests that consumers should pay based on cost-causality imposed on the network—kilowatt-hours consumed during peak periods are more dear than those consumed off peak. By the same token, kilowatt-hours generated and fed into the grid should receive the appropriate credit on the basis of their value to the network.

- Expand utilities' reach to include DERs with an expanded definition of service: This option seems attractive, especially in jurisdictions where there is competitive retailing or supply businesses. There are no insurmountable barriers for nonregulated retailers; suppliers; gentailers; or, for that matter, newcomers (call them energy service companies) to offer expanded services to customers, including energy efficiency investments, demand response, home energy management, and distributed generation. Such services can be offered in various forms, with the investments recovered on a lease option or other types of bilateral arrangements. Just as many solar rooftop PV installers lease the hardware or sell the output to consumers under a long-term arrangement, utilities could conceivably enter this business. It would be much easier in competitive retail markets than it would be in regulated ones. It is somewhat of a puzzle why the practice has not taken off to date, say in Texas or in Eastern Australian states, where there are major regulatory barriers.

- It may be too late: Those who subscribe to this option are of the opinion that the DER train has already left the station, or is about to leave soon, and response by the ESI may simply be too little, too late. This sentiment is captured in the following quote, which appeared in a *New York Times* article[17] on the debate on NEM: "We did not get in front of this disruption." Clark Gellings, a fellow at the Electric Power Research Institute, a nonprofit arm of the industry, said during a panel discussion at the annual utility convention last month, "It may be too late."

7. Conclusions

As the preceding discussion illustrates, DERs fit the classic definition of disruptive technologies. Their rapid rise is currently limited to a few places where existing tariffs are high and rising and/or where prevailing subsidies, regulations, or laws—such as NEM or feed-in-tariffs—make DERs commercially attractive. All indicators suggest that as time goes on, the cost-effectiveness of DERs would spread. The ESI's cost structure and its historical dependence on central generation delivering power through a massive transmission and distribution network are part of the reason. The falling costs of DERs are equally important. Energy efficiency has traditionally been cheaper than generation; self-generation options are getting cheaper, propelled by technological advancements, mass production, and new business models, such as leasing solar PVs rather than buying them outright, which has been a major obstacle to their penetration thus far.[18]

---

[16]This is an argument by solar advocates, who claim that consumers with solar PVs do little or no harm to utilities under the current NEM laws and in fact may be saving them by displacing expensive peaking generation. Moreover, there are savings due to the fact that much of what is self-generated on the customers' premises is locally used, reducing stress on the distribution network.

[17]Dianne Caldwell, On rooftops, a rival for utilities, *New York Times*, July 27, 2013.

[18]Since California regulators approved solar financing in 2007, the residential component of the business has reached $1 billion, growing from 10% of the market to nearly 70% by 2012. A similar decision in Oregon in 2011 has resulted in residential solar installation leasing growing from 25% to 52% in 2012.

The ESI appears ill prepared to face the challenge, hampered by a cumbersome regulatory process that does not allow fast response while discouraging innovations in bold service offerings, pricing, or business models. As with other disruptive technologies affecting other industries, the ball may no longer be in the ESI's court. This is not to say that the industry should throw in the towel; however, the time for decisive thinking and even more decisive action is now.

Energy Storage and the Need for Flexibility on the Grid

David Mohler[1], Daniel Sowder[2]

[1] *Vice President of Emerging Technology, Duke Energy,* [2] *Renewable Generation Development, Duke Energy*

The concept of "flexibility" is increasingly being recognized as a valuable attribute for the electric grid. Flexibility in the context of an electric power system is the ability to vary the performance characteristics of resources to maintain a balanced and efficient power system. This ability has value because of the critical need to instantaneously balance supply and demand to maintain grid stability. The balance between supply and demand is monitored on a bulk power system by measuring the system frequency.

Traditionally, flexibility has been achieved predominately by adjusting electric supply in response to varying electric load (demand). The supply-side adjustment occurs in different time domains. Hour-to-hour flexibility occurs by varying the output of generation resources and dispatching additional plants as needed. The shorter-duration, minute-to-minute imbalances between supply and demand are addressed by rapidly varying the output of resources through ancillary services, such as frequency response. Different-generation technologies provide varying degrees of flexibility at varying costs. Base load assets, such as nuclear generation, that are not designed for rapid output adjustments (or "ramping") provide hourly flexibility, whereas more flexible assets, such as natural gas turbines, provide a more rapid, minute-to-minute response for ancillary services. Energy storage technologies have the potential to provide second-to-second and even millisecond-to-millisecond response.

Flexibility on today's power grid is increasingly being discussed for two reasons. First, new technologies are becoming available that can provide a more effective source for grid flexibility relative to the resources used for flexibility today. Most of today's flexibility is provided by dedicated capacity from generators that require fuel to run. Newer technologies, such as energy storage, can provide the capacity and small amount of energy needed for flexibility without burning fuel. To the degree that new technologies can provide lower-cost and more rapid flexibility relative to today's resources, they can reduce the need to use generation assets for flexibility and allow them to be more focused on energy production. Also, many new fast response technologies can provide a response much faster than traditional assets, thus requiring less total capacity to be dedicated to second-to-second flexibility. Together, these benefits "free up" generation assets to use more of their capacity to produce energy, rather than being available for flexibility.

Second, another reason for an increased need for grid flexibility is related to the increasing connection of other new technologies that add volatility to the grid's supply and load balance. Today, this is primarily being driven by the connection of variable-generation resources, such as wind and solar. Absorbing generation sources, such as solar and wind, into a stable grid requires a higher capacity of offsetting flexibility to counteract the inherent flexibility of an intermittent resource. Because supply must instantaneously equal demand, the grid must be ready to respond with a flexible,

Renewable Energy Integration. http://dx.doi.org/10.1016/B978-0-12-407910-6.00023-5

285

dispatchable resource to counterbalance the rapid, unpredictable changes in generation that wind and solar frequently bring to the grid. More flexibility is required to enable the grid to absorb the increasing capacity of intermittent renewable generation that is expected.

Because new technologies undermine the traditional design assumptions of the grid, flexibility is increasingly important in other aspects of grid management beyond maintaining energy supply and demand balance. This includes aspects such as distribution system voltage management and system protection schemes. For example, the management of a distribution-circuit voltage profile is typically accomplished today by devices that respond to slowly varying load changes and single-direction current flow. Many devices are set with time delays of 45 s or more, which traditionally has been fast enough to respond to expected load changes. Distribution-connected solar undermines these design assumptions. A cloud-induced solar output fluctuation can cause current flow on the feeder to change direction and magnitude in just a few seconds. A higher degree of voltage-control flexibility is needed to ensure a feeder stays within acceptable standards.

The presence of new technologies that provide flexibility on the grid also provides new opportunities for advanced grid functions that were not previously possible and will increasingly be expected of efficient utilities. For example, a distribution system function, such as conservation voltage reduction (CVR) or the ability to finely control and vary the voltage and, therefore, power consumption of a feeder, requires the ability to rapidly adjust equipment settings, including the consumption and absorption of reactive power, real power, and other attributes. The demand for flexible resources will continue to grow as the value of advanced functions, such as CVR, is demonstrated.

A flexible grid is a more efficient grid today and is an enabler of the advanced grid capabilities and higher renewable energy penetration envisioned for the future.

Energy storage technologies come in many different forms, including electrochemical batteries of many different chemistries, capacitors, flywheels, pumped-hydro, and compressed air systems. The key commonality across these various energy storage technologies is the ability to draw electric energy from the grid on command and to discharge most of this energy back to the grid at a later time. Stated differently, energy storage enables supply and demand to be balanced even when the generation and consumption of energy do not occur at the same time. This ability to flexibly move energy across time is a tool that can be applied in many different applications on the electric grid. Some examples include balancing supply and demand on an hourly basis on the bulk power system, meeting frequency regulation needs with a second-to-second response, buffering power flows on a distribution feeder from rapid solar fluctuations, and storing energy downstream of capacity-constrained areas to avoid costly infrastructure upgrades.

Energy storage technologies that use solid-state power electronics, such as electrochemical battery systems, bring additional flexibility benefits. Four-quadrant inverters that are used to couple direct-current systems (e.g., electrochemical batteries) to the alternating-current grid have the ability to act as a load and supply of active power (kW) and act as a capacitor or inductor (reactive power). These inverters have the ability to change their output condition extremely rapidly relative to traditional generation assets on the grid. An inverter-based energy storage system can respond to an active and reactive power command virtually instantaneously.

These abilities to store energy and to rapidly provide any combination of real and reactive power to the grid (within the system's capacity) make energy storage a unique and valuable tool when applied to the grid. Many useful and diverse applications for this tool have been conceived and are beginning to be demonstrated and used. Although the following section describes several applications that are being used on the grid today, they represent just the beginning of fully taking advantage of energy storage's capabilities through advanced methods.

Examples of how flexibility, created by energy storage's unique ability to store energy and rapidly respond to real and reactive power commands, is creating value on the grid are described later.

Large, traditional spinning generators are limited to a certain "ramp rate," or rate of change of power output, based on their physical characteristics. As load varies from second to second, temporary imbalances between supply and demand are created while spinning generators overcome their inertia and vary their output to match load. Left unaddressed, these temporary imbalances between supply and demand will cause the system frequency to deviate from its nominal value. Frequency response is an ancillary service used to address this temporary imbalance by providing short-term, second-to-second energy to fill this gap until generation ramps meet the load. Traditionally, the most flexible generation resources are used to meet the frequency regulation command, issued as a real power command every few seconds and routinely varying between "ramp-up" and "ramp-down" signals. Constantly varying a generator's output can be stressful on some assets, and valuable generation capacity is displaced in the process.

Energy storage is well suited to provide frequency regulation because of its ability to rapidly ramp up and ramp down on command and rapidly cycle between charging and discharging. It can also respond to this need extremely fast, making it more effective in rapidly eliminating temporary supply-and-demand imbalances and, thus, lowering the overall capacity of frequency regulation needed to achieve a stable grid frequency. Because the system cycles between charging and discharging in response to a frequency regulation command, the net energy balance is near 0. Using energy storage for frequency regulation releases generation capacity that would otherwise be dedicated to frequency regulation and, because it can respond more rapidly, enables the power system to maintain a stable frequency with less total capacity dedicated to frequency regulation.

Flexibility from energy storage can also create value at other levels of the grid, such as the distribution system. Distribution systems have traditionally been designed under the assumption that current will flow in one direction from the substation source to the feeder loads. Distributed solar factors challenge this design assumption and often result in reverse current flow and rapid, unpredictable changes in current magnitude. Some possible symptoms of this condition include nonstandard voltage profiles along a feeder, ineffective equipment response (e.g., voltage regulators and capacitors), and ineffective feeder protection and restoration schemes.

The flexibility of energy storage can counterbalance the undesirable flexibility of distributed intermittent generation by charging and discharging in response to changes in power flows on a feeder. For example, when located at the output of a solar generator, energy storage can "smooth" the output by charging and discharging to limit the output rate of change. Thus, the distribution feeder does not experience the full magnitude of solar's output ramp rates. Energy storage can also be located along a distribution feeder to detect and mitigate reverse power flows, thus making feeder power flows look more like they were designed to even in the presence of rapidly fluctuating solar generation. This allows the grid to ensure quality distribution service as it was designed.

1. Energy storage as an integral part of the grid

The previously described examples of energy storage use cases describe single-functionality uses that utilize some portion of an energy storage system's dynamic capabilities to create value through a single mechanism. Realizing the full value of energy storage requires holistically optimizing its functionalities across the full spectrum of grid operations.

To do this, energy storage systems must first be enabled to create multiple and often simultaneous streams of value by using their full four-quadrant range of capabilities. For example, although some portion of a system is delivering active power, the remaining capacity can be used to deliver reactive power based on a different application. Therefore, multiple value streams can be created simultaneously by a single energy storage system. With sufficient control and optimization logic, an energy storage system can be dynamically switched between multiple functionalities, depending on which one creates the most value on the system at any given time. The ability of an energy storage system to maximize its value by dynamically optimizing and blending multiple possible objectives is determined by its control logic and degree of interoperability with the grid ecosystem.

Another critical component of holistic optimization is having the visibility and incentive to create value across the traditional divisions of the electric grid. Visibility includes having the data needed to optimize a storage system's operation within a particular aspect of grid operations. For example, for a distribution-connected battery to charge and discharge to relieve an upstream transmission constraint, enough data must cross between the transmission and distribution system to identify the constraint and provide the appropriate charge and discharge commands to the battery to effectively relieve the constraint. Thus, a battery connected to the distribution system is creating a benefit on the transmission system. Along with visibility, an incentive must exist to allow a resource to be compensated for a benefit it creates. In many cases, the generation, transmission, and distribution systems are owned and operated by different entities. Therefore, an incentive system must be in place to quantify and transfer value from one segment of the grid to a resource connected to another segment.

The need for visibility and incentive across grid "silos" is being addressed in various ways. Markets now exist that compensate resources for a valuable service regardless of their interconnection location. For example, independent system operators have established markets that compensate resources for various ancillary services that benefit the transmission grid, even when these resources are connected on adjacent distribution systems. Disaggregating ancillary services from other power system attributes and compensating resources that provide them through a market mechanism is an efficient way to enable resources that are connected at different points on the grid and under different owners to receive the visibility and incentive necessary to create value.

In some areas of the country, complex market mechanisms are not necessary because utilities are vertically integrated, meaning that generation, transmission, and distribution are all operated by a single entity. Thus, an incentive exists for any benefit created on the system to be realized anywhere on the system. This removes the need for complex market structures and the associated transaction costs needed to enable cross-"silo" value transfers.

2. An ecosystem of technologies enabling flexibility

Energy storage has established itself as a viable tool to be used to benefit the electric grid. To fully use this tool beyond the initial single-functionality applications, new methods and technologies are required to effectively deploy energy storage as part of an optimized grid. Rather than viewing energy storage as an individual technology, it must now be viewed as part of an optimized ecosystem of technologies that support one another to create a reliable electric grid.

Several key enabling technologies are needed to realize the full value of energy storage within this ecosystem. First, a robust and low-latency communications network is needed to give distributed energy storage devices the means to quickly gather information and commands from other devices across the system. Traditional supervisory control and data acquisition systems and other systems that rely on information being transferred back to a centralized system before being used are likely to introduce latencies that make distributed decision making and optimization ineffective. Thus, robust and distributed communications, including peer-to-peer communications, are needed.

Second, platforms that enable distributed decision making, or distributed intelligence, are needed. Decisions are often best made locally, particularly for functions that require fast decision making. Distributed intelligence also avoids unnecessarily incurring data transfer costs when there is no technical need to move data beyond a local environment. Field-deployed software platforms are needed that can receive and understand information from various sources, make decisions, and execute commands without relying on top-down control hierarchies.

Third, communications interoperability and built-in cyber security are necessary to enable devices to work together and avoid exposing the network to cyber-security risks.

The process of developing and demonstrating methods for holistically optimized, utility-embedded energy storage is just beginning, and important demonstration projects are emerging within multiple utilities. Duke Energy, a vertically integrated electric utility, has deployed more than six different energy storage systems as part of an effort to develop and demonstrate value creation using holistically optimized, utility-embedded energy storage. In addition to an emphasis on installing and operating multiple electrochemical battery systems, this work also required developing and deploying an interoperable and secure communications architecture that enables the energy storage devices to participate within the larger ecosystem of grid devices.

The following four descriptions of demonstrated energy storage applications provide examples of how energy storage, embedded within a vertically integrated utility structure and connected within an interoperable communications architecture, is demonstrating value. These examples represent just the beginning of deriving value from the flexibility offered by energy storage systems.

2.1 **Example 1**

Duke Energy's Rankin Energy Storage Project has demonstrated that an energy storage system coupled with a grid-interactive algorithm can help a distribution circuit better absorb the impacts of distributed solar (and thus lower the cost to absorb more distributed solar). This battery system, consisting of a 402/282-kWh sodium–nickel–chloride battery, is installed within a distribution substation located more than three miles from a large distribution-connected solar installation. Using circuit telemetry provided by relays and other equipment in the substation, the battery algorithm detects undesirably high rates of change in circuit loading that are caused by the rapid increase or decrease of solar output. The algorithm then issues charge and discharge commands to the energy storage system to counteract these undesirable "power swings." The substation equipment is thus effectively shielded from rapid load fluctuations. This allows the substation protection equipment to operate under the original design assumptions and prevents the substation voltage monitoring equipment (primarily a load tap changer) from responding unnecessarily to short-duration load changes caused by the solar site. Because of its placement at the substation, lack of need for direct solar output telemetry, and ability to make battery dispatch decisions locally without any long-range data back haul, the Rankin Energy Storage System demonstrates how energy storage can create value when optimized within the grid ecosystem using distributed intelligence. A demonstration is in progress to add a reactive power capability to the battery algorithm that will add an additional and simultaneous value stream to the system (Figure 1).

2.2 Example 2

Duke Energy's Marshall Energy Storage System is demonstrating the ability to create value across the traditional silos of the utility. This system, consisting of a 250/250-kWh lithium polymer battery, is connected to a 12 kV distribution circuit and was primarily designed for energy-shifting applications. The 250-kW, 3-h duration battery has been used to shift energy to create multiple utility benefits. These include distribution circuit peak shaving, transmission system peak shaving and constraint avoidance, and optimizing the generation portfolio to create fuel savings or reduced carbon emissions, all accomplished using a battery connected to a distribution feeder. The battery's ability to charge and discharge to accomplish energy shifting is clear; the challenge is to collect and use data from across the grid to determine the optimal energy-shifting priority at any given time. This includes a mixture of economic and operational data that must be optimized. Therefore, the ability to deliver visibility (through data) and incentive across traditional divisions within the grid is a key element to the Marshall Energy Storage System's value. This system is now being used to provide energy shifting while simultaneously providing output smoothing for an adjacent solar installation, thus demonstrating the bundling concept (Figure 2).

2.3 Example 3

Duke Energy's Community Energy Storage project is highlighting how the available value streams for an energy storage system are highly dependent on the location of the system. Located at the "edge of the grid," or near the customer premise, community energy storage (CES) systems are capable of creating unique value because of their proximity to the customer. With two 25/25-kWh CES units installed, multiple blended value streams that benefit both the utility and the customer are being

FIGURE 1

Rankin energy storage system, Mount Holly, NC.

FIGURE 2

Marshall energy storage system, Sherrills Ford, NC.

demonstrated. These values include energy shifting (grid benefit), automatic voltage regulation through reactive power dispatch (grid benefit), the ability to provide back-up power during a grid outage (customer benefit), and the ability to mitigate rapid power fluctuations caused by customer-connected solar (grid benefit). Each of these functionalities is accomplished through a high degree of local intelligence. The customer benefit of back-up power is made possible by interconnecting the battery with an appropriate amount of load enabled by its location near the customer site (Figure 3).

2.4 Example 4

A larger demonstration of blended, simultaneous value streams and blended customer/utility value is being done with Duke Energy's McAlpine energy storage system. This 200/500-kWh lithium–iron–phosphate battery is adjacent to a 50-kW solar installation and a critical load. When combined with some advanced switchgear and a local control system, these assets form a seamlessly islandable microgrid that is capable of disconnecting from the grid and providing reliable service to a critical load using all inverter-based technologies. When connected to the grid, the battery is being used to simultaneously shift energy and smooth the output of the solar

FIGURE 3

Community energy storage unit with transformer, Charlotte, NC.

FIGURE 4

McAlpine energy storage system, Charlotte, NC.

installation. But, when a grid outage occurs, the combined microgrid system responds by isolating itself from the main grid and providing seamless back-up power to the critical customer. Thus, an important customer value is realized while also fully using the microgrid assets to benefit the grid when they are available (Figure 4).

3. Conclusions

Significant progress has been made toward developing and demonstrating energy storage technologies that are capable of creating valuable flexibility on the electric grid. With these advanced flexibility tools in hand, utilities and grid operators are now able to focus on how they can be used to create the maximum value on the power system by holistically optimizing energy storage within an interoperable, secure, and autonomous grid ecosystem.

Variable Energy Resources in Island Power Systems

Renewables Integration on Islands 24

Toshiki Bruce Tsuchida
Principal, Cambridge, MA

1. Introduction

Renewable integration has been a hot topic in the electric industry for many years, but most studies have focused on large interconnected systems, such as those in the continental United States and Europe. Small island systems face different and more complex technical challenges when diversifying their generation mixes to include renewable resources with variable and intermittent output. There is a difference between using renewable fuels, such as biogas or refuse, in conventional power systems controlled and dispatched by the system operator and using variable and intermittent sources, such as from wind and solar power, whose output is not fully controllable by the operator. This chapter focuses on the latter, i.e., variable power sources, in island systems with peak loads below a few hundred MW. While the discussion and illustration may center around systems for physical islands located in tropical regions, such as the Caribbean and Hawaii, these systems do not necessarily have to be islands per se. The systems could be in isolated coastal or even inland areas such as rural Alaskan villages (there are over 180 electrically isolated communities in Alaska), military bases, mines, or any other microgrids.

Since the early days of electrification, power generation in island systems has mostly been fueled by oil because of the ease of transportation compared to other fuels, especially coal and natural gas. Exceptions include a few systems where hydropower or geothermal resources are available. Most of these oil-fired generators are reciprocating internal combustion engines (recip-gens) and some systems use combustion turbines and/or steam turbines along with these recip-gens. While the efficiency and reliability of these recip-gens have been steadily improving, the large increase in oil prices experienced lately has rendered the incremental cost of oil-fired power generation prohibitive for island economies.[1] The impact of oil price fluctuation on island systems is large because the transportation for fuel delivery (ships and barges for surface transportation to physical islands, trucks to remote inland locations, or, in extreme cases planes to isolated areas) also depends on oil. The recent oil price increase and the resulting high costs of generation have led to intense social and political pressures on island utilities to lower the cost of power and reduce dependency on imported fossil fuel.

Accompanying the high and volatile fuel costs are environmental concerns. On many islands, over two-thirds of greenhouse gas (GHG) emissions are thought to result from power generation. On a

---

[1]With recent crude oil prices at $100/barrel or higher, generation costs in excess of $200/MWh are observed for systems using fuel oil #6, or heavy fuel oil (HFO). Costs in excess of $400–$500/MWh are observed in systems using fuel oil #2, or light fuel oil (LFO). In comparison, the generation costs from fossil fuel observed in the continental United States range between $20/MWh and $100/MWh. This only accounts for generation; actual customers' utility bills are even higher.

Renewable Energy Integration. http://dx.doi.org/10.1016/B978-0-12-407910-6.00024-7

macro scale, increased GHG emissions contribute to climate change. The immediate impact of warmer temperature associated with climate change is higher electricity demand through increased usage of air conditioners (A/C) that could pull prices up. Climate change is also believed to cause rising sea levels, and increased frequency and severity of tropical storms, which are serious environmental threats to island communities. Fishing, an important industry for many islands, is also impacted significantly by climate change and associated environmental degradations. Islands are generally sensitive to GHG emissions and resulting climate change; though the direct impact on any island of its own GHG emission is negligible, the impact of global climate change on islands raises the awareness of its inhabitants regarding their own contribution to the problem and its solution.

Fortunately, many islands have great potential to exploit renewable resources, such as wind and solar resources. Utility-scale wind power has become recognized as one of the lowest-cost resources in recent years, leading many island communities to start exploring its potential. Solar energy, with the cost of photovoltaic (PV) cells falling precipitately by the day, has also been viewed as a viable option especially for tropical islands that are known for year-round high solar insolation. However, solar thermal systems with storage capability have an economic capacity of 50 MW or more today and are not considered to be a practical option for many smaller islands.

The short-run marginal costs of wind and PV plants are negligible compared to the fuel cost of traditional thermal generators. Avoided fuel costs, especially at today's high oil prices, can more than cover the capital costs of these renewable resources. Increasing generation from these indigenous, though variable, renewable resources to replace generation using imported fossil fuel, then addresses both the power cost and environmental concerns, and has become a popular political agenda. Examples of these political goals include: the Hawaiian Clean Energy Initiative aiming for 40% of delivered energy to come from renewable resources by 2030; the US Virgin Islands aiming to reduce fossil energy consumption by 60% by 2025; Aruba and Saint Lucia aiming at becoming 100% renewable by 2020; and Bermuda's GHG reduction goal that translates to generating 30% of electricity from renewable resources by 2020. While these policies developed by different island communities are not identical, most define successful renewable integration as utilizing the potential of these variable resources with minimum curtailment, resulting in a reduction of fossil fuel usage and a lowering of the overall cost of generation. Unlike the island jurisdictions within the United States where federal tax incentives or grants could support investments, projects on island nations need to be economically viable without subsidies. The high power cost on these islands is helping to make renewable energy economically competitive.

2. Lessons from renewable integration studies for larger interconnected systems

Renewable integration studies for interconnected systems have demonstrated the needs for system flexibility to accommodate the intermittency and variability of these resources. Study recommendations focus on enhanced ancillary services and transmission expansion that helps balancing the system while maximizing the renewable potential. The same ideas apply to island systems but without the various benefits afforded by transmission interconnections.

Interconnected transmission confers benefits of diversification of both supply and demand. A large number of generating resources with varying fuels interconnected across a broader region will minimize risk to a system associated with price volatility of any single fuel. Geographical diversity of renewable generation reduces the variability of these resources as a group. Similarly, loads across an expanded geographical area with diverse usage patterns exhibit lower variability when aggregated. Similar benefits are expected for forecast uncertainty of generation (particularly renewable resources) and loads. And, interconnected systems can share operating reserves with their neighbors. Interconnected transmission, as a combined effect, reduces overall net load fluctuation and deviations from forecasts, resulting in lower operational flexibility needs to balance the system while allowing the system operator to draw the required flexibility from a larger pool of resources.

Systems with no external connections cannot enjoy these transmission benefits. Island systems operate autonomously, and the continuous balancing of the system can only be addressed through ancillary services from internal resources. Therefore, island systems' planning is similar to resiliency planning discussed nowadays for interconnected systems. Statistics for island systems show that the minute-by-minute variation of renewable generation can be larger than that of load. The continuously variable power production from renewable resources becomes a major contributor to the net load variations that create operational challenges for different time frames. Stability issues occur on the milliseconds-to-seconds time frame. Frequency deviations caused by power production variability occur on the seconds-to-minutes time frame. Potential contingencies, such as the loss of generation or load, must be responded to within 10–15 min. Ramping limitations of existing generators could also lead to load-following issues on the minutes-to-hours time frame. When existing generators cannot follow the net load, renewable resources will be curtailed so that net load ramp is maintained within the existing generators' ramping limitations.

Island systems will fully and directly face these challenges caused by the intermittency and variability of renewable resources and the associated forecast uncertainty. Being autonomous, operating reserves are their only means of continuously balancing the system. The limited island resources do not allow for the luxury of unbundling operating reserves by type (e.g., spinning versus nonspinning reserves) like interconnected systems. Island system operators will use whatever resources are available as long as the required reserves can be provided within the required time frame. The scarcity of operating reserves is much more severe in island systems.

3. **Characteristics and challenges for island systems**

The unique characteristics of loads and generation in island systems further complicate the renewable integration challenge. Island loads are generally small with little diversity. The load profiles, especially for those located in tropic areas with mild weather, tend to be smooth and predictable with modest growth. For example, on islands where tourism is the main industry, the A/C load could account for nearly two-thirds of the island load. A/C load does not grow year by year unless the number of residential dwellings and/or hotels to accommodate tourists increase. There is little change in residential load or industrial and commercial load outside of the tourism industry on these islands. Load profiles for military bases and mines are based on the schedule and needs of the corresponding industry and share similar characteristics.

Generators for island systems were designed to serve these small and predictable loads. Many island generators, especially recip-gens fueled by heavy fuel oil (HFO), are inflexible. HFO, with its high viscosity, is known to cause issues such as start-up failures (due to the recip-gen pistons "sticking") and only provide limited ramping rates and operating range.[2] In some islands a large portion of the generators cannot provide regulation. Furthermore, island generators have smaller capacities for both operational and economic reasons. First, historically, smaller generators have been easier to transport and install. Generators along with the heavy construction machinery required for installation have been imported. The ease and associated cost of transportation and installation have been key factors for generator selection. Second, the small island loads have grown only incrementally, so only small units have been installed as load has grown. Third, system planners limit generator capacities to avoid exceeding the largest contingency in the system. Should a generator become the largest contingency, it could increase the contingency reserve requirements and lead to higher operating reserve needs. Therefore, many islands seeking proposals for renewable resources limit their renewable capacity.

As in any system, long-term planning for renewable integration on islands needs to account for short-term operational limitations determined by their specific load and generation characteristics. The bottleneck is usually the existing generation fleet's operational limitations unless additional means that provide or aid operational flexibility are being considered. Flexibility requirements of island systems are typically higher than those of interconnected systems.[3] For example, regulating reserves on the Maui system are driven by the wind plant output, and the quantity required varies according to wind conditions. Some islands tend to lose wind when it rains, and therefore the reserve requirements for renewables are even higher, with rain becoming a large contingency. Renewable resources that have nearly no inertia replacing thermal generators with inertia exacerbates fluctuations in frequency. As such, some island systems require developers of variable resources to also secure resources that provide ancillary services, such as batteries. From an operational perspective less renewable capacity is easier to integrate. However, the purpose of integrating renewable resources is to lower power costs by displacing generation from fossil fuel. Renewable generation quantity varies, even over the long term. Wind generation for a given island can differ between a strong and a weak wind year by over 60%. If renewables are to put a reliable cap on annual costs of generation, planners should consider a weak wind year and err on the side of building renewables in excess.

Combined with this operation versus planning dilemma is the difficulty of optimizing capacity and installation cost. With small loads, a single plant can account for a very large portion of an island resource mix. Installing ten 3 MW wind turbines on a system with a peak load of 80 MW will raise the wind penetration level (the amount of energy being served by renewable resources) from 0% to over 30% within the few months of installation.[4] This penetration level would be higher than the overall wind penetration of any interconnected system, but smaller installations will increase the per-MW cost dramatically. Mobilization costs of the heavy construction equipment and shipping costs of the turbines themselves significantly increase the per-unit capital costs required to install wind plants on islands as the number of units installed declines. Therefore, any economically sized wind plant could easily lead to high wind penetration with the variability and forecast uncertainty impact on the overall system being acute. Island system planners face multidimensional difficulties in determining the optimal renewable resource capacity.

---

[2]Gaseous fuels provides the greatest flexibility, and solid fuels provide the least flexibility for furnace/combustion control. Out of all liquid fuels, HFO, the heaviest commercial fuel obtained from crude oil, is the closest to solid fuel. HFO is the remnants of the crude oil refining process and requires preheating to be used as liquid fuel.

[3]Islands also require higher capacity reserve margins (i.e., excess generation capacity over peak load). For example, the Hawaiian systems require about twice as much margin when compared to the continental United States.

[4]This example assumes a 50% annual capacity factor of wind (average hourly output of 15 MW), and the average load to be 60% of the peak load (average hourly load of 48 MW), resulting in 31.25% of the hourly load being served by wind.

Issues unique to distribution networks further magnify this challenge. Island grids are smaller in scale than interconnected systems and operate at lower voltage levels, with many rated at 50 kV or below. In interconnected systems, this voltage level is considered part of the distribution network rather than the bulk transmission system. Distribution networks have distinct issues including reverse feeder flows, voltage fluctuation, and feeder imbalances caused by single phase PVs being connected to the three-phase system. On island systems where the distinction between transmission and distribution is not as clear as in larger interconnected systems, these distribution network issues are handled through ancillary services provided mainly using thermal generators.

In summary, long-term planning must consider the short-term operational limitations, that is, "How much flexibility does the system have to accommodate the generation from variable resources and how will this flexibility be provided through ancillary services?" The flipside of this question is, "How much additional flexibility can be added to the system and how?" System flexibility can come from three sources: existing generation; existing load; or new technology additions to the system, including storage technologies.

4. Ongoing efforts for island renewable integration

To increase renewable penetration and address policy agenda and goals, many island system operators have been studying innovative technology options, while recognizing the operational and planning challenges. Addressing operational flexibility, which impacts both operation and planning, has generally been the greatest challenge. On islands, operational flexibility, traditionally secured through operating reserves, has a direct correlation to the reduction in renewable curtailment and resulting economic benefits.

Figures 1–3 illustrate the impact of enhancing operational flexibility on an island system integrating 30% renewables, where the combined nameplate capacity of renewables (roughly 85% wind and 15% PV) is approximately two-thirds of the average island load. Figure 1 shows the hourly generation for an average day from traditional resources (dark gray area), renewable resources (checkered area), curtailment of renewables (light gray area), and the load profile of an average day (solid black line), assuming no operational flexibility enhancements (Reference Point). Details of wind and solar curtailment are shown within the lower dark gray area using the right hand axis. Figure 2 shows the same with operational flexibility enhancements provided using batteries/flywheels or other technologies (Enhanced Point), resulting in lower curtailments. Figure 3 compares the Reference and Enhanced Points and shows the economic benefits (reduction in short-run production cost from fuel savings) brought by reduction in renewable curtailments. These figures highlight the importance and significant benefits of increasing operational flexibility.

Operating flexibility can be secured by improving and fine-tuning existing generators and/or their operational procedures, using nontraditional methods of controlling or increasing loads, implementing new technologies, or reducing the need for operational flexibility itself.

4.1 Improving existing generators

Various island systems have sought ways to increase the flexibility of their existing generators with a focus on improving dynamic response through increasing ramp rates or unit responsiveness. One example of increasing responsiveness is replacing the generator exciter with permanent magnet

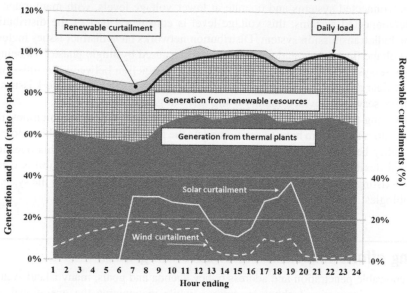

FIGURE 1

Impact of operational flexibility—Reference Point.

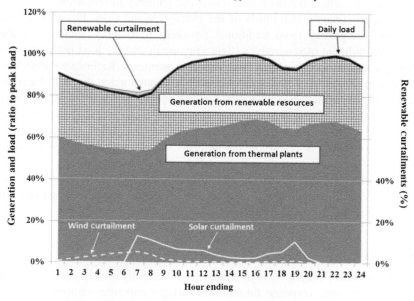

FIGURE 2

Impact of operational flexibility—Enhanced Point.

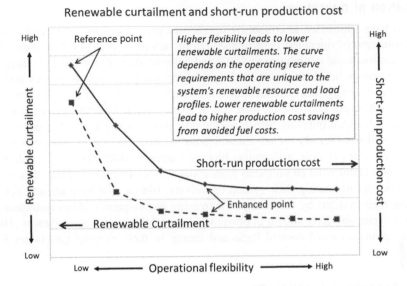

FIGURE 3

Renewable curtailment and short-run production cost.

exciters. Unlike exciters charged as parasitic load, permanent magnet exciters have improved generator response because they are not impacted by the system fluctuation. Means of widening the operational range of thermal generators by increasing maximum output or reducing minimum generation (min-gen) have also been explored. Lowering min-gen levels widens the generators' operational range and also reduces curtailment. Min-gen issues arise when the output of thermal units needed for providing energy or ancillary services cannot be backed down and is higher than net load. Under these circumstances, renewable resources are curtailed until net load becomes higher than the min-gen level. In some cases lower min-gen levels will even reduce the need for contingency reserves.

4.2 Diversifying fuel

Other islands have been exploring fuel switching, in particular, options that may lead to increased operational flexibility of thermal generators. These include introduction of higher grade fuel—the economic benefits of reduced renewable curtailment offsets the increased fuel costs. However, it has been observed that expanding the operational range or response rates by upgrading the fuel can lead to operational problems caused by difficulty in harmonizing the controls, even when using otherwise identical recip-gens. Nonconventional fuel options, such as biogas produced through a waste-to-energy process, or alternative fuel types, such as colloidal coal water mixture, have also been considered. Generally, these unconventional fuel types with lower carbon contents contribute to reducing fuel-oil consumption and GHG emission. However, some may encounter operational issues caused by the reduced response of recip-gens or by reduction in operating range caused by a new fuel's characteristics. The trade-off between the benefits of fuel diversity and GHG reduction, and the adverse impact of reduced operational flexibility, which hinders renewable integration, is still being studied.

4.3 Automation of operation

Several islands have succeeded in enhancing operational procedures to accommodate increased renewables. Automation of operation is recommended in various studies as a first step. The small and predictable island loads did not require sophisticated automated operation until the fluctuating output of renewable resources accelerated net load ramping to the point that operators could no longer respond manually. Adding to net load fluctuation is forecast uncertainty of these intermittent resources that could be much larger and volatile than that of load, further requiring automation. Automation allows for shorter dispatch intervals, which reduce operating reserve needs. Various studies indicate that automation will significantly reduce renewable curtailment, and the investment could be paid off within a few years. For example, a Hawaiian Electric Company (HECO) study concluded that automation could lower Oahu wind curtailments from about 15% to 4%. Recognizing this, several islands are in the process of automating their system operations. Islands that have already automated have been improving their systems by, for example, increasing the frequency of the Automatic Generation Control (AGC) cycle, and have shown improvements in stability and responsiveness. HECO's work included developing advanced control logic and tuning in their existing generators for improved response.

4.4 Demand response technology

Options to improve flexibility by controlling the load using advanced demand response technologies are also being explored. Thermal inertia is one of the simplest means of storing and moving energy. On tropical islands, this can be applied to controlling the on-and-off cycling of A/C loads or refrigeration facilities. Utility-controlled ice storage technology, which produces ice during the nighttime and uses it during the daytime for A/C, is another technology option of this type. When planned and implemented carefully, these sophisticated demand response technologies can provide operating reserves. Because the main challenge for renewable integration is the instantaneous balancing of generation and load, any load that can be controlled by the system operator becomes valuable. In the long term, policies should address how system operators can influence consumer demand for electricity. Visualizing real-time usage for end users, introducing real-time rates, and implementing automated real-time demand response may be a preferred option.[5] They are similar to the smart grid development.

4.5 Increasing load

There have also been attempts to increase overall island load to accommodate renewables. Higher load does not typically lower consumption of fossil fuel or GHG emissions. However, increasing loads can reduce the number of hours with min-gen issues and renewable curtailment. Utilizing this otherwise-curtailed renewable generation will lower the average cost of electricity, and higher load will increase the sales revenues for the electric utilities. One source of increased load whose cost may be worth incurring is the charging of electric vehicles (EV) discussed later.

Reducing load served by the system operator will complicate renewable integration but also require less generation and reduce both fuel consumption and GHG emissions. Load reduction will increase

---

[5]Limited variation of the marginal cost of generation leads to difficulty in tariff design, especially real-time rates or time-of-usage rates.

the number of hours with min-gen issues and may result in reduced operational flexibility because of fewer thermal generators being committed. This could result in increased renewable curtailments.

One commonly observed issue arises when policy dictates installation of generators that are not controlled by the system operator, either behind-the-fence generators or those of independent power producers.[6] These generators reduce the load served by the system operator and lower the utility's energy sales revenues. However, these generators also require greater operational flexibility because the system operator, as the last resort, needs to provide backup in case they become unavailable. The utility, receiving lower revenues than if it owned equivalent generation, will nonetheless assume greater responsibility and incur higher associated costs. Operating reserves, in particular, will become increasingly scarce relative to demand. Therefore, policies that lead to reducing load served by the system operator, including installing generators outside the system operator's control, can potentially increase the cost of renewable integration. Policies incenting rooftop solar PV installation will have similar impacts.[7] Installing rooftop water heaters that do not require backup from the system operator will not have these negative impacts. These lessons can improve the concepts of distributed generation currently discussed in the context of interconnected systems.

4.6 Storage and other new technologies

In conjunction with these various methods that utilize existing resources, many island systems have started looking into storage technologies. In general, storage can provide the means for power smoothing in the microsecond to second time frame, regulation in the seconds to minutes time frame, capacity for contingency reserves that typically is needed within 10–15 min, and load shifting across hours or days. Application within different time frames will require different storage technologies. The specific necessity for and application of storage technology further varies by system. With the advancement of inverters used in wind and PV technologies already providing power smoothing, island system operators tend to focus on storage technologies that address the need for regulation and provide capacity for contingency reserves (short-term storage) and allow shifting of load over several hours (long-term storage).

Short-term storage, such as batteries and flywheels, are used mainly to provide regulation and not energy. They are used to manage the intermittency and the rapid ramp changes of renewable energy sources that can cause deviations in frequency. Alternative technologies to short-term storage include dynamic resistors for system frequency control. Advanced wind turbine technologies involving inertia and frequency control are also being used. Inertial response from the wind generators can significantly improve the frequency response and help to avert underfrequency load shedding. Newer wind plants can provide fast response at a low cost for overfrequency events. These options are usually limited to renewable resources that are centrally controlled by the system operator. Short-term storage also buys enough time for other generators (governor response) to respond to any system changes while allowing the recip-gens to operate within a steady and optimal range.

---

[6]The effectiveness of competition, given the high capital costs, is unclear. If policies encourage installation of generators outside the system operator's control, the same policies should address the requirements, obligations, and compensation for the last resort of power.

[7]Rooftop PV policies raise issues of the rich being subsidized by the poor. Usually rooftop PVs are installed by the affluent. The associated system balancing costs to accommodate the rooftop PVs will be shared by all customers, including those not wealthy enough to install PVs.

Long-term storage shifts the load over several hours or days from periods of high generation costs to hours with surplus energy from low cost resources. Fixed schedule ice storage is a technology option available for islands. By reducing peak load and increasing minimum load, it creates a smoother load profile that reduces the cycling of the existing generators.

When installed alone, options for short-term storage provide larger economic benefits than long-term storage. For example, a study for Aruba showed a 5 MW short-term storage to have approximately 60% more production cost savings (as a result of reduced wind curtailment) compared to a 5 MW long-term storage. A Maui study showed that it required a 100 MWh battery system for long-term storage to yield results similar to using a 10 MW battery system for reserves. Short-term storage options lead to higher benefits for two reasons. First, short-term storage better addresses renewable resource variability/intermittency related issues—from inertia/stability and regulation to ramp issues. In contrast, long-term storage addresses more of the load following and demand cycle issues. Because renewable resources curtailment is mostly caused by the net load variability exceeding the operational flexibility of the system, short-term storage options that address this variability results in lower renewable curtailments and higher economic benefits. Secondly, most islands rely on a limited number of recip-gens that typically use the same fuel and have similar heat rates. The marginal costs of generation during peak hours and off-peak hours do not differ significantly. Therefore long-term storage options result in lower estimated economic benefits. However, it is important to recognize the operational value not captured by such estimates, which includes reductions in costs associated with cycling operation of thermal generators. The impact of cycling operation, which includes increased wear and tear, increased emissions, and higher heat rates, is oftentimes not accurately represented in estimates of short-term economic benefits. Also, once regulation requirements are met by short-term storage options, incremental long-term storage options provide larger economic benefits than does additional short-term storage options.

Utilizing EV, including plug-in hybrid cars and electric buses, is a technology option that is currently being favored in some islands. Power generation facilities have much higher fuel efficiencies than automobile engines. Therefore, EVs simultaneously reduce the usage of petroleum-based transportation fuel and GHG emissions even after accounting for the battery-charging cycle losses. EVs can also provide operating reserves while charging, contributing to increased operational flexibility of the system. The charging of these EVs, although still limited in capacity today, also increases electric load. Therefore, it has the further effects of reducing renewable curtailment and average generation cost while increasing revenues for utilities.

4.7 Role of policy

In tandem with the innovative engineering, policy plays a significant role. Most importantly, policies need to accommodate the scarcity of island nations' financial options. The small magnitude of island loads combined with the acute challenges discussed earlier tends to result in only marginally attractive investment opportunities. For example, unlike interconnected systems with access to broader markets, renewable resource investment on islands with limited load growth may have to compete against existing thermal generation capacity that has already paid off the capital costs. Therefore, investments, including overseas funds, need to be attracted through policies.

However, policies need to reflect holistic views and be adaptable over time because renewable energy projects and technology have lifetimes of 20 years or longer. In achieving a certain threshold of renewable energy penetration, the diminishing returns to the projects need to be considered. That is, the early projects will be economically viable but the later projects may not be, and as a result the renewable penetration level goal may not be met. Short-term policies that rush investments while the economics are favorable will end up in similar results.

Beyond attracting investment capital, policies also influence the choice of technology and therefore should be focused on its broader objectives rather than on the means of achieving them. For example, the combined benefit of EVs (reduction in fuel oil consumption and GHG emission for both generation and transportation) may not be realized if the policy is focused exclusively on adding renewable resources, such as rooftop PVs. Promoting EVs will require investment in infrastructure (charging stations, etc.) and is only possible with political support.

5. Conclusion

The key to successful renewable integration for islands lies in the continuous balancing of the system at an intense level, compared to interconnected systems. Islands are approaching this challenge by exploring many innovative technology options with policy support. Various studies have demonstrated that with these innovative technology options, renewable integration at a penetration level of 30% or even higher is quite possible, and several islands are leading this path. The challenges encountered by island systems when addressing renewable integration include a variety of issues that are treated separately from renewables in interconnected grids, ranging from smart grid, distributed generation, climate policy, and system resilience, to storage technologies. For the first time in history, island systems are being made a test bed for innovative technologies that can potentially shape the future of large interconnected systems.

Intentional Islanding of Distribution Network Operation with Mini Hydrogeneration

25

Glauco Nery Taranto, Tatiana M.L. Assis
Federal University of Rio de Janeiro/COPPE, Brazil

1. Introduction

The abundant presence of large river basins made Brazil a worldwide powerhouse of hydroelectricity. According to official numbers, the installed capacity of hydrogeneration in the country surpasses 70%, which, in a favorable precipitation year, can account for more than 90% of the country's annual electric energy needs. This natural characteristic of the country has been exploited in the past five decades, and much more is yet to come, because the Amazon River basin is just starting its generation enterprises in the far northern part of the country. The highly industrialized southern part, although largely hydro-based, faces an exhaustion of its hydrocapacity, and, in addition to waiting for the distant energy supply from the Amazon region, is looking for alternative renewable sources to maintain its long tradition of clean energy production. One such alternative is the generation coming from small hydroplants. In Brazil, because of its hilly terrain in large parts of the country, and government incentives, it is becoming a common scenario that the existence of mini hydroplants (of few megawatts) was able to supply small towns in rural areas.

This chapter focuses on a real case study of a mini hydro power plant connected to a 25-kV rural distribution feeder located at Rio de Janeiro in the southeastern part of Brazil.

2. Case study

2.1 System description

The system comprises two 6.6-MVA hydrogenerators, powered by horizontally mounted Francis turbines, connected to a 30-mile, 25-kV feeder with a peak load of approximately 7.5 MW. Figure 1 shows the one-line diagram of the feeder, where the location of the mini hydroplant is near its end.

2.2 Modeling aspects

Load-flow and transient analyses were performed in the case study. Both analyses considered three-phase phasor models for the network, the loads, the transformers, and the generators. The three-phase modeling is done in phase components, as explained in [1]. The mini hydrogenerators are

Renewable Energy Integration. http://dx.doi.org/10.1016/B978-0-12-407910-6.00025-9

FIGURE 1

Megawatt-scale microgrid.

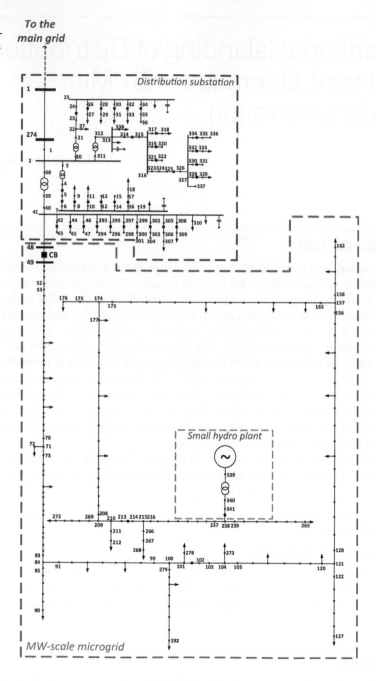

modeled as synchronous machines, and the hydroturbines are linearly modeled with typical non-minimum phase characteristics. The excitation system is represented with standard rotating brushless models.

3. Intentional islanding

With distributed generation (DG) increasing, the possibility of islanded operation becomes a reality. However, many distribution utilities want to avoid islanding because of personal safety concerns and loss of voltage and frequency control. Moreover, changes in the protection systems and a special scheme for safe reconnection to the main grid may be required.

Although most electric utilities do not look favorably on islanding, the pressure from the society and the government to improve reliability indices is forcing them to accept and to adapt their systems to allow islanded operation. In Brazil, intentional islanding is not yet mandatory, but it can be done if an agreement between the utility and the generation owner is established.

To put into practice intentional islanding, four main issues have to be addressed. The first one is the change in the control modes and in the protection systems that may be needed. The second one is the analysis of the microgrid formation, where the dynamic behavior of voltages and frequency has to be studied. The third aspect is the microgrid autonomous operation (i.e., its ability to maintain voltage and frequency profile within acceptable range). Finally, the reconnection of the microgrid to the main system has to be analyzed. Those aspects are evaluated in the next sections for the system described in Section 2.

3.1 Islanding detection and change in the control modes

When a problem occurs at the utility grid and the coupling circuit breaker is opened, the island is created. At this moment, two different strategies may be adopted. If the system is not prepared for isolated operation (i.e., the islanding is unintentional), all DGs in the microgrid must be shut down as quickly as possible. In this case, the islanding must be identifiable by the dispersed generators. If a communication system is available, such identification is simple because the circuit breaker status can be constantly monitored by the generators [2]. On the other hand, if no communication system exists, some islanding detection scheme must be installed such that the islanding can be recognized [2–4].

In the second strategy, the system is assumed to be prepared for intentional islanding. As in unintentional islanding, the separation from the utility grid must be identified by DGs, but for different reasons. Now, the islanding recognition is important such that the voltage and frequency control mode modifications and the necessary protection system adjustments can be performed. Again, if a communication system is available, the islanding recognition can be done simply by the circuit breaker status monitoring. If not, some islanding detection scheme must exist. A diagram with both strategies is illustrated in Figure 2.

Once the islanding is correctly detected, the necessary actions should be taken to guarantee that the microgrid will survive, and that it will be able to maintain an autonomous operation. In some cases, load or generation shedding may be required to preserve the voltage and frequency at the microgrid buses. In addition, control modes have to be changed because the voltage and the

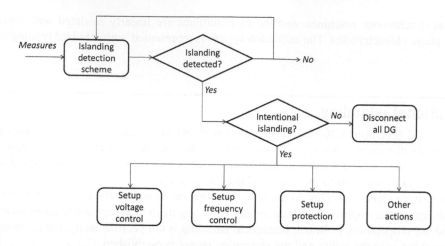

FIGURE 2

Intentional and unintentional islanding strategies.

frequency will not be controlled by the utility anymore. Typically, when those generators are connected to the main system, they do not participate in the voltage and frequency control, operating with constant power and constant power factor. Once islanded, they must provide voltage and frequency regulation.

The operation in isolated mode may also require changes in the protection schemes and their set points. For example, when the islanding is established, the short-circuit levels may vary drastically. Furthermore, other functions, such as directional and distance protection, may be necessary. So, the islanding detection will enable such modifications that can be done manually or in an automatic way. Depending on the balancing condition at the islanding formation, additional actions, such as load or generation shedding and capacitor switching, may be necessary. Evidently, as indicated in Figure 2, if the islanding is unintentional, the only action that should be taken is the disconnection of all DGs.

In Brazil, most electric utilities do not adopt islanding as a regular procedure. However, as indicated previously, because of the necessity to improve reliability indices, intentional islanding studies and practice are receiving more attention. In the study case described in Section 2, no communication system is available, so the islanding detection is based on observation of locally measured variables.

Islanding detection schemes usually used to shut down distributed generators when intentional islanding is not allowed are now used to detect system separation and make the necessary adjustments in the system so the isolated grid can operate in a safe way. Local islanding detection techniques can be divided into active and passive. Active techniques are based on small-disturbance application by the generator in the system. The detection is accomplished through observation of system impedance, short-circuit level, and reactive power error [2–4]. In general, active techniques are more effective and robust than the passive ones. However, because of lower cost and simplicity, passive techniques are more commonly used.

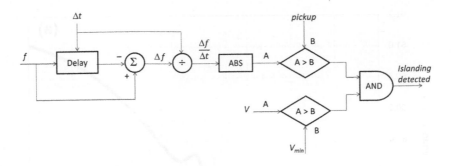

FIGURE 3

Islanding detection scheme using rate of change of frequency relay.

Two passive detection techniques were studied. The first one is based on the rate of change of frequency (ROCOF) relay, and the second one is known as vector surge relay. Both schemes have been tested and have presented similar performances. The main deficiency of these techniques is the existence of a nondetection zone when the generation and the load in the microgrid are balanced. In addition, undesirable DG tripping as the result of other network events may also occur. To solve those problems, more advanced techniques have been investigated [5].

To illustrate the detection techniques, Figure 3 presents the ROCOF relay implemented for simulation purposes. The local frequency is constantly monitored. A delay block is used to compute the frequency variation. Thus, the frequency rate of change may be computed, and it is compared with the pickup value. When the islanding occurs, a high frequency variation is expected, which will obviously depend on the microgrid net power flow. Moreover, to avoid incorrect islanding detection during short circuits, the ROCOF relay may have a voltage-restrained control. With this feature, the low voltage usually observed during faults will block the ROCOF relay operation.

Figure 4(a) shows the frequency for the islanding of the microgrid described in the previous section. The coupling circuit breaker is opened at $t = 1$ s. In this scenario, the total power generated by the mini hydroplant is 9 MW. On the other hand, the total load in the microgrid is 6.2 MW. Consequently, when the islanding occurs, the frequency increases because the mechanical power does not change immediately. The rate of change of frequency is shown in Figure 4(b). In this case, the rate is computed in a sliding window of 200 ms, which represents 12 cycles of 60 Hz. The pickup value is set to 0.6 Hz/s, and the islanding is easily detected, as indicated in Figure 4(b).

Figure 5 presents the system performance for a short-circuit simulation. One can see the frequency increasing (Figure 5(a)) and the pickup value of the ROCOF relay being exceeded (Figure 5(b)). This behavior results in an incorrect islanding detection. However, the voltage restrain control blocks the relay operation, as illustrated in Figure 6.

Figure 6(a) shows the voltage seen by the ROCOF relay during the islanding event. The voltage increases because the microgrid separation represents a loss of load from the mini hydro point of view. Instead, with the short-circuit occurrence, the voltage decreases, blocking the ROCOF relay operation. Figure 6 shows the minimal voltage value used to block the relay, which is equal to 0.8 pu. This value should be carefully set, because a short circuit far from the relay location may produce no significant decrease to properly block its operation.

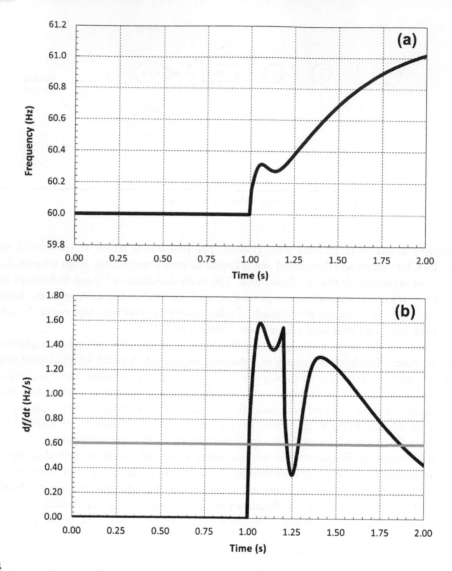

FIGURE 4

Frequency (a) and rate of frequency change (b) during islanding.

3.2 Islanding formation

This section presents results related to the island formation for the mini hydro studied. Table 1 shows the load and generation profiles analyzed. Those profiles were combined, producing nine different scenarios, as presented in Table 2. For each scenario, a short circuit at the utility bus is simulated, followed by the coupling circuit breaker opening after 100 ms. The islanding is considered to be

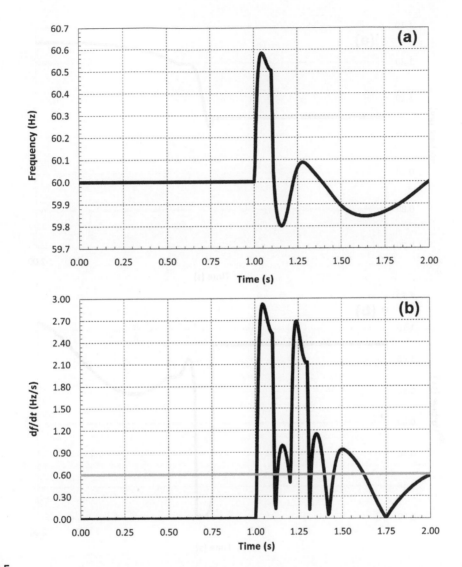

FIGURE 5

Frequency (a) and rate of frequency change (b) during a short-circuit.

successful if the minimum criteria of voltage and frequency performances are accomplished. Those criteria are established in the Brazilian grid codes published by the National Agency of Electrical Energy [6].

Table 2 indicates the active and reactive power flow from the utility grid to the microgrid for each scenario. The island formation result is also shown, stating when it was or was not successful. All scenarios succeeded, which means that both voltage and frequency requirements were observed.

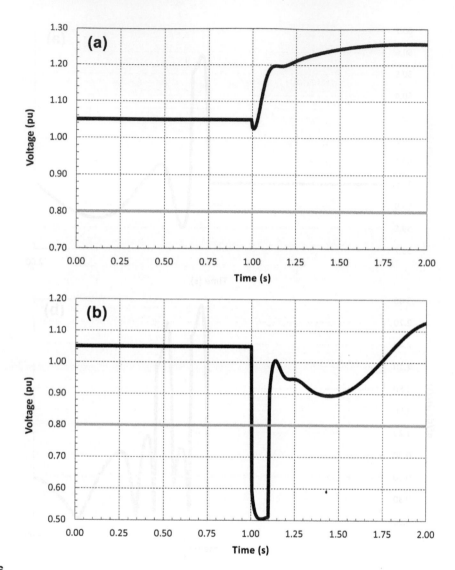

FIGURE 6

Voltage behavior during islanding (a) and a short-circuit (b).

Table 1 Microgrid Load and Generation Profiles

Load		Generation	
Heavy	9.3 MW	High	9 MW
Medium	6.2 MW	Intermediate	6 MW
Light	3.1 MW	Low	3 MW

Table 2 Microgrid Operating Scenarios

No.	Load	Generation	Flow from the Utility		Island Formation
			P (MW)	Q (Mvar)	
1	Light	Low	0.7	−0.2	Successful
2	Light	Intermediate	−2.0	1.2	Successful
3	Light	High	−4.4	3.0	Successful
4	Medium	Low	3.4	−1.5	Successful
5	Medium	Intermediate	0.4	0.2	Successful
6	Medium	High	−2.1	1.3	Successful
7	Heavy	Low	6.8	−0.6	Successful[a]
8	Heavy	Intermediate	3.6	−1.0	Successful[a]
9	Heavy	High	0.8	0.3	Successful[a]

[a]*Requires additional actions to reach steady-state voltage control.*

However, heavy load scenarios need additional voltage control actions after islanding. Those actions may include capacitor switching or the increase of a mini hydrovoltage reference. Both actions are required to maintain the steady-state voltage profile within acceptable limits. The capacitor switching and the change in the voltage reference may be either automatic or manual, depending on whether there is a communication link available.

Figure 7 illustrates the voltage performance for scenario 5. When the short circuit is applied at $t = 0.1$ s, the voltage decreases, as shown in Figure 7(b). After 100 ms, the protection system acts and the microgrid is separated from the main grid. Then, a transient overvoltage is observed, and after a few cycles, it stays within the acceptable steady-state limits.

Frequency performance is presented in Figure 8 for the same scenario. The acceptable steady-state limits are also shown. The transient oscillations are not cause for concern because the mini hydro-governors can readily control the frequency. When the short circuit is applied at $t = 0.1$ s, the frequency declines. The decrease is because of the increase in electrical power at the mini hydro, which occurs because of the high short-circuit current associated with the low X/R relationship observed in such a rural feeder.

It is important to observe that when the microgrid is connected to the utility grid, the mini hydrogenerators operate with constant power control. So, once the island is formed and detected, their control is switched to islanded mode, in which an isochronous operation applies, because only one plant supplies the microgrid. In the case of voltage control, no switching is necessary because the generators already operate in a voltage control mode.

The analyses developed for the actual microgrid have indicated that the islanding success is strongly dependent on the power flow through the coupling circuit breaker at the islanding formation. As a result, if this power flow can be measured and made available to the mini hydrosite, the islanding success may be predicted and the correct action can be taken. However, if no information about the microgrid net power flow exists, the decision has to be taken based on the mini hydrogeneration and the estimated load, which can be established as a function of the time. Also, a scheme of load/generation shedding may be implemented to guarantee the microgrid survival in all probable scenarios.

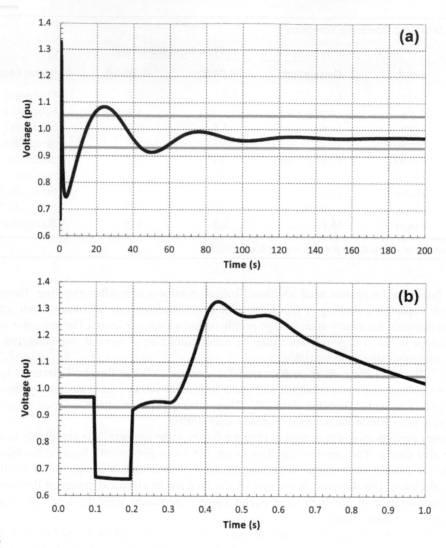

FIGURE 7

Voltage during islanding: Scenario #5 (a) and the detail of initial voltage drop (b).

For the cases of successful islanding, the autonomous operation of the microgrid must be evaluated. This topic is addressed in the next section.

3.3 Microgrid autonomous operation

The microgrid autonomous operation analyses have considered the occurrence of short circuits and usual load variation as well. Three- and single-phase short circuits at all microgrid buses were

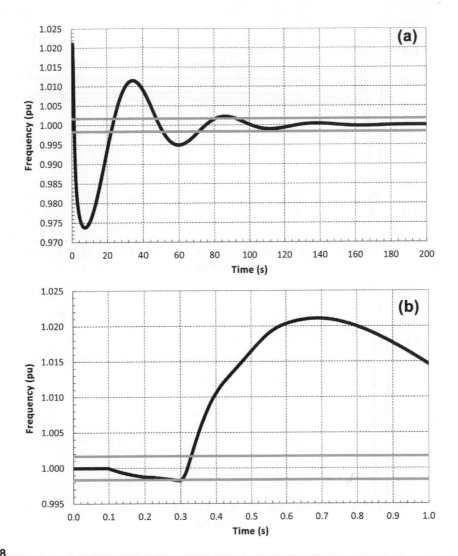

FIGURE 8

Frequency during islanding: Scenario #5 (a) and the detail of initial frequency drop (b).

simulated. The mini hydro has responded well for all of them, considering the presence of protection devices that can properly isolate the fault. However, in the actual network, the quantity of such devices is limited and they do not cover all points. Consequently, the occurrence of faults at some locations would result in the mini hydro shutdown because its protection system would be responsible to isolate the problem.

Figure 9(a) shows the positive sequence voltages at different buses of the microgrid, considering a single-phase fault. The system is stable, and the voltages recover after fault clearing. In Figure 9(b),

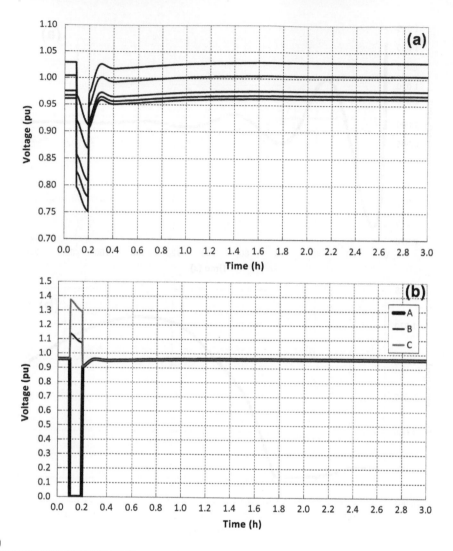

FIGURE 9

Positive sequence voltages during a single-phase short-circuit (a) and phase voltages at the faulted bus (b).

one can see the voltage in each phase at the short-circuited bus. The phase voltage at phase "A" goes to 0 during the fault. On the other hand, there is an overvoltage in the healthy phases. The overvoltage level is associated with the grounding level of the system.

The load variation study has considered the load profile over 24 h. Figure 10(a) presents the load curve for the entire microgrid. The correspondent voltage profile is shown in Figure 10(b). To guarantee that the voltages will be within acceptable limits, some capacitors must be switched over during the day. The active and reactive power generated by the mini hydro is presented in Figure 11.

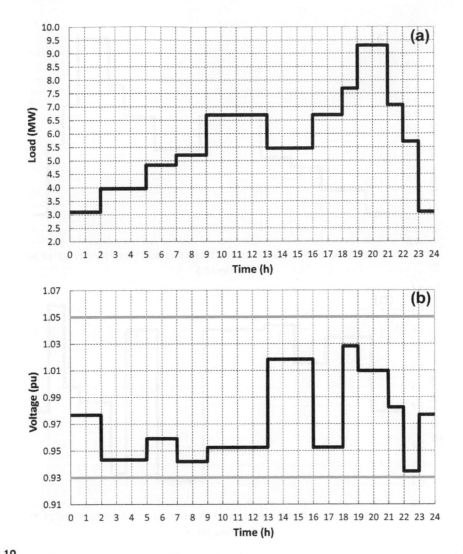

FIGURE 10

Load (a) and voltage (b) curve over 24 hours.

3.4 Reconnection

When the problem that has caused the islanding is solved, the microgrid can be reconnected to the main system. Because the analyzed microgrid is fed by mini hydrogenerators that operate with synchronous machines, the reconnection has to be performed carefully to avoid excessive torsional effort.

When the coupling circuit breaker is closed, transient power oscillations will naturally appear, which may reduce the machine's expected life time. The quantification of the machine damage, along

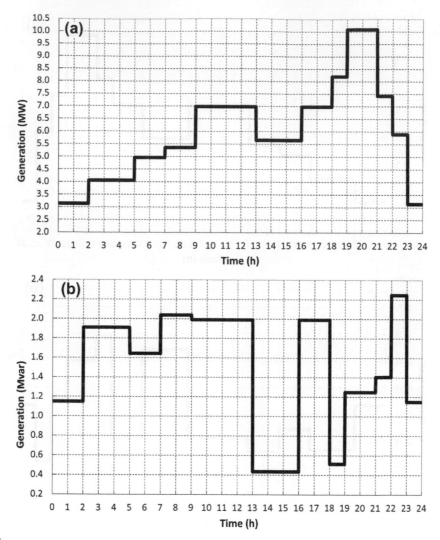

FIGURE 11

Active (a) and reactive (b) power generated over 24 hours.

with switching events, can be evaluated through the behavior of electric power, as proposed in [7]. The criterion, originally established for steam turbine systems, states that, after switching, the maximum acceptable change in the electric power is ±50% of the generator MVA capacity. If the electric power stays within these limits, the loss of life associated with the switching event can be acceptable.

To illustrate the importance of controlled reconnection, two simulations were performed considering the medium load condition. In the first one, the circuit breaker is randomly closed (forced reconnection). In the second case, the reconnection is supervised by a synchronism check relay [8] and safe reconnection conditions are achieved after the application of a 10% step in the generator's voltage reference. The following figures compare the results obtained in each situation.

Figure 12 presents the voltage magnitudes at both sides of the coupling circuit breaker. When the forced reconnection is considered (Figure 12(a)), the breaker is closed at $t = 1$ s. On the other hand, when the synchronism check relay is used (Figure 12(b)), the reconnection occurs at $t = 4.3$ s, after the step application in the reference voltage. In both cases, there is no significant different between voltage magnitudes.

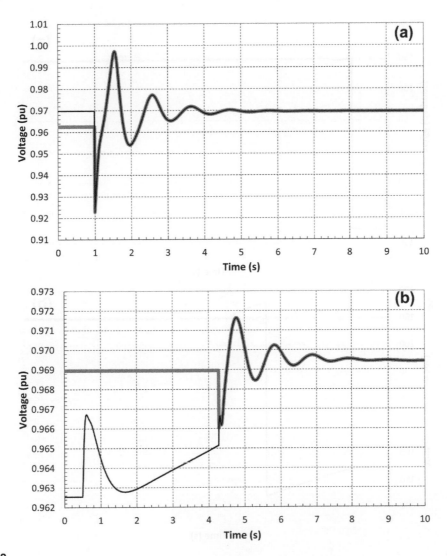

FIGURE 12

Voltage magnitudes evaluation during forced (a) and controlled (b) reconnection.

Figure 13 presents the voltage angles at both sides of the coupling circuit breaker. When the forced reconnection is considered (Figure 13(a)), the breaker is closed when the phase shift is 84°. Instead, when the synchronism check relay is used (Figure 13(b)), the breaker is closed when the phase shift is approximately 9°. The smaller phase displacement was achieved after the application of the voltage reference step.

Figure 14 shows the electrical power behavior with the torsional criterion. When the forced reconnection is considered (Figure 14(a)), the criterion is violated. On the other hand, when the

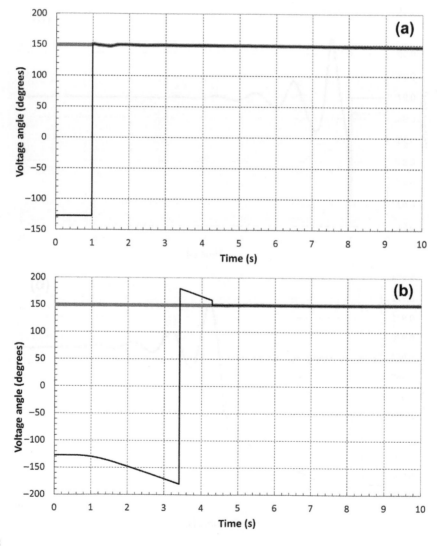

FIGURE 13

Voltage angles evaluation during forced (a) and controlled (b) reconnection.

synchronism check relay is used (Figure 14(b)), the reconnection is smooth and the mechanical stress in the generators is drastically reduced.

In the simulation presented herein, a manual action was required to reduce the mechanical stress at the mini hydrogenerators. However, an automatic reconnection scheme may be used, as the one proposed in [9]. The method allows soft resynchronization, but remote sensing of voltage and frequency is required. The remote measurements are used to produce additional signals for both voltage and frequency regulators of the mini hydro.

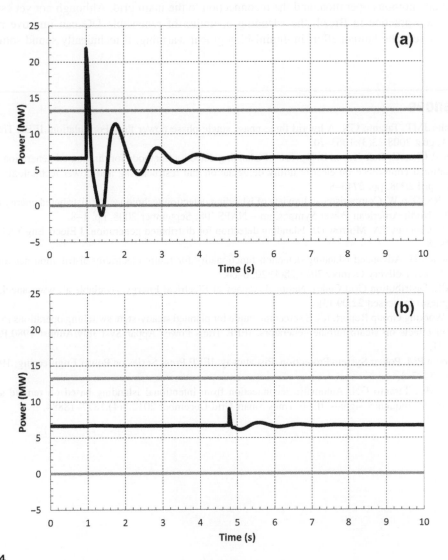

FIGURE 14

Torsional effort evaluation during forced (a) and controlled (b) reconnection.

4. Conclusions

This case study showed an intentional islanding operation of a mini hydro power plant connected to a medium-voltage distribution feeder in an actual rural area in the southeastern part of Brazil. The investigations presented herein highlighted important stages for a successful intentional islanding operation. Those were the detection of the islanding, the necessary change in the control modes of voltage and frequency regulation, the islanding formation with load/generation balancing, the islanding autonomous operation, and the reconnection to the main grid. Although not yet established as a common practice in Brazil, the islanding operation of renewable DG can improve reliability indices and help the human effort in diminishing global warming, if technically sound solutions are applied.

References

[1] Marinho JMT, Taranto GN. A hybrid three-phase single-phase power flow Formulation. IEEE Trans Power Syst August 2008;23(3):1063–70.

[2] Mahat P, Chen Z, Bak-Jensen B. Review of islanding detection methods for distributed generation. In: Third international conference on electric utility deregulation and restructuring and power technologies – DRPT 2008; April 2008. pp. 2743–8.

[3] Kunte RS, Gao W. Comparison and review of islanding detection techniques for distributed energy resources. In: 40th North American Power Symposium – NAPS '08; September 2008. pp. 1–8.

[4] Ding X, Crossley PA, Morrow DJ. Islanding detection for distributed generation. J Electr Eng Technol 2007; 2(1):1928.

[5] Laaksonen H. Advanced islanding detection functionality for future electricity distribution networks. IEEE Trans Power Delivery October 2013;28(4):2056–64.

[6] ANEEL, Distribution Grid Codes, National Agency of Electrical Energy. Available at: www.aneel.org.br (in Portuguese) [accessed 22.09.13].

[7] IEEE Working Group Report. IEEE screening guide for planned steady-state switching operations to minimize harmful effects on steam turbine-generators. IEEE Trans Power Appar Syst July/August 1980;PAS-99(4): 1519–21.

[8] Anderson PM. Power System Protection. Piscataway: IEEE Press Series on Power Engineering; 1999, ISBN 0-7803-3427-2.

[9] Assis TML, Taranto GN. Automatic reconnection from intentional islanding based on remote sensing of voltage and frequency signals. IEEE Trans Smart Grid December 2012;3(4):1877–18884.

Solar, Tidal and Wave Energy Integration

CHAPTER

Economic and Reliability Benefits of Large-Scale Solar Plants

26

Udi Helman

Independent consultant, San Francisco, California, USA

1. Introduction

Large-scale solar plants[1] are interconnecting rapidly to electric power systems around the world, driven by renewable energy policies intended to stimulate cost reductions in these technologies and reduce greenhouse gas emissions. Whereas countries such as Germany have achieved high penetration of distributed solar generation, in regions with high solar insolation, such as Spain and the southwestern United States, there has also been rapid expansion in larger solar projects interconnected to the high-voltage transmission network, with installed capacity measured from 2 MW up to several hundred MW.[2]

In the first phase of renewable generation expansion, the primary metric used to evaluate policy effectiveness and solar technology competitiveness was the change in the levelized cost of energy (LCOE). However, as electric power systems around the world achieve renewable penetrations of 10–20% of annual energy and higher, more detailed quantification of economic benefits and "net costs" is becoming more important as a means to rank alternative investments in renewable resources as well as in any other resources needed to ensure operational feasibility and long-term reliability (see, e.g., [1,2]). This chapter is focused on analysis of the economic benefits of such large solar projects, typically measured as the avoided costs of operations and capacity from fossil-fuel generation, and provides a critical survey of the research in the United States and some other countries. A survey of trends in solar technology LCOE is beyond the scope of this chapter, but it can be found in many publications.[3] The chapter first describes the categories of large-scale solar technologies, summarizes the framework for solar valuation, and reviews some of the modeling methods. The results of solar valuation studies to date are then surveyed.

There are two general categories of large-scale solar technologies: solar photovoltaic (PV) and concentrating solar power (CSP). PV is the lower-cost solar resource and the leader in terms of actual and forecast installed capacity globally [3]. Nevertheless, there are many large CSP plants in commercial operation, primarily in Spain and the United States, including several with integrated thermal energy storage (CSP-TES). The literature to date suggests that storage will be increasingly important to continued solar expansion in some regions experiencing high solar penetration, which

---

[1]Large solar projects are also called "utility-scale."
[2]There is no standard size definition of a large or utility-scale solar plant [3]. Rate plants of 2 MW and greater as utility-scale.
[3]For PV and CSP costs, see, e.g., [3,6,29].

could spur the electrical storage market and improve the comparative valuation of CSP-TES.[4] However, other than CSP-TES, the chapter does not consider valuation of other types of storage, whether co-located with large-scale PV or elsewhere on the power system (see, e.g., [4,5]), nor does the chapter explicitly consider the comparative costs and benefits of other resources that can support integration of solar resources, such as flexible conventional generation, demand response, or transmission upgrades. There is also a growing literature on those topics.

2. Technology categories and production characteristics

As noted, solar resources fall into two general categories for purposes of economic valuation: PV and CSP. Within each category, each of these technologies can be further divided into (1) resources that are not on economic dispatch,[5] and (2) resources that are on economic dispatch because of active power management and/or integrated storage, subject to energy limitations and operational constraints. For the purposes of economic valuation, the PV and CSP resources in the first subcategory are modeled as fixed production profiles, typically on a minute time-step or aggregations up to one hour. In contrast, the resources in the second subcategory, which include CSP-TES, will have production profiles that reflect the co-optimization of their energy, ancillary service, and possibly capacity benefits within the power system being modeled. Within these two categories, there are some differences between solar technologies, as discussed next.

Regardless of technology type, the key characteristic of solar energy without storage is that its production is limited to sunlight hours. Common metrics for describing solar production are the capacity factor[6] and annual penetration by energy for particular power systems (as percentage of total load). However, for operational evaluation and capacity valuation, the key metric is the quantity of solar production during the sunlight hours, the proportion of load during those hours, and the resulting "net load" shapes, including effects on system ramps. For example, the mix of PV and CSP expected for California's 33% renewable portfolio standards (RPS) by 2020 has a capacity factor of approximately 15–30% (depending on location) and equals approximately 11% of annual energy, but it composes on average between 25% and 42% of the energy on the power system during the midday and early afternoon hours, depending on the month (in addition to other renewable energy during those hours).

2.1 PV

The PV technologies discussed in this chapter are projects that do not integrate storage; hence, their valuation is conducted primarily on the basis of the shape of their forecast production profiles, their spatial distribution, and the capabilities of their inverters. Although individual smaller PV projects can have extreme variability on days with transient clouds, spatial distribution greatly smooths the aggregate solar production profile, and most long-term studies reviewed here assume that solar resources are widely distributed. Larger plants may actually experience more production variability than smaller, more widely distributed plants, but there has been little analysis of how this might affect

---

[4]As of this writing, several regions with high actual or forecast solar penetration have established energy storage procurement mandates. In California, the storage mandate allows electrical storage and types of thermal storage, including CSP-TES.
[5]Economic dispatch refers to optimization of production from generation and nongeneration resources to minimize system production costs or on a bid-basis to clear a wholesale power market. Renewable generation not on economic dispatch is sometimes called "as-available" or "must-take."
[6]Capacity factor is the plant's average production as a percentage of maximum production.

system operations, and, conversely, larger plants are also more amenable to control by system operators. The key technological distinction in the economic modeling literature is between fixed-tilt and tracking PV systems. On clear days, fixed-tilt plants oriented to maximize production, have production profiles that generally slope more gradually toward the daily peak PV production; hence, they contribute less to the predictable diurnal power system ramps caused by solar production. The orientation of fixed PV can also be modified to determine when its daily production peaks. Plants with single- or double-axis tracking generate production shapes with more rapid ramps up and down at the start and end of daily operations; hence, they have a more "rectangular" production shape. Tracking can also be used to reduce production. These differences are then reflected in estimated energy and capacity benefits, as discussed in the next subsection.

2.2 CSP

CSP plants use mirrors or other reflective surfaces to heat a working fluid, which then heats steam to operate a conventional generation power block with a steam turbine. Unlike PV plants, CSP plants need high direct normal insolation (DNI) to achieve desirable operating efficiencies; hence, they are, with a few exceptions, only built in such regions. There are several different CSP plant designs, which are surveyed in [6] and other sources.[7] The studies reviewed in this chapter evaluate the economic benefits of the two primary commercialized designs: the parabolic trough and the tower. In terms of production profiles, for different types of CSP plants without thermal storage (or hybridization with other fuels), production is differentiated primarily in terms of seasonal capability. The positioning of parabolic troughs is optimized to maximize production during the summer months; power towers with tracking heliostats are better able to shape production smoothly across the year. Both types of plants tend to ramp up and down fairly rapidly when there is sufficient DNI, partly as a function of other operational characteristics, such as some degree of natural gas augmentation to manage heat transfer fluid temperatures for start-up or during transient cloud conditions. During periods of transient clouds, CSP designs also provide some degree of inertia.

Thermal energy storage can be directly or indirectly integrated into any of the CSP designs, and the CSP-TES design may affect the operational flexibility of the plant. A key feature of CSP-TES is that in current designs, the storage system charges entirely from the solar field; hence, the plant appears to the system operator as a conventional, dispatchable thermal generator with limited energy, which must be forecast daily. Thus, the simulation models used for economic valuation of either of these design approaches must be able to simulate the hourly conversion of potential thermal energy from the solar field into the state of charge in the energy storage system, and then for operations of the steam turbine to jointly provide energy and ancillary services. In addition, depending on the simulation model, the plant representation can include a detailed range of operational attributes of the power block [7].

3. Overview of valuation methods

In regions where large solar plants are being constructed under utility requirements to meet RPS, the economic valuation of alternative project offers is typically an explicit requirement for resource

---

[7]CSP can also be hybridized with other fuels, typically natural gas or coal but also biofuels, to support plant operations. The hybrid approaches are not reviewed in this chapter.

Table 1 Electric Power Products and Services

	Existing Products	New Products and Modifications for Renewable Integration
Energy	• Hourly schedules, unit commitment (pre-day-ahead, day-ahead) • 5–15 min economic dispatch (real-time)	• Ramping reserves
Ancillary services	• Frequency regulation • Spinning and nonspinning reserves (10 min) • Supplemental reserves (>10 min)	• Regulation pay for performance • Increased regulation procurement
	• Frequency response (primary frequency control) • Voltage control • Blackstart	• Frequency response reserves • Inertial response reserves
Capacity	• System megawatts • Locational megawatts	• Operational attributes ("flexible capacity")

procurement in the period measured. For regional or utility planners and procurement analysts, the basic cost-benefit equation to determine the net cost of a large solar plant is as follows[8]:

$$Net\ cost = levelized\ cost\ of\ solar\ energy\ (or\ contract\ cost) + transmission\ cost$$
$$+ integration\ cost - energy\ benefits - ancillary\ service\ benefits - capacity\ benefits$$

The net cost is generally reported in terms of dollars per megawatt-hour (or other applicable currency), with the megawatt-hour referring to the total production of the individual or aggregated solar plant being evaluated in the period measured. Table 1 shows the conventional subdivisions of these benefit categories into specific products or services.[9] Energy and ancillary service costs and benefits also include reductions in emissions costs (typically modeled as emissions taxes) as well as changes in the startup costs and variable operations and maintenance of fossil-fuel generators.

In most regions experiencing significant growth in solar penetration, the valuation of a particular solar project or portfolio then further requires multiyear, scenario-based estimates of the value for each of these components because they will be highly dependent on system conditions and the evolving resource mix. These scenario analyses, as discussed further below (e.g., [2,7,11]), include variables that have not typically been considered historically in long-term valuation of generation resources, such as rapidly changing penetration levels of different renewable resources and associated renewable integration requirements. As a result, new modeling approaches have been required that link planning and operational models. The geographical scope of the analysis also plays a major role in the production characteristics of spatially distributed solar projects of different capacities. Further

---

[8]The California Public Utilities Commission has embodied this equation in the market valuation requirements for utility procurement under the state's RPS. In the research literature, there are several papers that attempt to comprehensively conduct such valuation (see, e.g., [2,9]).

[9]For a useful survey of product definitions in different regions, see [30].

complicating evaluation, some of these costs and benefits may be highly nonlinear as a function of solar penetration; for example, when reaching the limits of particular power systems to absorb additional solar energy without further significant investments in integration capabilities. These limits can be difficult to identify accurately in large, regional power systems, and given the many integration solutions that may be under consideration or in early stages of implementation (see, e.g., [8]).

Table 2 summarizes the primary modeling methods used for economic valuation or to provide inputs into economic valuation. A few of the studies reviewed jointly analyze several of the different categories of economic benefits within a single model. Other studies may use different models, sometimes linked, to provide more comprehensive benefit estimates. Further comment on methodology is offered in the following subsections.

3.1 Calculation of economic value

As noted, the studies surveyed here focus on a subset of the economic benefits of solar energy—the avoided costs of energy, ancillary services, and capacity from conventional, fossil-fueled generation (or any added costs in the case of solar integration).

For calculations of energy and ancillary service benefits, the modeling methods in the papers reviewed are of a few basic types. The first method calculates the optimal market revenues of the solar plant, such that plant revenues $= \max z' = \sum_{i,t} p_t^i q_t^i$, subject to operational and network constraints (if applicable), where z' is the variable being maximized, p is the market price for service i in interval t, and q is the quantity of the service provided. For a variable energy resource, the plant's operations are not being optimized; hence, the q_t for energy is a fixed profile whereas for CSP-TES it is a variable to be optimized hourly for energy and ancillary services. Prices are either (1) fixed and determined exogenously to the model, using historical market prices (see, e.g., [15]) or simulated future market prices; or (2) determined endogenously within the model by clearing supply and demand (see, e.g., [2]).[10]

In the second common method for evaluating alternative scenarios, energy and ancillary service value are calculated using a dispatch model, which is solved for the least-cost solution to meet demand plus reserves, such that total production costs $= \min z'' = \sum_{i,t} c_t^i q_t^i$, where c represents the variable operational costs. As noted above, q is fixed or variable depending on whether the plant is dispatchable.[11] These production cost studies (see, e.g., [7,11,33]) calculate the change in total production cost (z'') between scenarios with different types of solar technologies providing the marginal projects and required to provide equivalent annual energy. This allows for comparison of the effect of different solar production profiles (and other renewable or proxy resource profiles) on production costs by individual component.

As discussed in Section 4.1, there are several different methods for calculating capacity credits. However, the capacity credit (megawatts) that generally results from the valuation method is

---

[10]A further extension is to model estimates of the value of lost load to approximate a market in which all revenues are earned through energy and ancillary services, with new investment in capacity taking place when scarcity prices signal the need for new capacity. For a description of such a model as applied to renewable resource valuation under long-run equilibrium assumptions, see [2].

[11]The model can be extended to include startup costs, which would be represented separately as an integer variable associated with each particular resource.

Table 2 Summary of Key Modeling Applications and Selected Studies of Solar Projects

Type of Model	Key Application	Brief Description	Selected Solar Studies
Portfolio planning/capacity expansion models	Optimal resource portfolios under uncertainty.	Evaluation of many resource alternatives, generally with simplified consideration of operational characteristics. Linked to production simulation models to verify operational feasibility of resource portfolios.	[2,9,10]
Capacity valuation models	Capacity credits and capacity value.	See Table 3	See Table 3
Production cost models	Energy, regulating reserves, spinning and nonspinning reserves, load-following. Capacity credits using approximations of high-risk hours.	Hourly and subhourly economic dispatch with co-optimization of energy and ancillary services on time frames of days to years.	[7,11,12,13,33]
Plant-level dispatch models	Energy, regulating reserves, spinning and nonspinning reserves, load-following. Capacity credits using approximations of high LOLP hours.	Models of individual resources producing as "must take" or dispatched against exogenous fixed prices, historical or forecast, to evaluate economic benefits. A flexible modeling framework for evaluating detailed design parameters for CSP and CSP-TES.	[14,15,16]
Statistical models of operational requirements for solar integration	Subhourly requirements for regulating reserves, load-following reserves.	Variants on these models have been developed to estimate subhourly reserve procurement under different wind and solar portfolios. Can be linked to production simulation models to evaluate operational capabilities of particular power systems.	[12,17,32,33]

Type of Model	Key Application	Brief Description	Selected Solar Studies
Frequency control on primary and secondary frequency control time frames	Requirements for interconnection standards and frequency responsive reserves.	Models of power system frequency, with estimates of potential frequency control contributions of wind and solar resources (as well as other eligible resources).	[25]

Table 2 Summary of Key Modeling Applications and Selected Studies of Solar Projects—cont'd

multiplied by the annualized cost of new capacity (dollars per megawatt) as derived from industry surveys or from current capacity market prices.

3.2 Methods for scenario development and baselines for measurement of benefits

In addition to how they measure economic benefits, the studies reviewed here use several different methods for developing scenarios and baselines for measuring economic benefits, which can affect the comparative results for particular solar technologies. Earlier studies typically modeled individual solar plants against historical market prices, and they estimated capacity values as if these were marginal plants on the existing system rather than components of an expanding portfolio (see, e.g., [16]). More recent studies that evaluate scenarios with different renewable penetrations have developed two key methodological approaches. First, they measure the value of a marginal solar plant added to the portfolio to distinguish the average value of the portfolio from the marginal value of the added plant. Second, they assume that each marginal plant being compared as an alternative portfolio addition has equal annual energy (megawatt-hour; see, e.g., [2,33]). The equal energy assumption has a major effect on the production shape of alternative solar technologies—PV, CSP, and CSP-TES—primarily by adjusting the installed capacity to ensure the same total energy production.[12] For additional comparisons, several studies also model other baseline scenarios, such as cases with zero or low renewable penetration, and proxy resources, such as a "flat block" of equal energy (see, e.g., [2,33] on the interpretation of the flat block, see also [18]).

4. Survey of research results

This section reviews methods and results from studies of solar valuation, primarily in the western United States, but also some other regions.

---

[12]These approaches can also be used to evaluate different designs within the same solar technology type—for example [7], to evaluate different CSP designs, varied by solar multiple and storage capacity, but each adjusted to provide equal energy.

4.1 Capacity benefits

Solar capacity credits refer to the rating, conventionally denominated as a percentage (%) of the plant's installed capacity (in megawatts), which can be counted toward the resources needed to prevent loss of load on a near-term (e.g., resource adequacy)[13] or long-term resource planning basis [8]. Capacity value (dollars per megawatt) refers to the capacity costs avoided by the solar project from alternative existing or new capacity resources, often denominated in the periods used for capacity procurement (e.g., dollars per megawatt-month or -year).[14]

Models to calculate solar capacity credits fall into two general categories [8]: (1) those using statistical simulations to calculate solar generation's contribution to maintaining the loss-of-load expectation (LOLE) to within a reliability standard such as one event in 10 years, and (2) those that use approximations to simplify the analysis, either of system capabilities or by focusing on actual or simulated solar production in hours modeled as having high risk of loss of load. As shown in Table 3, there are several variants of each type of method, with further description in the studies cited. Moreover, other factors may significantly affect the result, such as the number of weather years used, the geographic scope of the load shape being modeled, and the net load shape associated with the portfolio of wind and solar resources being modeled.

At zero or low penetrations, incremental solar resources in regions with high solar insolation generally obtain high capacity credits and value because of the positive correlation of their production with summer peak loads. Nevertheless, locations with lower solar insolation may derive a high capacity credit if this correlation is especially high. Figure 1 summarizes several study estimates of PV capacity credits at different locations and different penetration levels; the figure is adapted from [9] and includes additional study results.[15] The studies of incremental additions of solar generation at low penetration, furthest left on the x-axis, results in the highest capacity credit in each case. Figure 2 then summarizes the capacity benefits (dollars per megawatt-hour) calculated in some of the same studies and a few others. A few studies have compared different valuation methodologies using the same data. For example, [19] evaluate different methods for calculating PV capacity value at locations in the western US. Their results, which are not shown in the figures below, suggest a range of 52–93% for the capacity credits of different PV technologies at different locations[16] with interannual variation of approximately 16%. Single-axis tracking PV has an average of approximately 26% improvement in the ELCC rating when compared with fixed tilt. Double-axis tracking adds an average of a 3% improvement compared with single-axis tracking.

---

[13]The resource adequacy or capacity obligations imposed on load-serving entities in many regional power systems in the United States are typically 1 to 3 year-ahead accounting mechanisms requirements that require showings a few months before the calendar year or season/month.

[14]Capacity costs obviously vary between regions. For example, of the studies in Figure 2, for new gas combustion turbines [11], use $147/kW for Colorado, while [33] use $212/kW and [2] use $200/kW for California.

[15]Ref. [9]; Figure 8, p. 24. Mills and Wiser obtained some of these estimates by converting capacity penetration to energy penetration using load factor assumptions for the locations; see discussion in the original figure. Note that the figure in this chapter excludes some of the results in Mills and Wiser because the original paper or presentation was not able to be reviewed. The author is grateful to Andrew Mills for sharing the data points in Mills and Wiser's original figure.

[16]However, [19] models PV production in all of the locations evaluated against the load shape in the entire western United States, whereas some of the locations modeled have different load shapes and may further encounter transmission constraints that limit delivery as capacity resources to other locations. Hence, as they note, their results are somewhat inconsistent with those from other studies of the same locations and should be interpreted given those limitations.

Table 3 Solar Capacity Rating and Valuation Methods Applicable to High Penetration Studies

Method	Description	Selected U.S. Regional Studies
Statistical methods		
Equivalent load-carrying capability (ELCC)	The capacity of an incremental variable generator or aggregate portfolio required to meet incremental load and meet the desired LOLE	[8,19,20,21]
Equivalent conventional power (ECP)	The solar generator compared to a conventional generator derated by an expected forced outage rate.	[19]
Equivalent firm power (EFP)	The solar generator compared to an ideal generator (not derated by an expected forced outage rate).	[19]
Approximation methods without system simulations		
Garver's approximation and multistate generator extensions	Risk function to calculate incremental generator ELCC for a particular system	[19,22,23]
Capacity-factor based approximation	Average solar production during a subset of high-risk hours (high load, high net load, high loss-of-load probability (LOLP), etc.)	[8,15,16,23]
Approximation methods coupled with power system simulations		
Production cost or market price simulation	Average solar production (MWh) during high-risk hours, either high price or high load/net load	[2,11,33]
PV, photovoltaic; CSP, concentrating solar power; TES, thermal energy storage.		

For CSP, [15,16] find a range by location in the southwestern United States, from 46% to 95% of nameplate capacity. CSP-TES plants obtain higher capacity credits as storage capacity increases, and they reach approximately 100% capacity credit when 4–5 hour of thermal storage have been added [14–16].[17]

As solar penetration increases, the capacity credits for incremental PV and CSP without storage are reduced as solar production shifts the "net load" peak hours into the evening hours. These results are

---

[17]Studies that compare the "equal energy" portfolios of PV, CSP, and CSP-TES adjust the installed capacity of each technology to reflect its capacity factor (e.g., [2,11,33]). As such, in low solar penetration scenarios, they can award a similar or higher capacity value to the PV or CSP without storage than the value awarded to CSP-TES. This is because, as modeled, the individual or aggregated plants without storage have a higher installed capacity (megawatts) than the CSP-TES plant or portfolio, and the "net load" peak shift has not yet become significant. This phenomenon can be seen at the low penetration levels in Figure 2.

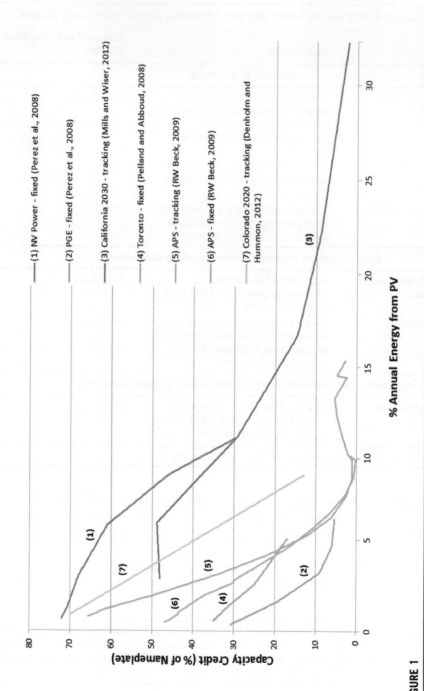

FIGURE 1

Capacity credits (% of nameplate MW) of marginal photovoltaic additions to existing portfolios from selected studies of increasing solar penetration.

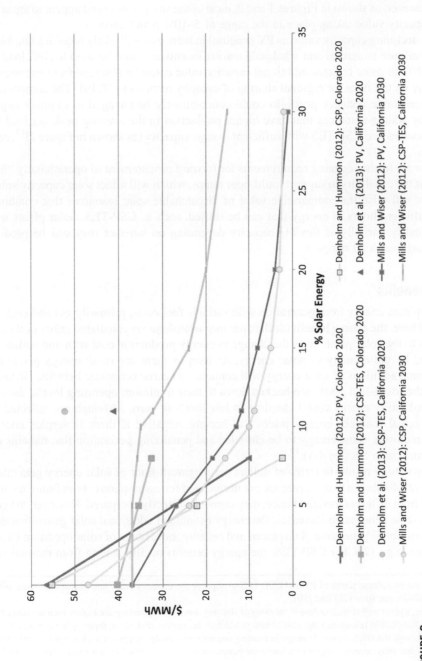

FIGURE 2

Capacity benefits ($/MWh) of solar resources from selected studies of increasing solar penetration.

sensitive to the power system being modeled, and even for studies of the same system under similar conditions they may vary depending on the model used, suggesting a need for further model evaluation (e.g., [2,33]). However, as shown in Figures 1 and 2, most of the studies surveyed appear to show major reductions in capacity value taking place in the range of 5–10% penetration.

To offset the declining capacity value as PV penetration increases will likely require a combination of solutions. Retail rate structures and wholesale market incentives could be used to shift load to the sunlight hours [10]. In some regions, additional capacity value could be obtained from exported solar energy depending on the rules for regional sharing of capacity resources [18,19]. The composition of the expanding renewable resource portfolio could concomitantly be changed to improve aggregate capacity value by adding resources that have higher production in the evening peak net load hours, such as wind resources and CSP-TES with sufficient storage capacity (as shown in Figure 2)[18], or other types of storage.

Some regions are further defining requirements for forward procurement of operationally "flexible capacity" such as the capability to support multi-hour ramps, which will affect solar capacity valuation (e.g., [24]). This will raise the comparative value of dispatchable solar resources that combine fast ramping with sufficient hours of energy that can be shifted, such as CSP-TES. Solar plants without storage could qualify partially as flexible capacity depending on whether they can be predictably curtailed to manage net load ramps.

4.2 Energy benefits

Solar energy displaces energy from generation with variable fuel costs, primarily gas and coal. In the studies reviewed here, the energy benefit (dollars per megawatt-hour) is calculated either as the energy market revenues to the solar plant, or as the change in energy production cost with and without solar resource, divided by the quantity of solar energy. As long as there are fossil-energy plants on the margin, incremental additions of solar energy will continue to accrue economic benefits. However, in hours in which the fossil-fuel plants are backed down to their minimum operating levels, the market prices (or avoided utility energy costs) can drop to low levels or zero. Although not reflected in the simulations, in actual markets, energy prices can become negative if there is surplus energy, or overgeneration, requiring some energy to be curtailed and penalizing generation that remains online (see further discussion in Section 4.4).[19]

Figure 3 shows study results in terms of dollars per megawatt-hour of solar energy generated. As noted above, the results shown in the figure are not normalized for assumptions about future natural gas prices and result from different models; hence, they cannot be directly compared. However, the general finding is that as solar penetration increases, the energy benefits of additional solar generation decline because higher cost fossil generation is displaced and because more hours of solar operation have very low or zero prices (e.g., [2]). For CSP-TES, the energy benefits of dispatching from thermal storage

---

[18]For discussion on the complementarity of PV and CSP-TES to improve the capacity value and operational feasibility of an aggregate solar portfolio, see Refs [28] and [10].

[19]As a general matter, a plant will remain online if the costs of shutting down and restarting are higher than the cost of paying the negative prices. Renewable resources may also obtain production incentives that allow them to bid negatively down to those payments. Although the effect of wind energy in causing negative wholesale energy prices has been observed in many regions, solar energy has only recently begun to achieve the penetration levels in a few locations to contribute to this effect during sunlight hours.

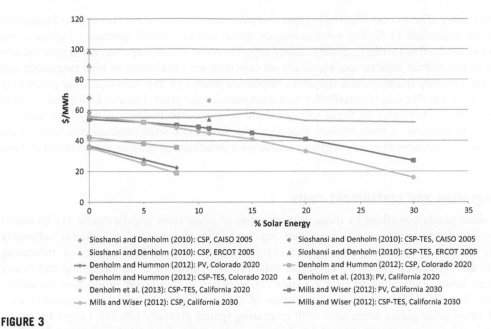

FIGURE 3

Energy benefits ($/MWh) of marginal solar resources from selected studies of increasing solar penetration.

decline much less rapidly at higher penetrations because of the ability to shift energy to the remaining highest energy value intervals ([2,11,33]).

4.3 Ancillary service benefits

In regions with sufficient solar insolation, large solar plants are likely to eventually provide some ancillary services simply because they will provide much of the energy on the system during the sunlight hours. CSP-TES is currently the only solar technology that has been modeled as providing ancillary services because it can be economically dispatched. Because of fast ramp rates, such plants are well suited to being operated at minimum load overnight and used to provide spinning reserves [16]. They can also potentially provide frequency regulating reserves [7]. Models of CSP-TES that co-optimize energy and ancillary service production have shown a range of ancillary service value for particular locations, between approximately $1.50/MWh [2] and more than $10/MWh ([16,33]), as a function of storage capacity and the scenario and year evaluated.

At higher penetrations, solar plants without storage may either be required administratively to provide some inertial response and frequency responsive reserves under evolving grid codes,[20] or

---

[20]In some European countries and Texas, wind farms are already expected to self-provide frequency responsive reserves (see, e.g., [31]).

opportunities may arise to cost-effectively these and other ancillary services if lower-cost integration solutions are exhausted [3,5]. For solar resources, these services would primarily include some smoothing of production variability during cloudy days, provision of ramping reserves to slow the rate of change of the diurnal solar ramps, especially on days with net load ramps of high magnitude and duration, and possibly an inertial and frequency responsive reserve [8,25]. Although solar power may continue to have a higher cost of curtailment than wind plants, solar plants can also be located closer to loads where the market value for these services could be higher (particularly if they have local requirements). In addition, in many regions with high solar insolation, wind production is lower in the late afternoon; hence, operators may require variable solar generators to provide ramp control at those times [8].

4.4 Integration and curtailment costs

Variable solar plants contribute to three primary types of integration requirements: (1) increased frequency regulating reserves, (2) increased net load-following needed to address subhourly variability caused by transient clouds, and (3) the intrahourly net load ramps of increasing magnitude and frequency caused by aggregate solar production ramps at the beginning and end of the sunlight hours. Individual small PV plants can experience significant production variability during transient clouds. However, in the absence of transmission congestion, the aggregate variability of small solar plants diminishes with increasing spatial diversity [26,27]. Larger PV plants may experience more production variability because of transient clouds than more distributed resources, but may be able to manage that variability because their size buffers the changes in aggregate plant production, and active power management can be used to further manage the resulting plant ramps. However, increasing penetration of variable solar plants can only increase the magnitude of the aggregate intrahourly net load ramps.

To date, most utility resource planners, at least in the western United States, appear to use estimates of variable solar integration costs in the range of $1.25–11.00/MWh primarily based on "rules of thumb" because of lack of operating experience [9]. Simulation studies to date have generally found costs within this range for 20% – 30% solar penetration scenarios (e.g., [12]).

As discussed previously, curtailment cost refers here to the lost contracted revenues when a variable energy resource is required to reduce production without a fully compensating market payment. If solar generators, or the utilities that had contracted for the power, were to bid curtailment into wholesale markets for compensation, then the bid curve would reflect the vintage of the projects, with newer projects having a much lower cost of energy [3]. The studies surveyed here that simulate increasing solar penetration invariably find that curtailment of variable solar energy increases with penetration beyond certain levels, with a more significant increase in curtailment often appearing in the range of 15–20% solar penetration. The simulation studies of 33% RPS in California in 2022, with approximately 11–15% distributed and large-scale solar energy (without storage) depending on the scenario, have not found any significant curtailment of renewable energy to date, although they do find an increase in exports during sunlight hours.[21] However,

---

[21]Selected results for the California 33% RPS simulations can be found at http://www.cpuc.ca.gov/PUC/energy/Procurement/LTPP/ltpp_history.htm. See also [33].

preliminary studies of solar penetration in California greater than that range do find significant potential increases in curtailment (e.g., [28]).

An open question, given the results discussed above, is the quantity of solar that can be utilized in very high renewable penetration scenarios for particular regions. The National Renewable Energy Laboratory (NREL) [10] has conducted the most detailed such study currently available, which models 50–80% renewable penetration scenarios in the entire United States. This study notably anticipates much greater penetration of wind compared with solar, with solar nationally ranging between 3% and 22% depending on the scenario, although with higher penetrations in the southwestern U.S. states. In the high solar penetration case, PV provides a little over 10% of annual renewable energy (half distributed, half utility-scale) whereas CSP-TES provides up to approximately 14%, primarily to support operational flexibility. This level of penetration will also require a significant redesign in how electricity markets and institutions currently operate and how costs are allocated.

4.5 Total economic benefits

The total economic benefit of a large solar project, or an aggregation of solar projects, is defined here as the sum of energy, ancillary service, and capacity benefits, net of any integration and curtailment costs, over the period being evaluated. As noted above, the summation of the components of the benefits calculation may be determined using a single comprehensive power system model, or it can be derived from a set of linked models or constructed from different sources. Figure 4 shows the declining total economic benefit of marginal solar projects as solar penetration increases (and hence increasing net cost, if costs do not continue to decline). This is driven primarily by the change in capacity value. There is also the potential for increasing the curtailment of variable solar resources at higher penetrations, which will increase the costs per megawatt-hour of energy produced.

5. Conclusions

As large-scale solar projects expand on power systems around the world, calculation of economic and reliability benefits will become more important in regional planning, utility procurement, and project development. Several modeling methods have been demonstrated in the research literature and in commercial settings, ranging from plant-level market value analysis to simulation of regional power systems under different scenarios and at different levels of spatial and temporal aggregation. This research will increasingly be informed by actual operational experience with large solar plants.

A key finding in the literature to date is the declining capacity value of solar resources without storage as solar penetration increases. This measurement is specific to particular power systems and needs further simulation and empirical verification. However, across studies to date, significant reductions in solar capacity value are found in the range of 5–10% solar penetration by annual energy, which is the penetration expected in regions such as the southwestern United States by 2020, if not earlier. Much further analysis is needed to understand how solar energy is managed within regional power systems as over-generation conditions potentially emerge in particular locations during solar production hours. Although in the models to date the energy benefits of marginal solar plants appear quite stable until very high penetrations, there is the potential for increasing curtailment, and most of the regional studies to date use simplified network representations for large regions that may not fully capture operational and

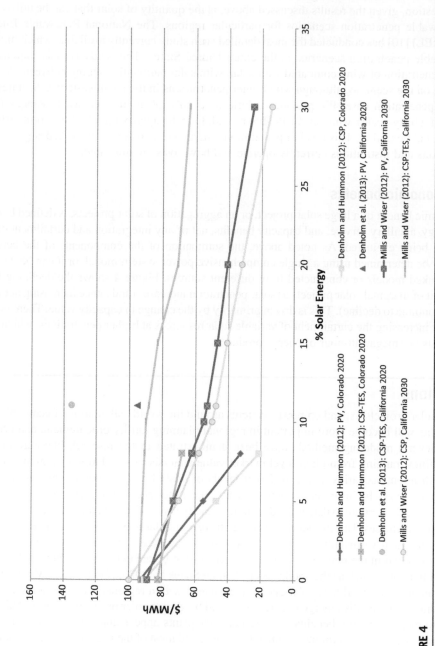

FIGURE 4

Total economic benefits ($/MWh) of marginal solar resources from selected studies of increasing solar penetration. PV, photovoltaic; CSP, concentrating solar power; TES, thermal energy storage.

transmission constraints. These results point to the need for regional coordination, market design changes, and operational enhancements to improve operational flexibility, but they also point to the potential value of the storage used to capture solar energy that would otherwise be curtailed.

Trends in PV prices as well as its scalability and wide applicability have resulted in a significant diminishment in the forecast for CSP projects. The literature surveyed here suggests that CSP-TES would become more economically viable in high solar penetration scenarios because its operational flexibility allows it to provide sustained economic benefits when compared with alternative solar resources without storage. However, in utility procurement and regional planning, CSP-TES will compete with other solutions – load management, other flexible resources and transmission upgrades that can provide operational flexibility or that can change the shape of the net load curve.

Acknowledgments

Part of this chapter was initially developed as a survey of the valuation of CSP-TES plants for the CSP industry. The author thanks Paul Denholm, Joe Desmond, David Jacobowitz, Andrew Mills, Tex Wilkins, and the other participants and reviewers of that earlier study.

References

[1] Joskow PL. Comparing the costs of intermittent and dispatchable electricity generating technologies. Am Econ Rev May 2011;101(3):238–41. http://dx.doi.org/10.1257/aer.100.3.238.

[2] Mills A, Wiser R. Changes in the economic value of variable generation at high penetration levels: pilot case study of California. Lawrence Berkeley National Laboratory; June 2012b. 5445E. See, http://eetd.lbl.gov/ea/emp/reports/lbnl-5445e.pdf.

[3] Bolinger M, Weaver S. Utility-scale solar 2012: an empirical analysis of project cost, performance, and pricing trends in the United States. LBNL06408E. Lawrence Berkeley National Laboratory; September 2013. http://emp.lbl.gov/sites/all/files/lbnl-6408e_0.pdf.

[4] Denholm P, Margolis RM. Evaluating the limits of solar photovoltaics (PV) in electric power systems utilizing energy storage and other enabling technologies. Energy Policy September 2007;35(9): 4424–33.

[5] Denholm P, Ela E, Kirby B, Milligan M. The role of energy storage with renewable electricity generation; 2011. Technical Report, NREL/TP-6A2–A47187, January 2010.

[6] International Energy Agency (IEA). Technology roadmap: concentrating solar power. OECD/IEA; 2010. See http://www.iea.org/publications/freepublications/publication/csp_roadmap.pdf.

[7] Jorgenson J, Denholm P, Mehos M, Turchi C. Estimating the Performance and economic value of multiple concentrating solar power technologies in a production cost model; December 2013. Technical Report, NREL/TP-6A20–58645.

[8] North American Electricity Reliability Corporation (NERC) and California ISO (CAISO). 2013 Special reliability assessment: maintaining bulk power system reliability while integrating variable energy resources – CAISO approach, a joint report produced by the North American electric reliability corporation and the California independent system operator corporation. Available at: http://www.nerc.com/pa/RAPA/ra/Reliability%20Assessments%20DL/NERC-CAISO_VG_Assessment_Final.pdf; November 2013.

[9] Mills A, Wiser R. An evaluation of solar valuation methods used in utility planning and procurement processes. Environmental Energy Technologies Division, Lawrence Berkeley National Laboratory; December 2012a. LBNL-5933E.

[10] National Renewable Energy Laboratory (NREL). Renewable electricity futures study, vols. 1–4; 2012. See http://www.nrel.gov/analysis/re_futures/.

[11] Denholm P, Hummon M. Simulating the value of concentrating solar power with thermal energy storage in a commercial production cost model. National Renewable Energy Laboratory; November 2012. Technical Report, NREL/TP-6A20–56731.

[12] Navigant, Sandia National Laboratories, and Pacific Northwest National Laboratory. Large-scale PV integration study prepared for NV Energy; July 30, 2011. See http://www.navigant.com/~/media/Site/Insights/NVE_PV_Integration_Report_Energy.ashx.

[13] Denholm P, Margolis R, Milford J. Production cost modeling for high levels of photovoltaics penetration. National Renewable Energy Laboratory; 2008. See http://www1.eere.energy.gov/solar/pdfs/42305.pdf.

[14] Sioshansi R, Denholm P. The value of concentrating solar power and thermal energy storage. IEEE Trans Sust Energ 2010;1(3):173–83.

[15] Madaeni SH, Sioshansi R, Denholm P. How thermal energy storage enhances the economic viability of concentrating solar power. Proc IEEE February 2012a;100(2):335–47.

[16] Madaeni SH, Sioshansi R, Denholm P. Estimating the capacity value of, concentrating solar power plants: a case study of the southwestern United States. IEEE Trans Power Syst May 2012b;27(2):1116–24. http://dx.doi.org/10.1109/TPWRS.2011.2179071.

[17] Ibanez E, Brinkman G, Hummon M, Lew D. A solar reserve methodology for renewable energy integration studies based on sub-hourly variability analysis. NREL pre-print, Available at: http://www.nrel.gov/docs/fy12osti/56169.pdf; November 2012.

[18] Milligan M, Ela E, Hodge B-M, Kirby B, Lew D, Clark C, et al. Cost-causation and integration cost analysis for variable generation; June 2011. Technical Report, NREL/TP-5500–51860.

[19] Madaeni SH, Sioshansi R, Denholm P. Comparing capacity value estimation techniques for photovoltaic solar power. IEEE J Phot January 2013;3(1):407–15.

[20] Beck RW. Distributed renewable energy operating impacts and valuation study. Arizona Public Service; 2009. http://www.solarfuturearizona.com/.

[21] Perez R, Taylor M, Hoff T, Ross JP. Reaching consensus in the definition of photovoltaics capacity credit in the USA: a practical application of satellite-derived solar resource data. IEEE J Sel Top Appl Earth Observ Remote Sens 2008;1(1):28–33.

[22] Garver LL. Effective load carrying capability of generating units. IEEE Trans Power Appar Syst August 1966;PAS-85:910–9.

[23] Pelland S, Abboud I. Comparing photovoltaic capacity value metrics: a case, study for the City of Toronto. Prog Phot Res Appl 2008;16(8):715–24. http://dx.doi.org/10.1002/pip.864.

[24] Lannoye E, Milligan M, Adams J, Tuohy A, Chandler H, Flynn D, et al. Integration of variable generation: capacity value and evaluation of flexibility. IEEE Power Engineering Society (PES); 2012. Summer Meeting, 2012.

[25] GE Energy Consulting and California ISO (CAISO). Frequency response study, final draft; November 2011. Prepared by Miller NW, Shao M, Venkataraman S. See http://www.caiso.com/Documents/Report-FrequencyResponseStudy.pdf.

[26] Mills A, Wiser R. Implications of wide-area geographic diversity for short-term variability of solar power. Tech. Rep. LBNL- 3884E. Berkeley (CA): Lawrence Berkeley National Laboratory; September 2010.

[27] Mills A, Alstrom M, Brower M, Ellis A, George R, Ho_ T, et al. Understanding variability and uncertainty of photovoltaics for integration with the electric power system. Tech. Rep. LBNL-2855E. Berkeley (CA): Lawrence Berkeley National Laboratory; 2009. http://eetd.lbl.gov/EA/emp/reports/lbnl-2855e.pdf.

[28] Denholm P, Mehos M. Enabling greater penetration of solar power via the use of CSP with thermal energy storage. National Renewable Energy Laboratory; November 2011. Technical Report, NREL/TP-6A20–52978. See http://www.nrel.gov/csp/pdfs/52978.pdf.

[29] International Renewable Energy Agency (IRENA). June 2012. Concentrating solar power, renewable energy technologies: cost analysis series, Volume 1: Power Sector, issue 2/5, IRENA Working Paper.

[30] Ela E, Milligan M, Kirby B. Operating reserves and variable generation. National Renewable Energy Laboratory; August 2011. Technical Report, NREL/TP-5500–51978.

[31] GE Energy and Exeter Associates. Task report: review of industry practice and experience in the integration of wind and solar generation, PJM renewable integration study; November 2012. Prepared for: PJM Interconnection, LLC. Prepared by: Exeter Associates, Inc. and GE Energy.

[32] California ISO (CAISO) and DNV-KEMA. Final report for assessment of visibility and control options for distributed energy resources; June 21, 2012. See http://www.caiso.com/Documents/FinalReport-Assessment-Visibility-ControlOptions-DistributedEnergyResources.pdf.

[33] Denholm P, Wan Y-H, Hummon M, Mehos M. An Analysis of Concentrating Solar Power with Thermal Energy Storage in a California 33% Renewable Scenario. National Renewable Energy Laboratory, Technical Report, NREL/TP-6A20-58186; March 2013.

State of the Art and Future Outlook of Integrating Wave and Tidal Energy

27

Timothy R. Mundon[1], Jarett Goldsmith[2]
[1] *Oscilla Power, Inc,* [2] *DNV GL - Energy*

1. Introduction

Wave and tidal energy resources exist around the world in quantities that can make a significant contribution to clean electricity generation targets. The ongoing challenge posed is how to extract this energy at an economically feasible cost. At present, while there is much development in this emerging sector, there remains a scarcity of commercial systems operating anywhere in the world.

It is common for those without experience in these particular fields to have the misconception that wave and tidal energy are equivalent. This is not the case, and it is important for any text that discusses these resources to clarify this in advance. While tidal and wave resources are indeed primary areas of interest within the realm of marine renewables, these resources and the technology needed to utilize them are completely different.

With that distinction made, it can be noted that both wave and tidal energy are variable resources, and like other variable renewables the ability to predict their energy content is a critical component in increasing the value of available power. Fortunately, the ability to predict the potential output of these generation sources exists, even for relatively long time horizons. This is a key advantage that can help the adoption of these emerging technologies into the energy mix.

This chapter will discuss wave energy and tidal energy as separate areas and will cover the key features of each resource, including the basic principles involved in generating electric power. It will then highlight the variability of the resources and explain how we can use modern tools to predict the power and energy output successfully.

1.1 Common challenges

It is well known that the cost and complexity of marine operations are dramatically higher than equivalent land-based operations. As a result, the economics of marine projects tend to be significantly more sensitive to external factors that influence installation and operation and maintenance (O&M) costs. Experience from the oil and gas industry has taught that while it is technically feasible to build structures to survive and operate in the extremities of the marine environment, it is not done without expense proportional to the challenge. Bearing in mind the inherent challenges in developing marine projects and offshore operations, a number of common challenges are important to understand before we approach the resources themselves.

Renewable Energy Integration. http://dx.doi.org/10.1016/B978-0-12-407910-6.00027-2

1.1.1 Electrical challenges

At the present time, wave and tidal devices have been connected to the grid primarily as individual units or in very small numbers, and the subsequent information presented here can consider only these circumstances. As commercial generation is expected to occur in arrays of units, some variability issues are expected to become less of a challenge as very short time frame variability (~10 s) is smoothed out by the spatial distribution of devices within the resource. Conversely, as project sizes increase, electrical aggregation becomes a challenge in its own right. On land, this can be solved by relatively straightforward implementation of switchgear, transformers, and power electronics. However, the same solutions are more difficult to implement at sea and require thought given to the location of electrical subsystems and protection from saltwater, while retaining the ability for service and maintenance. Present solutions range from locating equipment on fixed offshore platforms, as is the standard approach for offshore wind projects today, to custom floating or subsea substations, although given the currently small number of deployed devices it is not clear what electrical infrastructure solution will become typical.

1.1.2 Grid integration

Resource intermittency and variability have historically been considered key drawbacks of renewable energy sources, reducing their integration value. The Ocean Energy Systems Implementing Agreement, an intergovernmental collaboration between countries operating under the framework established by the International Energy Agency, identified Grid Integration as a key issue and established Annex 3 to the implementing agreement to cover research in the topic. Some notable reports produced through the Annex that discuss the subject specifically look at wave and tidal energy and can be found in Refs [1,2].

1.1.3 Regulatory, environmental and other common issues

While this chapter concentrates on particular technical aspects of power generation from wave and tidal resources, it does not address other challenges that may make the adoption of these resources difficult. These issues can be nationally or regionally specific and encompass topics such as site permitting, environmental regulations, competing marine spatial uses, and the electrical power market.

2. Tidal energy

Solar and lunar gravitational forces, combined with the rotation of the Earth, generate periodic changes in sea level known as the tides. This rise and fall of ocean waters can be amplified by basin resonances and coastline topography to create very large surface elevation changes in specific geographic locations. Most coastal sites throughout the world experience two high and two low tides each day (semi-diurnal tides), although some places experience just one high and low tide per day (diurnal tides) and others are characterized by a combination of diurnal and semi-diurnal oscillations (mixed tides). The difference in sea level height between a high and low tide at a given location is called the tidal range, and it can vary each day depending on the location of the sun and the moon, and globally depending on the coastal location. The vertical rise and fall of water is accompanied by an incoming (flood) or outgoing (ebb) horizontal flow of water in bays, harbors, estuaries, etc., called a tidal current (or tidal stream). These can be exceptionally strong in certain areas where large tidal ranges are further constrained by geography, such as in the Bay of Fundy in Canada, which experiences tidal ranges over 15 m and flow velocities that can range from zero at slack tide to greater than 5 m/s at peak in certain places. This results in what can be considered a locally concentrated energy resource: while on a broad geographic scale the average or total resource can be quite low, the nature of the resource is such that it is typically concentrated in certain areas. For example, the United States as a whole has a limited tidal resource estimated at 60 TWh annually, but this energy is concentrated in only a few locations such as Alaska, Maine, and Washington, providing the potential for significant local generation in those places [3].

2.1 Energy extraction

There are two general approaches to tidal energy extraction, a barrage approach (also called tidal range because it utilizes the potential energy created by the difference in sea level between high and low tides) and a hydrokinetic approach (also called tidal stream, which captures kinetic energy from the horizontal flow of tidal currents). The barrage approach is based on conventional hydropower principles and requires a structure (e.g. a dam) to impound a large tidal body of water, such as an estuary. As the tidal height varies outside of the impounded area, water is discharged either into or out of the enclosed area through conventional hydro turbines housed in the structure, creating electrical power as the water moves from one side of the dam to the other. Only a handful of projects exist that use this principle worldwide, notably including two large ($\sim$250 MW) projects in France and Korea. The inevitable environmental consequences of such schemes mean that there are significant challenges to permit a project of this nature.

A less environmentally intrusive solution is the hydrokinetic approach, which converts the kinetic energy of flowing water into electricity in the same way that wind turbines convert the kinetic energy of flowing air (wind). This method does not require any impoundment structure or dam and uses devices that are designed to extract energy from moving water without a requirement for a significant hydrostatic head. The hydrokinetic approach is gaining considerable support and development around the world because the environmental impacts are anticipated to be relatively minor. Devices are typically representative of wind turbines and use the same principles of energy extraction, with designs modified to suit the much higher density medium of saltwater. When hydrokinetic systems are used in a tidal environment they are often referred to as tidal turbines, tidal in-stream energy converters, or tidal/marine current turbines. A horizontal-axis, three-bladed, open, axial flow turbine, as shown in Figure 1, is the preferred technology concept currently being pursued by the large original equipment manufacturers (OEMs); however, other designs exist that use different approaches, such as ducted turbines, cross flow turbines, drag turbines, and reciprocating hydrofoils.

The principles of energy extraction from a moving fluid are well known, but several notable factors need to be considered when designing a tidal turbine:

- The density of water is roughly 850 times that of air, leading to a smaller requisite capture area compared to a wind turbine, but with higher associated loads.
- Power density increases with the cube of the flow velocity: The power density of a 3 m/s flow is more than triple that of a 2 m/s flow.
- Alternating flow direction: tidal flows reverse a number of times per day.

FIGURE 1

Typical three-bladed, open, axial-flow tidal turbine.

Alstom Ocean Energy.

- Highly limited accessibility: installation operations can typically only be done in low current speeds, providing accessibility that can be as short as 30–45 min per slack tide.
- Significant flow shear and turbulence levels: the geography and bathymetry of tidal stream sites often produce flows with a range of shear profiles and eddies on a variety of spatial and time scales that impact loading and performance.

So far only a handful of companies have demonstrated megawatt scale (1 MW + rated capacity), grid-connected devices, and have plans for utility-scale deployments. Multi-megawatt tidal arrays of these devices are currently in the planning stages in a number of locations around the world.

2.2 Project power rating and capacity factor

A tidal turbine power curve is quite similar to a standard wind turbine, especially where the blade pitch is altered to regulate power production above the rated flow speed. However, pitching blades are generally more difficult to implement in a tidal turbine compared to a wind turbine, driving device costs up, and so an alternative being pursued by a number of developers is to employ fixed pitch blades. Some fixed pitch blade concepts use a control method that caps the power by speeding up the rotor rpm beyond the rated flow speed.

2.3 Tidal resource and energy prediction

Since the tidal resource is driven primarily by the orbits of the sun and moon, predicting the future tidal currents and surface elevations at a particular point is achieved with reasonable accuracy using well-established harmonic analysis methods with site-specific measurements. A good coverage of harmonic analysis and prediction can be found in Refs [4,5], and also many textbooks.

For tidal barrage projects, a hydrodynamic model is required to model the interaction of the dam with the site-specific tidal harmonics and also to include any hydrological effects from rivers and storm surges.

For hydrokinetic tidal projects, the power density of a flow will primarily vary with the cube of the water velocity,[1] and therefore the tidal stream velocity needs to be predicted very accurately to determine the turbine energy output. Harmonic analysis can be used to predict the ($\sim$hourly) temporal changes in the resource, but it is not an adequate predictor of shorter time frames or the spatial variability of the resource [6]. The tidal stream velocity is a function of the channel bathymetry for some distance upstream and downstream of the turbine, which will introduce non-uniform flow dynamics [7] into the channel velocity. The result of this is that the generated power will vary spatially across the channel and thus needs to be considered at each turbine location. In addition to the deterministic components discussed, there are also additional components of the tidal current generated by turbulence, and meteorological effects (waves, storm surges, riverine/hydrological contributors, etc.).

Limits to the predictability of tidal currents are discussed in Ref. [8], which found that at a typical tidal energy site both current speed and kinetic power density can still be well described by harmonic analysis. Furthermore, the aforementioned challenges in predicting power output in hydrokinetic turbines using harmonic analysis are significant mainly before the turbines are installed. Once turbines are operating and power curves are validated, measured flow and power production data can be used to progressively refine the tidal prediction models for the site, providing highly accurate energy output predictions.

Natural geographic phase variations in the tidal resource may provide a way to reduce the grid integration impact by strategically placing a portfolio of generation stations along (or around, in the case of an island) a coast [9]. If properly distributed, the peaks at some sites may coincide with the troughs of others to produce a more regular output. At a reduced geographic scale, it may also prove possible to use tidal phase variations across a single tidal site to reduce generation intermittency for a large array [10].

3. Wave energy

The typical definition of wave energy is the kinetic and potential energy contained in water waves that propagate across the ocean surface. Ocean surface waves, as they are called, are generated by wind blowing across the ocean. They can propagate with minimal energy loss and will combine and continue to gain energy from the wind over long open ocean stretches. Although the specific interaction mechanisms between the wind and the surface of the ocean are complex, ocean surface wave formation is primarily influenced by the speed of the wind, the duration it blows, and the 'fetch' (distance of open water over which the wind blows).

---

[1]Power in flow given by $P_{(kW)} = \frac{1}{2}\rho A v^3$, ρ = fluid density, A = swept area, v = velocity.

It is possible to consider waves as a concentrated form of solar energy since it is solar heating that generates winds, which then generate ocean waves. If we follow this energy flux, we can see that the average power density experiences a spatial concentration, increasing from around 0.1–0.3 kW/m^2 for solar to typical values of 0.5–0.8 kW/m^2 for wind [11] and annual average wave power values of around 10–100 kW/m of wave crest length depending on the location in the world.

Wave energy converters (WECs) are the devices that convert the kinetic and potential energy of ocean waves into a more useful form, typically electricity. Extraction of this energy at useful scales has to date proven challenging, with devices currently emerging being the result of a significant body of knowledge and many cycles of research and development [12,13]. Serious academic attention started to be directed at this problem in the early 1970s, although realistic devices that can provide low energy costs while retaining physical robustness are only just starting to emerge. At present there are a number of grid-connected devices installed in high-energy environments, representing the pre-commercial prototypes of devices that are aimed for buildout into utility-scale arrays in the next several years.

Much of the challenge in capturing the energy in waves is created by the fact that instead of a linear particle flow, as in tides or currents, water particles in ocean waves move in elliptical paths around a fixed point (the apparent progression of waves across the surface is due to the energy *transfer* between adjacent particles). As waves move from deep water and start to approach the shore, this particle motion moves from circular to increasingly elliptical and the energy content reduces from the frictional losses with the seabed.

Ocean waves can be thought of as a superimposition of regular waves with many different heights, periods and directions, creating a particular 'sea state' with a common period and height. The sea states vary over time as conditions change. As a result, extraction methods tend to be based on oscillating systems, tuned to provide efficient generation within a band of wave periods. One simple way to think about the performance is to consider the wave height as the energy carrier, while the wave period is an efficiency 'tuning' parameter.

Many different concepts of energy extraction have been developed, taking advantage of various degrees of freedom (i.e. heave, surge, pitch, sway, roll, and yaw), and each WEC concept tends to be suited to a particular regime, such as offshore (>30 m water depth), near shore (~10–20 m depth), and coastline. A variety of different power take-off methods have also been developed within the WECs to convert the slow linear or rotational motions of the devices into electricity or other forms of energy (e.g. pressurized saltwater for desalination), including pneumatic, hydraulic, and electromechanical direct drive systems. An excellent coverage of the principles of ocean wave extraction can be found in Ref. [14].

3.1 Terminology

In short time frames of the order of seconds to minutes, the water surface elevation at a stationary point in space will vary considerably. Figure 2 shows a typical water surface level variation over a few minutes and the relative power contained within. This short-term water surface level is impossible to predict accurately for distances more than a very short distance away from a measuring location. However, if a statistical approach over a slightly longer time period is taken (of the order of tens of minutes), then it can be seen that they can be very well described and predicted.

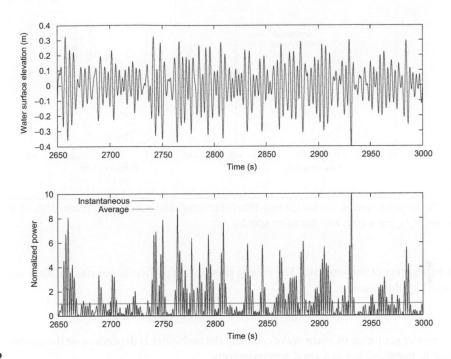

FIGURE 2

Short-term water surface elevation and normalized power content for a site near the Isle of Shoals, NH, USA.

3.2 Sea state

The sea state is a statistical description of the water surface elevation at a particular point in the ocean, and is typically given as a plot of wave energy against period as shown in Figure 3. This is known as a wave spectrum and can be used to derive the majority of key wave parameters such as significant wave height (H_s) and peak period (T_p). The spectral shape will vary geographically, but common examples are JONSWAP [15] or Bretschneider [16].

To relate this concept in terms of water surface elevation, it can be seen that for a particular value of H_s, the individual wave heights will fit a particular distribution (approximately Raleigh), meaning that we can predict the maximum height of a single wave in a sea state as roughly $1.8 \times H_s$. A detailed coverage of wave statistics can be found in most oceanography or ocean engineering textbooks (see Ref. [17] as an example).

3.3 Energy production

While the short-term stochastic wave environment is important to consider during design, the spectral wave parameters are a much better representation of the resource. Therefore, if we assume that adequate short-term smoothing and power conditioning is incorporated into a wave energy converter, the power output can be described and estimated with reasonable levels of accuracy based on the

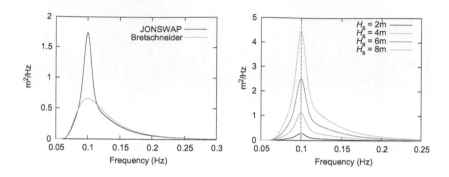

FIGURE 3

Typical spectra for ocean waves. On the left two different spectral distributions are shown, while on the right differing levels of H_s are shown with the same spectra.

statistical parameters of the sea state. The energy present in a particular sea state in *deep water* can be approximated by the equation:

$$P_{(kW/m)} = (0.3 \rightarrow 0.5) \times T_e H_s^2$$

where P = power per meter of linear wave crest and the multiplier is dependent on the spectral shape at the location in question; 0.4 is a good approximation.

Just as wind (or tidal) turbine manufacturers have power performance curves that are used to determine the power output at a given wind (or current) speed, a WEC developer will provide performance data for particular sea states. However, from the above equation, it can be seen that the power is dependent on two parameters, H_s and T_e,[2] the consequence being that a matrix rather than a curve must be used when describing the output of a particular wave energy converter.

As an example, Figure 4 shows performance data for a notional commercial-scale WEC based on the 'Pelamis' P1 device, taken from a study by the Scottish Executive [18] (Section 3.3.3). However, it should be noted that many factors influence the power production of a device that are not fully represented by H_s and T_p, including sensitivity of the WEC to wave direction, and spectral shape [19].

3.4 Variability and predictability of wave energy

Although the short-term water surface elevation changes rapidly, it can be seen that the statistically represented sea state as a whole varies much more slowly over much longer periods, generally hours for significant differences. Figure 5 shows the sea state variations in power density for a location in the northern hemisphere over the course of a full year. The seasonality is clearly apparent, with a significant difference in available energy between summer and winter. Of course, given this variability, sea state forecasting is clearly very important to be able to predict device output and manage grid integration [20,21].

---

[2]Although T_p is provided as a common wave statistic, T_e (energy period) is typically used when dealing with wave energy. This is the mean wave period with respect to the spectral energy distribution.

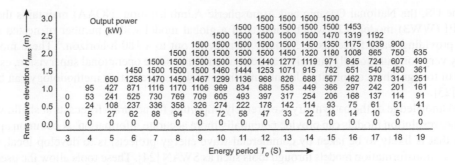

FIGURE 4

An example of a power matrix for a generic 1.5 MW wave energy device.

After leaving their generation zone as swell, ocean waves can travel in deep water over large distances with very little energy loss. As such, the consideration of complete ocean and global weather patterns allows for reliable sea state predictions further into the future than conventional wind or solar forecasts. Modeling the wave climate requires large and complex numerical models that are fed by a wide global network of sensors. Fortunately, accurate forecasting is essential for commercial marine traffic and offshore operations and as such, the models and infrastructure already exist, with good quality forecasts being made available publicly. More localized, high-resolution models can be used to refine these regional forecasts to a specific area to provide improved spatial resolution.

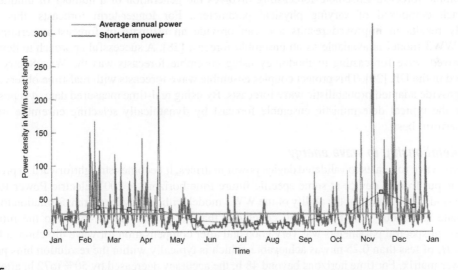

FIGURE 5

Annual variation in wave power density. The blue line shows the incident power averaged over hourly intervals, the green boxes show the monthly averages, and the red line shows the annual average. (For interpretation of the references to color in this figure legend, the reader is referred to the online version of this book.)

In the US, the National Oceanic and Atmospheric Administration (NOAA) maintains the Wave-Watch III (WW3) model [22], which consists of a global model with a number of nested regional models providing wave forecasts in 3-hourly increments out to a 180-h horizon.[3] These models are typically very computationally intensive, and can also require some operational supervision, especially in areas of topographic complexity. Useful guidelines on wave forecasting methodology can be found in Ref. [23].

Good knowledge of the bathymetry is extremely important for accurate forecasts at shallower sites, and higher resolution models can provide potentially more accurate information. A common approach, and one that is likely to be taken by commercial wave energy projects, is to develop local, shallow water wave transformation models through tools such as SWAN [24]. These tools allow the user to take the open ocean forecast conditions (where such global forecasts are the most accurate) and incorporate a much higher resolution local bathymetry to generate the most accurate sea state predictions. This will be of particular importance for wave projects that are closer to shore [25,26].

3.5 Improving forecast accuracy

Forecasts may be improved once site measurement buoys are in place, by enhancing the frequency and accuracy of forecasts, through assimilation of localized buoy data into the forecast model. Statistical models such as regression analysis can be effective for short-term forecasts ($\sim$1–5 h) [27] and may also be combined with advanced predictive techniques to further refine the short-term forecasts of energy output. These techniques could include genetic algorithms, advanced pattern recognition, and intelligent model combinations, amongst others.

An adaptation of existing forecasting methods that can improve on the single-model approach is the ensemble forecast. Ensemble forecasting involves the generation of a number of unique model runs, each composed of varying physical parameters. For longer-term forecasts, this process generally results in improved results and can provide an estimate of forecast uncertainty. The NOAA WW3 model is available as an ensemble forecast [28]. A successful approach to developing an improved wave forecasting methodology using ensemble forecasts was the WaveSentry project employed in the UK [29]. This project couples ensemble wave forecasts with real-time observed wave data to provide adapted probabilistic wave forecasts. By using real-time measured data, it is possible to improve the typical deterministic ensemble forecast by dynamically selecting ensemble members which perform best.

3.5.1 Relationship to wave energy

By linking wave forecasts to validated device power matrices, it becomes straightforward to predict the energy output from a device at some specific future time horizon. A 2007 Electric Power Research Institute study [30] looked at the ability of the WW3 model without any local transformation to predict the climate at various time horizons. It concluded that for periods up to 48 h into the future, the prediction of spectral parameters was very good and after compensation for wave direction, a forecast error for H_s of less than 0.25 m was achievable, which is typically within the resolution bins provided on a power matrix. For time horizons beyond 48 h, the accuracy decreased by 50% to 72 h, after which it stayed relatively constant before deteriorating rapidly past 136 h.

---

[3]The frequency and resolution of the WW3 forecast is primarily defined by the computational requirements of the numerical model.

3.6 Direction and spectral sensitivity

Although it can be seen that the wave spectral parameters can be predicted accurately up to 48 h in advance, depending upon the device, other parameters not fully represented in the power matrix may be important when determining the device power output, such as wave direction, spreading, and spectral shape. Although these parameters can be predicted well, the sensitivity of the device may not be represented in the power matrix and hence may be an additional factor that will contribute to power uncertainty. Spectral and directional sensitivity will be unique to each WEC concept.

4. Summary

Wave and tidal energy have many advantages as a renewable resource, and as a result there is significant current interest in developing viable extraction technologies. Importantly, both of these resources offer the ability for prediction of power output that has the potential to be significantly more reliable than that currently offered for other variable renewable energy technologies.

Tidal power projects, including hydrokinetic installations, have the ability to provide accurate, long time-horizon predictions of energy output. Although the output from tidal turbines will vary from rated capacity during peak currents to no generation during slack water, once power curves are validated and performance data are established, power production can be accurately forecasted over both short and long time horizons. This forecast ability is very beneficial for the planning and long-term scheduling of the grid operator and may significantly elevate the value of this resource over other variable renewables.

In the case of wave energy, the high inertia of the resource provides an enhanced predictability in comparison to weather-based resources like wind and solar, and reduces the potential for unexpected or sudden ramping of generation stations at the site. By developing on existing methodology, using local numerical wave transformation models, and employing state-of-the-art wave forecasting routines, there is clearly the ability to develop reliable power predictions into the 48–136 h time horizon. As such, it is expected that the quality of forecasting for wave farms can achieve, and even surpass, that found in the now well-established wind industry.

References

[1] Khan J, Bhuyan G, Moshref A. Potential opportunities and differences associated with integration of ocean wave and marine current energy plants in comparison to wind energy. A report for the IEA-OES Annex III. Available from: www.iea-oceans.org; 2009.

[2] Santos M, Salcedo F, Ben Haim D, Mendia JL, Ricci P, Villate JL, et al. Integrating Wave and Tidal CurrentPower: Case Studies through Modelling and Simulation. A report prepared jointly by Tecnalia (Spain),- Powertech Labs (Canada) and HMRC (Ireland) for the OES-IA. Available: www.iea-oceans.org; 2011.

[3] Haas KA. Assessment of energy production potential from tidal streams in the United States. Georgia Tech Research Corporation; 2011.

[4] Foreman MG, Crawford WR, Marsden RF. De-tiding: theory and practice. Coast Estuar Stud 1995;47:203–39.

[5] Codiga DL. Unified tidal analysis and prediction using the UTide Matlab functions. Graduate School of Oceanography, University of Rhode Island; 2011.

[6] Kutney T, Karsten R, Polagye B. Priorities for reducing tidal energy resource uncertainty. In: European wave and tidal energy conference, Aalborg, Denmark; 2013.

[7] Stiven T, Couch SJ, Iyer AS. Assessing the impact of ADCP resolution and sampling rate on tidal current energy project economics. In: OCEANS, 2011 IEEE-Spain; 2011. pp. 1–10.

[8] Polagye B, Epler J, Thomson J. Limits to the predictability of tidal current energy. In: OCEANS conference 2010, Seattle, WA; September 2010.

[9] Iyer AS, Couch SJ, Harrison GP, Wallace AR. Variability and phasing of tidal current energy around the United Kingdom. Renew Energy Mar. 2013;51:343–57.

[10] Polagye B, Thomson J. Implications of tidal phasing for power generation at a tidal energy site. In: Proceedings of the 1st Marine Energy technology symposium, Washington D.C; April 2013.

[11] Falnes J. A review of wave-energy extraction. Mar Struct 2007;20:185–201. Elsevier.

[12] Evans DV. Power from water waves. Annu Rev Fluid Mech 1981;13(1):157–87.

[13] Cruz J. Ocean wave energy: current status and future perspectives. Springer; 2007.

[14] Falnes J. Ocean waves and oscillating systems. Cambridge University Press; 2004.

[15] Hasselmann K, Barnett TP, Bouws E, Carlson H, Cartwright DE, Enke K, et al. Measurements of wind-wave growth and swell decay during the Joint North Sea Wave Project (JONSWAP); 1973.

[16] Bretschneider CL. Wave variability and wave spectra for wind-generated gravity waves. Washington DC, Beach Erosion Board: US Army Corps of Engineers; 1959.

[17] Holthuijsen LH. Waves in oceanic and coastal waters. Cambridge University Press; 2007.

[18] Boheme T, Taylor J, Wallace AR, Bialaek J. Academic study: matching renewable electricity generation with demand: full report. Scottish Executive; 2006.

[19] Portilla J, Sosa J, Cavaleri L. Wave energy resources: wave climate and exploitation. Renew Energy September 2013;57:594–605.

[20] Carballo R, Iglesias G. A methodology to determine the power performance of wave energy converters at a particular coastal location. Energy Convers Manag September 2012;61:8–18.

[21] Kavanagh D, Keane A, Flynn D. Capacity value of wave power. IEEE Trans Power Syst February 2013; 28(1):412–20.

[22] Tolman HL. The 2002 release of WAVEWATCH III. In: 7th international workshop on wave hindcasting and forecasting; 2002. pp. 188–97.

[23] World Meteorological Organization. Guide to wave analysis and forecasting. Geneva, Switzerland: Secretariat of the World Meteorological Organization; 1998.

[24] Booij N, Holthuijsen LH, Ris RC. The 'SWAN' wave model for shallow water. Coast Eng Proc 1996;1(25).

[25] Cooper W, Saulter A, Hodgetts P. Guidelines for the use of metocean data through the life cycle of a marine renewable energy development. London: CIRIA; 2008.

[26] García-Medina G, Özkan-Haller H, Ruggiero P, Oskamp J. An Inner-Shelf Wave Forecasting System for the U.S. Pacific Northwest. Weather and Forecasting June 2013;28(3):681–703.

[27] Reikard G, Rogers WE. Forecasting ocean waves: comparing a physics-based model with statistical models. Coast Eng May 2011;58(5):409–16.

[28] Chen HS. Ensemble prediction of ocean waves at NCEP. In: Proceedings of 28th ocean engineering conference in Taiwan, NSYSU; 2006.

[29] Tozer N. WaveSentry – probabilistic wave forecasting with data integration. HR Wallingford Report; September 2013.

[30] Bedard R. Feasibility of using Wavewatch III for Days-Ahead output forecasting for grid connected wave energy projects in Washington and Oregon. EPRI, EPRI-WP-012; February 2008.

CHAPTER

German Renewable Energy Sources Pathway in the New Century

28

Matthias Müller-Mienack

1. Introduction

Looking at the climate goals of the European Union, the key factors are the so-called 20-20-20 targets: the aimed reduction of greenhouse gas emission by at least 20% up to 2020 compared with 1990, the increase of renewable energy share to 20%, and the increase of energy efficiency by 20%. More ambitious emission targets are stated by the German government, by the Federal Ministry for the Environment, Nature Conservation and Nuclear Safety.

The overall German decarbonization target is (compared with 1990 values) the reduction of German greenhouse gas emission by 40% in 2020 emission, by 55% in 2030, by 70% in 2040, and by 80–95% in 2050. Referring to the German electricity sector, the concrete targets regarding the renewable electricity generation and the reduction of electricity consumption are shown in Figure 1.

But what about the German status quo of usage of renewable energy sources (RES)? As a result of the German Renewable Energy Act and dedicated RES incentive scheme, significant progress has been achieved as shown in Figure 2. The total installed RES power by end of 2012 was already 76 GW, of which photovoltaic (PV) power was 33 GW, wind power was 31 GW, and biomass

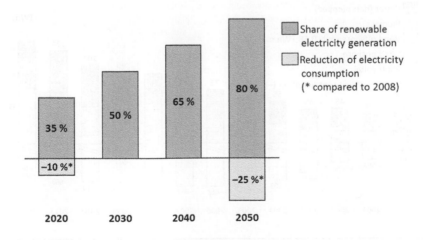

FIGURE 1

Electricity targets of official German Energy Concept 2020 [1].

Renewable Energy Integration. http://dx.doi.org/10.1016/B978-0-12-407910-6.00028-4

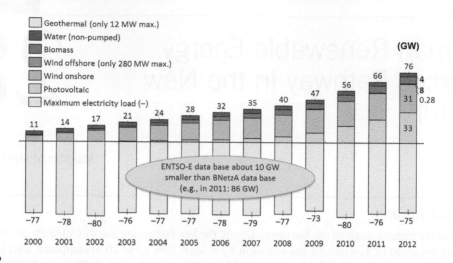

FIGURE 2

Installed renewable energy sources power in Germany versus peak load development in gigawatts [2].

power was 8 GW. In same year 2012, the German peak load was close to 80 GW (i.e. the same level as).

Looking at the real energy contribution by the RES, the EU goal of 20% share was already topped in Germany by end of 2012. Figure 3 shows the infeed development of different RES types compared with the yearly gross electricity consumption in Germany. For 2012, the ratio of RES energy contribution to total consumption already had reached a share of 22.8%.

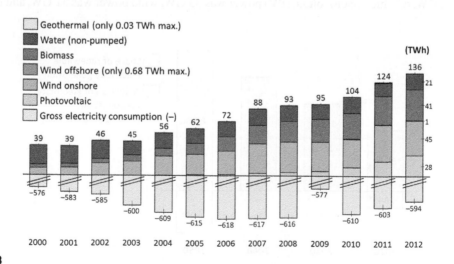

FIGURE 3

Infeeded renewable energy sources energy in Germany versus gross electricity consumption in terawatt-hours [3].

2. Increasing challenges of RES integration into the German electricity system

To understand the challenges caused by the ongoing decarbonization of the electricity supply in Germany, the special geographic situation has to be considered more in detail. As a consequence of the massive RES development in Germany, the neighboring transmission grids are affected as well, especially by RES loop flows stressing these foreign networks.

Figure 4 shows the four German transmission system operator (TSOs) as well as the discrepancy between the regional yield of renewable energies and the location of German load centers. In the northern part of Germany, significant wind development takes places; however, here only few industries respectively only few areas with heavy electricity loads can be found. In the southern part of Germany both are present: strong PV development and big industry load centers. In the control area of 50 Hertz as TSO for the eastern part of Germany, a share of 36% of RES energy contribution to the electricity consumption was reached for the year 2012. Thus, the first challenge that can be recognized in Figure 4 is the need for massive grid extension, which mainly is related to the north-south direction, to meet the German RES targets as displayed in Figure 1. Long-lasting permitting procedures of public authorities and the significant lack of public acceptance in the grid extension areas means several years will be required to build new extra-high-voltage transmission lines. The latest official grid studies have shown a need for grid extension measures of about €62 billion (meaning 62 000 000 000 €) over the next decade, of which €22 billion is needed for the onshore transmission grid investments until 2023 [5], €22 billion will be needed up to 2023 for the grid connection of the offshore wind farms in the German part of the Northern and Baltic Seas [6], and €18 billion will be needed up to 2020 for the reinforcement and extension of the distribution grids [7].

The second challenge is related to the daily operation of the German electricity system, especially for the four German TSOs responsible for the stability of the electricity system. For example, Figure 5 exemplarily shows a squall line week around the end of January 2013, and Figure 6 shows strong PV infeed in Germany on June 5, 2013. Both graphs split the infeed into the single contributions of the control areas of the four German TSOs, and the EPEX electricity spot price curve for the same time frames is displayed in parallel.

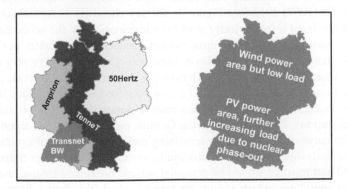

FIGURE 4

The four German TSOs and the RES distribution among Germany [4]. PV, photovoltaic.

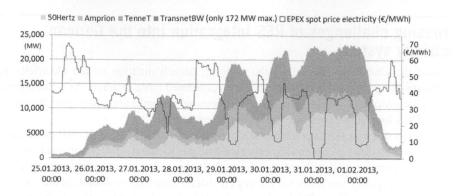

FIGURE 5

Squall line in Germany and electricity spot price curve at the end of January 2013 [8,9].

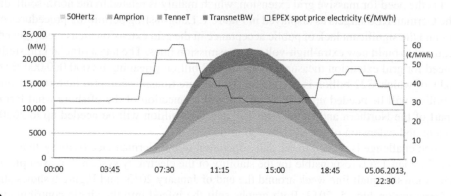

FIGURE 6

Photovoltaic infeed and electricity spot price curve on June 5, 2013 [8,9].

Looking at the high and fluctuating RES infeed values compared with the maximum electricity load in Germany of 75 GW in 2012 (see Figure 2), the day-ahead (D-1) forecast deviations for wind and PV infeed that has legal infeed priority is of high importance for the scheduled planning of the conventional power stations. The forecast error has been reduced significantly in Germany in recent years. Nevertheless, there remain uncertainties at some points in time that are very critical for system stability. For instance on January 19, 2012, in Germany, a D-1 wind infeed forecast deviation of 7 GW compared with real-time infeed was registered (i.e., a wind infeed surplus of about 70% occured). This clearly exceeds the available contracted German reserve power that still is provided mainly by the conventional power stations. Thus, sometimes the German electricity system depends on the reserves from outside of Germany—that is, from the remaining part of the European synchronous network.

Considering the daily operation of the electricity system, overloads on transmission lines are mainly RES related and must be solved by remedial actions organized by the responsible TSO. As a first step, special switching measure must be taken to push the power flows to other lines. If insufficient or not possible, so-called redispatch measures must be taken—that is, the infeed of conventional power stations and pumped storage power stations located closely to this congestion needs to be adapted to further decrease the loading of the congested line. Related to the control area of the TSO 50 Hertz, in 2012, redispatch measures were necessary on 262 days with a total volume of 2.8 TWh [10].

Another challenge related to the daily operation occurs when the switching and redispatch measures are not sufficient to secure system stability in certain times. Then, the TSOs are mandated and obliged to regulate generation and loads. In the first step, because of the legal priority given to RES, the conventional power stations need to be curtailed, but the provision of required ancillary services must be considered (especially reserve power, reactive power for voltage stability, short-circuit power, black start capability for grid restoration). If this measure is insufficient, in a second step, RES infeed must be curtailed by the TSOs and relevant distribution system operators DSOs, respectively. For instance in 2012, the TSO 50 Hertz had to curtail RES on 77 days, which amounted to 120 GWh (without consideration of the curtailment measures taken by the DSOs). Because of the unbundling of the electricity supply chain in Germany, a main problem with RES curtailment management is the TSO's lack of real-time RES infeed information (i.e. observability and dedicated controllability). This lack of information is due to the fact that the majority of RES is decentralized and therefore connected to the DSO grids. Despite the given legal regulation for the DSO to provide data about RES infeed, it will take several years before the TSOs have the complete picture about what is happening at the DSO level. Obviously, the smartgrid rollout road map for Germany will push forward this issue [10].

The fourth challenge is the ongoing nuclear phase out in Germany: 8.3 GW of installed nuclear power was shut down in 2011 after the Fukushima accident in Japan; the remaining 12.1 GW will be phased out by 2022. As the affected nuclear power stations mainly are located in the southern part of Germany (see Figure 4), in periods with high wind power infeed in the northern part of Germany, the phase-out caused increasing overloads and thus redispatch became necessary, especially on the north-south directed lines. Moreover, especially in the southern part of Germany, significant voltage problems are the consequence; and in the winter, electricity shortages also were observed.

In terms of energy market design, there is still not enough market for flexibility products in Germany to address the operational challenges. The significant influence of wind power and PV infeed on the electricity supply is shown by the response of electricity spot price curves in Figures 5 and 6. In the classic peak price time at noon, the PV infeed increasingly reduces the peak price. As a consequence, the discussion on large-scale decommissioning of fast and flexible gas-fired stations already has started. On the other hand, during off-peak times, the nuclear phase-out increases the off-peak prices. Here, the fast and flexible pumped storage stations are affected as the decreasing price spread is making it difficult to justify the business case.

3. Future outlook

To decarbonize the electricity system requires special risks to be solved. So, in the beginning of PV deployment, the value of 50.2 Hz frequency was defined as the infeed threshold. Upon reaching this value, all PV units had to be stopped to contribute to system stability. But considering the significantly growing PV generation, the implemented infeed stop at 50.2 Hz turned out to be a risk factor for system stability. To lose 20 GW of PV infeed in a very short time theoretically would mean a sudden frequency drop of about 1 Hz, taking into account the strength of whole synchronous European

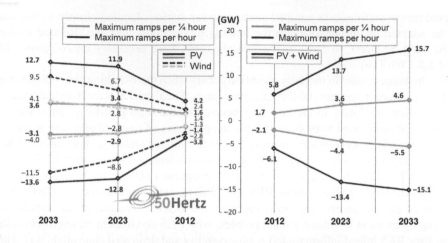

FIGURE 7

Power ramps by German photovoltaic and wind power—status quo and future outlook [11].

electricity system. As a result, all new PV devices now need to follow a new regime, but the older German PV devices continue to present a pan-European stability risk, which becomes clear when looking at the frequency peaks of 2011 (50.16 Hz) or 2012 (50.13 Hz).

On the basis of the general observation of what is happening in the German electricity sector (including this example of a PV-related challenge), it is clear that there are not enough flexibility products in the market to cope with the grid operational challenges.

Nevertheless, the need for flexibility will grow steadily on the pathway toward further decarbonization of Germany's electricity sector. For instance, looking at the power ramps caused by fluctuating RES infeed, more than 4 GW infeed change per hour by PV infeed already has been observed in Germany and almost 3 GW has been observed as PV infeed change per quarter hour (see Figure 7). Looking toward 2033, more than 15 GW infeed change per hour by PV and wind power is assumed. Without massive deployment of demand response and storage in future, these power ramp numbers will be unmanageable from the current view not only for Germany but also for the entire synchronous European system.

Finally, as main message for Germany, apart from the massive need for grid reinforcement and extension, to push forward with the decarbonization of the electricity system, the exceptional treatment of RES units eventually must be replaced by increasingly requiring the same ancillary services of RES units as the conventional power stations currently provide. Moreover, the energy market design must be reviewed and provide sufficient mechanisms for flexibility products. With such measures, Germany could remain a leader in the context of decarbonization of electricity supply.

References

[1] GridLab, German Federal Ministry for the Environment, Nature Conservation and Nuclear Safety.
[2] GridLab, German Federal Ministry for the Environment, Nature Conservation and Nuclear Safety, Erneuerbare Energien 2013, February 2013; UCTE/ENTSO-E "System Adequacy Retrospect" reports, 2000–2011.

[3] GridLab, German Federal Ministry for the Environment, Nature Conservation and Nuclear Safety, Entwicklung der erneuerbaren Energien in Deutschland im Jahr 2012, February 2013.

[4] 50Hertz, GridLab.

[5] The 4 German TSOs. Grid development plan 2013. Second draft for consultation; July 17, 2013.

[6] The 4 German TSOs. Offshore grid development plan 2013. Second draft for consultation; June 24, 2013.

[7] German Energy Agency (dena). dena-Verteilnetzstudie; December 11, 2012.

[8] EEX, www.transparency.eex.com, August 5, 2013.

[9] EEX, http://www.eex.com, August 5, 2013

[10] 50Hertz, Almanach 2012.

[11] 50Hertz Transmission GmbH, RES power ramp update 2013.

Enabling and Disruptive Technologies for Renewable Integration

Control of Power Systems with High Penetration Variable Generation

29

Christopher L. DeMarco[1], Chaitanya A. Baone[2]

[1] *Department of Electrical and Computer Engineering, University of Wisconsin-Madison, Madison, WI 53706, USA,*
[2] *Electric Power Systems Lab, Power Conversion and Delivery, GE Global Research, Niskayuna, NY 12309, USA*

1. Introduction and motivation

Throughout many parts of the world, installed capacity and production levels for renewable electric generation are growing rapidly. Among renewable sources, both wind turbines and photovoltaics present interesting new challenges to grid operations as they come to represent a larger share of the generator mix. Traditional synchronous generators driven by steam, water, or gas turbine prime movers have long dominated electricity production, and in many ways their characteristics implicitly underlie the philosophies of primary and secondary control practice for grid frequency and active power. This chapter will consider the impact on control design of potentially very different characteristics from the new power delivery technologies of renewables.

If increased penetration of renewable generation brings a new class of control actuators to the power grid, in a complementary fashion, increasing penetration of phasor measurement units (PMUs) brings a new class of sensor technologies. A key aspect of PMU technology relative to traditional measurements is their much higher sampling rate and bandwidth. PMUs typically report at 30 or 60 times per second, versus older systems that sampled only once per several seconds. PMUs thereby offer a high dimension vector of wide-area signals, with bandwidth that captures the behavior most significant to power system electromechanical dynamics. These measurements open new opportunities for dynamic state observation and improvements in control.

While important progress to exploit these advances has been made in prior research and in vendors' controller technology implementations, many have been limited by design choices that seek to force renewable generation to behave much like traditional synchronous machines. Many such works propose supplemental controls that try to mimic the inertial characteristics inherent to traditional generators [1,2]. The work here approaches renewables' control design differently, seeking to answer: (1) what are the objectives associated with primary control that are necessary to maintain stable, secure operation of the power grid; and (2) what control action can renewables and storage provide to meet these objectives, while remaining within their operating limits?

2. The case for advanced control methodologies

Future grid control design requires techniques that respect differing bandwidth and operational limits available from the diverse technologies of renewables and storage. To this end, the tools of optimal

Renewable Energy Integration. http://dx.doi.org/10.1016/B978-0-12-407910-6.00029-6

control [3], and more recent advances based on convex optimization [4], hold great promise for active power/frequency control designs that better facilitate high penetration renewables. To place such approaches in context, recall the history of optimal control in the power systems literature of the 1970s and early 1980s. Beginning from Elgerd's classic 1970 paper [5], a number of authors demonstrated the potential for improved performance via application of optimal control designs to generator governors and AGC [6,7]. The drawbacks in these approaches were almost entirely in practical limitations on implementation. Most methods required dynamic observation of full system states, and were beyond the reach of 1970s and 1980s grid sensor technologies. Today, high bandwidth PMU measurements, coupled with tremendous advances in real-time computational power, make it attractive to revisit optimal control for renewable generation.

3. The roles of inertial response, primary control, and secondary control: past and future

Large electric power grids are complex dynamical systems, displaying electromechanical coupling on a nearly continental scale. Electrical frequencies are among the key output variables reflecting electromechanical behavior. At locations with synchronous generation attached, the inherent physics of the synchronous machine dictate that electrical frequency of the generator voltage and mechanical rotational speed of the generator are locked in fixed proportion to each other, so that variation of frequency directly reflects deviations in rotational speed away from the desired steady state. Moreover, the nature of AC power transmission is such that a synchronous region is in exact equilibrium only if electrical frequency is equal at every node in the network. In these basic observations lie two key requirements of grid control: (1) in credible disturbance scenarios that cause individual generator frequencies/speeds to deviate, all should asymptotically return in a stable fashion to an equal value; (2) this shared equilibrium frequency should be regulated to a tight band about its desired 60 Hz value.

Among available wind generation technologies, variable speed wind turbines utilizing doubly fed induction generators (DFIGs) have gained wide acceptance, and use of fully powered electronically coupled machines is growing. In DFIGs and other more recent designs, the turbine's mechanical inertia is not fully coupled to its electrical terminal characteristics, unlike a traditional turbine/synchronous machine generator set. If such a source substitutes for synchronous generation, it reduces inertial response in the grid. Several studies have looked at the consequences of reduced system inertia due to increasing penetration of wind generation [8–10].

Large-scale energy storage technologies such as batteries, high-speed flywheels, and electro-chemical capacitors are currently being researched, developed, and deployed for grid applications [11]. Storage applications can be loosely divided into power applications and energy management applications, which are differentiated based on the storage/discharge durations. Our focus here will be on short-duration storage technologies that can be used to provide control power during transients and assist in primary frequency control. Storage technologies that suit such applications, typically grid scale batteries, can respond at a much faster rate than the mechanical action of traditional governor controls, or of blade pitch wind turbine speed control mechanisms.

For control action available from renewable generation and electrical storage technologies, complementary saturation and bandwidth limits are quite natural. To represent these, we consider two broad classes of actuators: (1) low bandwidth, slow actuators with broad saturation limits (e.g. power control available by varying blade pitch in wind generators, or from traditional governor controls in synchronous generators), and (2) high bandwidth, fast actuators with narrow saturation limits (e.g. power control available from battery or flywheel energy storage). With such characterizations of multiple actuators, one may formulate the problem in the context of multi-input control objective, that of stable regulation of grid frequency during disturbances. Utilizing the broad bandwidth, wide-area measurements available from PMUs, it becomes feasible to consider state feedback control, in which system state is estimated via a dynamic observer.

4. Frequency regulation in power systems

Frequency control in power systems is often undertaken at two levels and time frames, primary control and secondary control [12]. The term *primary frequency control* is traditionally reserved for local, automatic control action that adjusts the active power output of the generating units in direct response to measured frequency variations [13], and with the added goal of restoring balance between load and generation. Its design should be chosen to achieve its regulation objective in a stable fashion, following credible generation or load disturbances. Such control is commonly exercised through continuously acting turbine governors, with operating time scale of seconds. The direct feedback term (the droop) is independent of any centralized control signal, using only the local measurement of rotational speed or terminal electrical frequency. While the control designs to be considered here will extend beyond simple governor droop, the focus will be on primary generation (and storage) control.

4.1 Frequency regulation with emerging grid technologies

Until recently, wind energy sources and other renewables were typically not asked to contribute to primary control. While it is accepted operating practice that only a subset of generators need exercise governor control, any plant not contributing to primary control is in a sense a "free rider" on the governor action provided by others [14]. However, even when not exercising governor control, traditional synchronous machines inherently contribute to moderating the time rate of change of frequency, just by virtue of inertia. In [15], the authors examine possible degradation of grid performance when DFIG wind turbines displace synchronous generators; to offset this degradation, they propose a supplementary control that yields an inertia-like response in DFIG generators. Using rotational speed/electrical frequency error as the measured output, an inertia-like control feeds back the derivative of this output. Hence, inertia-like control is nothing more than derivative feedback, with droop control being proportional feedback. Textbook results of classical control theory suggest that simple proportional/derivative feedback is unlikely to yield good performance in high order, lightly damped systems such as the power grid. It is not surprising therefore that concerns have been raised regarding effectiveness of inertia-emulating control for wind turbines [14].

4.2 Western electricity coordinating council wind turbine generator system model

To examine the impact of control designs for new actuator technologies, it is useful to have a credible model of the actuator of interest. For representation of DFIG wind turbine generators, an industry-standard dynamic model has been developed, the "Type-3 Wind Turbine Generator (WTG) Western

Electricity Coordinating Council (WECC) generic model," detailed in [16]. Two coupled rotating masses represent the mechanical behavior of the turbine-generator system, allowing torsional stress on the coupling between the two rotating masses to be considered. The equations governing this wind turbine model are given by

$$2H_t\Delta\dot{\omega}_t = T_{\text{mech}} - K\delta_{\text{tg}} - D_{\text{shaft}}\Delta\omega_{\text{tg}} \tag{1}$$

$$2H_g\Delta\dot{\omega}_g = -T_{\text{elec}} + K\delta_{\text{tg}} + D_{\text{shaft}}\Delta\omega_{\text{tg}}. \tag{2}$$

Here H_t, H_g are the turbine and generator inertia constants, respectively, $\dot{\omega}_t$, $\dot{\omega}_g$ are the time derivatives of turbine and generator rotational speeds, respectively, T_{mech} is the turbine side torque, T_{elec} is the generator side torque, ω_{tg} is the difference in the turbine and generator rotational speeds, δ_{tg} is the difference in the turbine and generator angles, respectively, K is the shaft stiffness, and D_{shaft} is the shaft damping constant. All quantities are expressed as per unit (p.u.) values. To measure torsional stress on the drivetrain shaft, we consider $(T_{\text{mech}} - T_{\text{elec}})$, which can be expressed as a function of state by adding Eqns (1) and (2)

$$2H_t\Delta\dot{\omega}_t + 2H_g\Delta\dot{\omega}_g = (T_{\text{mech}} - T_{\text{elec}}). \tag{3}$$

The time derivatives $\dot{\omega}_t$ and $\dot{\omega}_g$ can be written as a linear combination of the states of the overall system, which includes the power system, the storage devices, and the WTG systems.

4.3 Supplemental primary control: model of power delivery from battery storage

Battery energy storage systems with power electronic grid coupling can have very high bandwidth in their linearized transfer characteristic, describing commanded-to-grid-delivered electric power. This has the potential to complement the low bandwidth inherent in the mechanical action of wind turbine blade pitch control. However, from the standpoint of grid applications, many practically sized battery systems are very limited in their maximum energy and power. Batteries therefore represent a class of actuator having narrow saturation limits on their available control action, but broad bandwidth. Work in [17] develops linearized models for lithium-ion battery terminal behavior. When examined in the frequency domain, these yield a high-pass filter characteristic, confirming our premise regarding the qualitative frequency domain behavior of a battery.

4.4 Model of wide area electromechanical dynamics for control design

The state space formulation of the power system model representing wide area electrodynamics is assembled as below. For simplicity, we omit details of the disturbance model that represents fast time scale variations in wind input power; the interested reader is referred to [18]. The electromechanical behavior of a traditional synchronous generator can be captured by standard swing equation models [13]. The frequency ω and angle δ are related by definition

$$\Delta\dot{\delta} = \Delta\omega \tag{4}$$

The equation governing the generator dynamics is given by

$$M\Delta\dot{\omega} = \Delta P_m - \Delta P_e - D\Delta\omega \tag{5}$$

where M is the normalized rotational inertia of the generator, D is the damping constant, ΔP_m is the mechanical shaft power input to the generator, and ΔP_e is electrical power output of the generator.

The state equations governing the various subsystems (synchronous generators, wind generators, and energy storage devices) can be written together with the network power balance equations in the form

$$\Delta \dot{x}_{\text{sys}} = A_{\text{sys}} \Delta x_{\text{sys}} + B_{\text{grid}} \Delta y_{\text{grid}} + B_{\text{sys}} \Delta u_{\text{sys}} + E_w \Delta w \tag{6}$$

$$0 = C_{\text{grid}} \Delta x_{\text{sys}} + D_{\text{grid}} \Delta y_{\text{grid}} \tag{7}$$

where, x_{sys} are the overall system dynamic states, y_{grid} are the algebraic variables of the power flow, and u_{sys} are the inputs (battery inputs). Barring degenerate operating conditions that yield a singular Jacobian, the algebraic variables can be solved in terms of the dynamic variables using Eqn (7) and substituted in Eqn (6) to form a standard state equation.

5. Optimal control design

In this section, we briefly review a linear quadratic (LQ) regulator–based design methodology as described in the research monograph [19], which is explicitly developed for applications in which the control action is subject to hard saturation limits. The presentation below exploits the saturation-bandwidth characteristics relevant to the case study scenario of wind turbine pitch control supplemented by battery storage. We then describe a distributed observer-based control scheme utilizing one local and one remote PMU measurement at each location.

In the optimal control design method of the monograph [19], the objective is output regulation against disturbances (e.g. wind power variations) characterized by an exosystem, where the exosystem is explicitly constructed as part of the overall model. Consider a standard linear state space description for the small signal behavior of the system of interest:

$$\dot{x} = Ax + B\sigma(u) + E_w w \tag{8}$$

$$\dot{w} = Sw \tag{9}$$

$$y = Cx + D_{yw} w. \tag{10}$$

Equation (8) describes the plant with state $x \in \mathbb{R}^n$, and control input $u \in \mathbb{R}^m$, subject to the effect of an exogenous disturbance represented by $E_w w$, where $w \in \mathbb{R}^s$ is the state of an exosystem. Equation (9) describes the state space realization of the autonomous exosystem. The output is $y \in \mathbb{R}^p$, and σ is a normalized vector-valued saturation function indicating the saturation limits on the input.

Conditions for solvability of the output regulation problem are:

1. All the eigenvalues of A lie in the closed left-half plane, and the pair (A, B) is stabilizable.
2. There exist matrices Π and Γ such that they solve the regulator equation

$$\Pi S = A\Pi + B\Gamma + E_w \tag{11}$$

$$0 = C\Pi \tag{12}$$

3. There exists a $\delta > 0$ and a time $T \geq 0$ such that $\left\| \Gamma w \right\|_{\infty, T} \leq 1 - \delta$ for any *allowable* initial condition on the exosystem.

The desired matrix P_ε is obtained by solution to the algebraic Riccati equation (ARE), given by

$$P_\varepsilon A + A^T P_\varepsilon - P_\varepsilon B B^T P_\varepsilon + Q_\varepsilon = 0 \qquad (13)$$

where, $Q : (0, 1] \to \mathbb{R}^{n \times n}$ is a continuously differentiable matrix-valued function such that $Q_\varepsilon > 0, \frac{dQ_\varepsilon}{d\varepsilon} > 0$ for any $\varepsilon \in (0, 1]$, and $\lim_{\varepsilon \to 0} Q_\varepsilon = 0_{n \times n}$.

The solution of Eqn (12) yields a unique positive definite P_ε that is continuously differentiable with respect to ε, is monotonically increasing with ε, and $\lim_{\varepsilon \to 0} P_\varepsilon = 0$.

From this solution, one then constructs the state feedback gain matrix F_ε as

$$F_\varepsilon = -B^T P_\varepsilon. \qquad (14)$$

where, in the terminology of [19], ε is the low gain parameter. Feedback using F_ε yields an asymptotically stable undisturbed system for any $\varepsilon \in (0, 1]$. The feedback control law is given by

$$u = F_\varepsilon x + [\Gamma - F_\varepsilon \Pi] w. \qquad (15)$$

6. Distributed control design for practical implementation

In a large-scale power system, with actuators and measurements spread over distances of many hundreds of square kilometers, a real-time full-state feedback control scheme that communicates measurements and commands to every controller is clearly impractical. Therefore, here we propose a distributed observer-based scheme using minimal communication; the observer is allowed to use any local measurement signal but is restricted to one remote measurement signal.

The controller design consists of two components: observer design and state feedback design. For the feedback design, the low gain design method described in the previous subsection is adopted. The partitioning of the Q matrix is done by looking at the most controllable modes among a subset of critical modes "assigned" to a given actuator (i.e. a given generator or storage unit is responsible for improving damping of only a small number of modes). This feedback control is fed by the local observer that dynamically estimates system states.

For the observer, the dynamic equation corresponding to input channel i is given by

$$\Delta \dot{\hat{x}}^i_{comp} = \overline{A}_{comp} \Delta \hat{x}^i_{comp} + \overline{b}^i_{comp} \Delta \overline{u}_i + \overline{E}_w \Delta w + L^i \left(\Delta \hat{y}^i - \overline{C}^i_{comp} \Delta \hat{x}^i_{comp} \right) \qquad (16)$$

Here i denotes the observer/actuator location index, x_{comp} denotes the states estimated by the observer, A_{comp} denotes the state matrix of the composite system (plant and actuators together), b_{comp} denotes the ith column of the input matrix B_{comp}, U_i denotes the input to be fed to the actuator, L_i denotes the observer gain matrix, and C_{comp} denotes the output matrix. The matrix L_i is designed according to the method described in [18]. The dynamics of the system are governed by

$$\Delta \dot{\overline{x}}_{comp} = \overline{A}_{comp} \Delta \overline{x}_{comp} + \sum_{i=1}^{k} \left(\overline{b}^i_{comp} \Delta \overline{u}_i \right) + \overline{E}_w \Delta w. \qquad (17)$$

Here k denotes the number of actuators in the system, and $\bar{x}_{\text{comp}}$ denotes the states of the composite system. The control law U_i is given by

$$\Delta \bar{u}_i = F_e^i \Delta \hat{x}_{comp}^i + \left(\Gamma^i - F_e^i \Pi^i \right) \Delta w \tag{18}$$

where, F_e^i, Γ^i, and Π^i are the state feedback and solutions of the regulator equations, respectively, for a particular location, indexed by i.

7. Case study results: multiobjective evaluation of optimal control performance

To provide a case study, we employ a modification of the standard IEEE 14 bus system to illustrate the effectiveness of the control methods for frequency regulation, and reduction in torsional stress on the wind turbine drivetrain. For this case study, two of the five synchronous generators are replaced by wind generators. A disturbance input is created as periodic wind power variation at each of the wind generators, modeled using an exosystem state equation [18]. To assess performance, we monitor the weighted sum of frequencies from each of the traditional synchronous generator locations (Note: this signal is used for assessment only, not in the design, as it would undermine distributed control). The resulting plot of frequency without any supplemental control from a storage device is shown in Figure 1. As seen in the figure, frequency deviations have a steady state amplitude of about 0.9 Hz, representing the effect of the disturbance input. The resulting turbine and rotor torque difference (torsional stress) for one of the wind generators is shown in Figure 2.

Next, we demonstrate the performance of a distributed observer-based control method on this test system. We have one observer-based controller feeding a battery at a traditional synchronous generator location. Since the actuator associated with the wind generation system is already part of the WECC

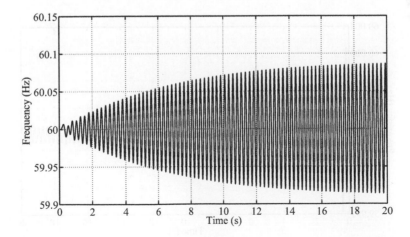

FIGURE 1

Frequency variation, no control case—it displays periodic variation due to the periodic input disturbance.

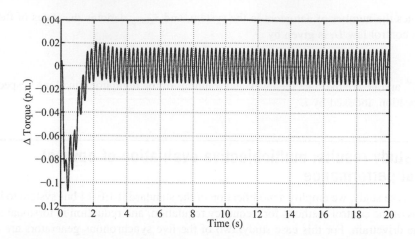

FIGURE 2

Drivetrain stress indicated by torque difference ($T_{mech} - T_{elec}$), no control case.

WTG model, we avoid adding any supplemental feedback to the wind machine. It acts only with its standard blade pitch control, as represented in the WECC model. The design focuses on the battery controller, with objectives: (1) force frequency deviations toward zero, and (2) reduce the stress on the drivetrain of the wind system. The control design is performed on a low dimensional subspace associated with a subset of lightly damped, oscillatory modes in the system. The local generator frequency and angle measurements, and one remote measurement of another generator frequency, are used by the observer to form the state estimates.

We demonstrate the distributed control scheme, with the dual objective of frequency regulation and drivetrain stress reduction. The drivetrain stress objective is included by having the turbine and rotor

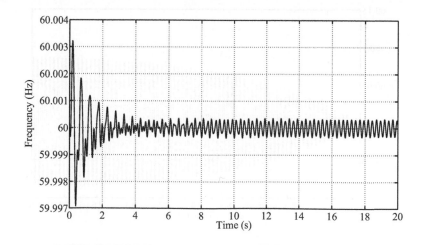

FIGURE 3

Frequency behavior with feedback design—control objective includes torque difference.

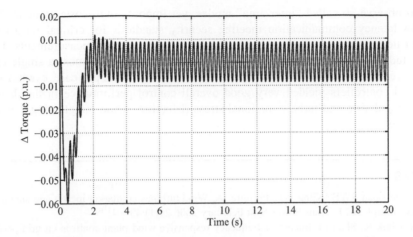

FIGURE 4

Torque difference ($T_{mech} - T_{elec}$) behavior with feedback design—control objective includes torque difference.

torque difference constructed as an explicit state in the control design, and contributing to the quadratic objective function. The resulting plot of frequency is shown in Figure 3. Clearly, the controller is able to regulate the frequency to within acceptable band of regulation about its nominal 60 Hz value. The resulting plot of turbine and rotor torque difference is shown in Figure 4. Comparing this with the torque difference for the case without any control in Figure 2, it can be seen that this design also significantly reduces the drivetrain stress.

To ensure that the design is respecting battery limits, the peak per unit power produced/absorbed by the battery is evaluated, and found to be less than 2×10^{-3} p.u. The system MVA base for this example is 100 MVA, so this per unit figure corresponds to 200 KW. For a battery installation with power capability of 2 MW (typical of test installations today), the amount used for frequency regulation would represent approximately 10% of maximum capability.

8. Conclusions

This chapter and its accompanying case study have considered a distributed design for grid frequency control using an industry-standard wind-powered electric-generation model, supplemented by a modest amount of controllable storage. The optimal control design addresses dual goals: the system-wide objective of contributing to grid frequency regulation, and the local, "equipment-centric" goal of minimizing mechanical stress in the wind turbines. The nature of the design also tailors it to the electric power application in other ways. First, control effort is focused on behavior within a subspace associated with a small number of lightly damped electromechanical modes of interest. This greatly reduces communication and computation needs, with each local controller having an observer that estimates state behavior only for the small number of modes "needing" control. In the power systems context, use of geographically remote measurements is sometimes necessary to observe so-called "inter-area" modes, yet imposes high costs associated with long-distance, low-latency communication meeting security standards for critical infrastructure. Our local control design employs the minimum number of such remote measurements. In the simulation case study examined, with an overall state dimension of 36, at most a single remote measurement proved adequate for excellent performance. In the example case examined, the local observer-based controllers yielded very good overall control performance in the dual objectives of frequency regulation and turbine torsional stress reduction, while maintaining actuation within saturation limits.

References

[1] Morren J, de Haan SWH, Kling WL, Ferreira JA. Wind turbines emulating inertia and supporting primary frequency control. IEEE Trans Power Syst February 2006;21(1):433–4.
[2] Miller N, Clark K, Shao M. Impact of frequency responsive wind plant controls on grid performance. In: National renewable energy laboratory workshop on active power control from wind power; January 27, 2011. Boulder, CO.
[3] Anderson BDO, Moore JB. Optimal control linear quadratic methods. Englewood Cliffs (NJ): Prentice Hall; 1990.
[4] Boyd S, El Ghaoui L, Feron E, Balakrishnan V. Linear matrix inequalities in control and system theory. Philadelphia (PA): Society for Industrial and Applied Mathematics (SIAM); 1994.
[5] Fosha CE, Elgerd OI. The megawatt frequency control problem: a new approach via optimal control theory. IEEE Trans Power App Syst April 1970;PAS-89(4):563–77.
[6] Tacker EC, Lee CC, Reddoch TW, Tan TO, Julich PM. Optimal control of interconnected electric energy systems: a new formulation. Proc IEEE 1972;60(10):1239–41.
[7] Mukai H, Singh J, Spare JH, Zaborszky JA. A reevaluation of the normal state control of the power system using computer control and system theory, Part-III: tracking the dispatch targets with unit control. IEEE Trans Power App Syst January 1981;PAS-100(1):309–17.
[8] Eto JH, Undrill J, Mackin P, Daschmans R, Williams B, Haney B, et al. Use of frequency response metrics to assess the planning and operating requirements for reliable integration of variable renewable generation. Federal Energy Regulatory Commission report LBNL–4142E; December 2010.
[9] Vittal V, McCalley J, Ajjarapu V, Shanbhag U. Impact of increased DFIG wind penetration on power systems and markets. PSERC Publication; October 09–10, 2009.
[10] Doherty R, Mullane A, Nolan G, Burke DJ, Bryson A, O'Malley M. An assessment of the impact of wind generation on system frequency control. IEEE Trans Power Syst February 2010;25(1):452–60.
[11] DOE report Electric power industry needs for grid-scale storage applications; December 2010.
[12] Jaleeli N, VanSlyck LS, Ewart DN, Fink LH, Hoffmann AG. Understanding automatic generation control. IEEE Trans Power Syst August 1992;7(3):1106–22.
[13] Wood AJ, Wollenberg BF. Power generation, operation and control. New York (USA): Wiley; 1996.
[14] Undrill J. Power and frequency control as it relates to wind-powered generation. Federal Energy Regulatory Commission report LBNL–4143E; December 2010.
[15] Lalor G, Mullane A, O'Malley M. Frequency control and wind turbine technologies. IEEE Trans Power Syst November 2005;20(4):1905–13.
[16] Generic Type-3 wind turbine-generator model for grid studies. WECC Wind Generator Modeling Group; September 2006. Version 1.1.
[17] Johnson VH, Pesaran AA, Sack T. Temperature-dependent battery models for high-power lithium-ion batteries. In: Presented at the 17th electric vehicle symposium, Montreal, Canada; October 2000.

[18] Baone CA, DeMarco CL. From each according to its ability: distributed grid regulation with bandwidth and saturation limits in wind generation and battery storage. IEEE Trans Control Syst Technol 2013;21(2): 384–94.

[19] Saberi A, Sannuti P, Stoorvogel A. Control of linear systems with regulation and input constraints. London (UK): Springer-Verlag; 2000.

Enhancing Situation Awareness in Power Systems: Overcoming Uncertainty and Variability with Renewable Resources

30

Mica R. Endsley[1], Erik S. Connors[2]

[1] *United States Air Force,* [2] *SA, Technologies, Inc.*

1. Introduction

Electricity blackouts can cost nations billions of dollars while severely disrupting the day-to-day lives of ordinary people. These costly impacts have been seen in recent years following blackouts across large geographic regions in the United States, Europe, Australia, India, and other parts of the world. The August 14, 2003, power outages in North America, for example, affected an estimated 50 million people across eight U.S. states and the province of Ontario. The total estimated cost in the United States alone ranged between $4 and $10 billion [1]. The joint U.S.–Canada Power System Outage Task Force identified four major causes of the 2003 North American blackout in their 2004 final report. Prominent among these causes was "inadequate situation awareness." The Joint Task Force remarked that training deficiencies, ineffective communications, and inadequate reliability tools and backup capabilities all contributed to a lack of situation awareness (SA) for the operators involved. Similarly, the power outage experienced in Europe on November 6, 2006, left 15 million EU citizens without power for several hours. A report by the German Utility EON on the incidents and causes of this blackout listed lack of coordination and miscommunication—characteristics of operations that directly affect *team situation awareness*—as causal factors. Other blackout incidents, such as Scandinavia and London in 2003, and Australia in 2006, have been related to decreased SA, which resulted from operators' poor mental models about modes of operations and lack of knowledge regarding the health of transmission and distribution (T&D) equipment. Most recently, the 2012 blackout in India, where inadequate SA was attributed as a major factor, affected more than 600 million people.

Examining the problem more broadly shows that many of these major outages exhibit negative impacts both directly and indirectly attributable to a lack of SA, including the inability of system operators and coordinators to visualize events across the entire system. Reliable, sustained operation of today's highly interconnected T&D systems cannot be accomplished safely and effectively without high levels of SA. Robust and efficient operations require that system operators have the right information at the right time and that this information be provided in an effective manner so that they can fully understand the state of a complex and dynamically changing cyber and physical systems, allowing them to project future changes and respond in a timely manner.

In addition, as the industry advances toward energy-friendly smarter grids, compensating for the complications and challenges that variable generation imposes on operator SA within

T&D control centers becomes increasingly important. Supporting the overall accuracy of opera-tors' decision making and performance in the face of rapidly increasing system complexity associated with renewable energy sources is critical to the overall safety and reliability of T&D operations.

Renewable generation sources, such as wind and solar, introduce a number of new challenges because of their intermittent nature. These power sources do not generate energy on demand or per easily predictable schedules, unlike traditional sources such as hydroelectric, nuclear, or coal-powered generation plants. Wind energy, for example, tends to increase at night, generating the most energy when system load is lowest, and it can be quite variable in both timing and intensity. Solar power has different timing, but it is accompanied by similar variability and reliability challenges.

Operators often struggle to incorporate these relatively unpredictable and variable energy sources into daily operations that strive to be scheduled, projected, and exact. Although generation output for these sources can be forecasted in advance, poor reliability of those forecasts is a common problem. Additional challenges with renewable power integration include understanding its effects on power system operating cost, power quality, power imbalances, power system dynamics, and transmission planning [2]. These characteristics of renewable resource generation increase the levels of uncertainty and variability in power grid operations, significantly degrading operator SA and simultaneously creating an increased inherent need for operating flexibility.

1.1 What is situation awareness? Why is it important?

Throughout this discussion, the prevailing factor for ensuring optimal operator performance and de-cision making in the face of the aforementioned industry trends remains SA—but what exactly is SA? One of the earliest and most widely used definitions of SA describes it as the "perception of the elements in the environment within a volume of time and space, the comprehension of their meaning and the projection of their status in the near future," [3]; p. 36. Based on this definition, SA includes three levels: (1) perception, (2) comprehension, and (3) projection (see Figure 1).

Level 1 SA, perception, involves the sensory detection of significant environmental cues. For example, operators need to be able to see relevant data or hear an alarm sound. In the field, the use of other senses to gather information (e.g., the smell of burning wire) also may be pertinent. SA, however, is far more than simply perceiving a bunch of data. Understanding the meaning or significance of that information in relation to one's goals, referred to as Level 2 SA, comprehension, is also important. This

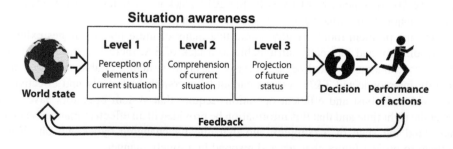

FIGURE 1

Situation awareness is the key to optimal operator performance and decision making.

process includes developing a comprehensive picture of the state of the system based on integrating the various pieces of data collected by the system. For example, operators with good Level 2 SA are able to understand that a particular voltage value is "over the limit," or comprehend the severity of a new system outage on other parts of the system. Level 2 SA involves understanding the "so what?" of a piece of data or system reading.

Projection, the highest level of SA (Level 3 SA), consists of extrapolating current system information forward in time to determine likely future states of the operating environment. In the power system industry, an example of projection is the prediction that under a set of credible contingencies (e.g., a transmission line outage or a wind plant outage), the system will remain within operating limits. In addition to the use of automated real-time contingency analysis (RTCA) tools, experienced operators also actively merge what they know about the current situation with their mental models of the system to continuously predict what is likely to happen next—for example, projecting the impact on the system of removing an element from service, a forecasted wind ramp, or future system demands over the course of the day. The higher levels of SA allow operators to function in a timely and effective manner, even with complex and challenging tasks.

Although the term "situation awareness" and much of the research on SA originated within the aviation domain, SA as a construct is studied widely and exists as a basis of performance across many different domains such as air traffic control, transportation, military operations, satellite monitoring and management, oil and gas operations, and weather forecasting. The individual elements of SA can vary widely from one domain to the next, but the *importance* of SA as a foundation for decision making and performance applies to every field of endeavor in which critical decisions are made. Designing systems to assist individuals in developing and maintaining SA facilitates decision-making activities and improves both operator and team performance.

1.2 Challenges for SA

Developing and maintaining SA can be a demanding process. The difficulty occurs in the interaction between the human information–processing capabilities of power system operators and the design of the technologies with which they interface in the control room. On the basis of a detailed cognitive model of SA and a review of 40; years of research in the field, Endsley et al. [4] summarized these difficulties, labeling them "SA demons." System designers need to be cognizant of these SA demons to avoid them in the design of energy management systems and support tools, as they can undermine operators' ability to develop and maintain SA during grid operations. Eight categories of SA demons have been identified and each presents its own set of challenges to developing and maintaining SA.

1. *Attentional narrowing*—Humans tend to concentrate their attention on specific information to the exclusion of other data that may be indicative of an escalating problem. This attentional narrowing can lead to operational errors when attention is not directed to critical cues associated with competing, important events.

2. *Requisite memory trap*—Short-term memory is limited and easily disrupted, especially under high workload conditions. When system designs require operators to hold information in short-term memory, it greatly increases the chance of an operational error.

3. *Workload, fatigue, and other stressors*—These factors all act to reduce already limited short-term memory and disrupt information acquisition. Long hours, night shifts, high workload peaks, as

well as personal and job-related anxiety can all contribute further to inappropriate attentional narrowing and limiting cognitive processing of information.

4. *Data overload*—The volume and rate of change of data in many systems can outpace operators' abilities to keep up with the state of the system. Data flow in utility control rooms is extremely high, system components and interactions can be complex, and most displays in the industry present only basic data (e.g., only a voltage level, rather than showing its proximity to a limit).

5. *Misplaced salience*—The overuse of prominent visual features such as bright colors, movement, and flashing lights can overwhelm and misdirect operators' attention. In this environment, critical information has to compete for attention with data that is much less relevant, creating poorer awareness and understanding of the most important information and events.

6. *Complexity creep*—The more complex the system, the harder it is for operators to develop accurate situation comprehension and projection, making the higher levels of SA take longer and be less accurate Over time, power systems have become increasingly complex, and variable energy resources will significantly increase this trend.

7. *Errant mental models*—Without good mental models of how a system operates, it is easy to misinterpret data based on how a different part of the system works. The more modes an automated system has, for example, the easier it is for its operator to misinterpret what the system is doing, leading to poor decision making.

8. *Out-of-the-loop syndrome*—Highly automated systems can leave operators with low awareness of the state of the system and increase the risk of an operator being unable to intervene when a problem arises.

A significant number of these "demons" that act to undermine operator SA were identified during reviews of T&D control rooms across several major power companies in the United States [5,6]. *Data overload* is foremost among these. In the typical control room, operators rely on between 6 and 10 different software applications to do their jobs, which are spread across multiple computer monitors and large screen displays. Operators regularly scan through thousands of pages of Supervisory Control and Data Acquisition (SCADA) data tables and trend graphs. Station one-line diagrams typically span several screens and are densely populated with element data, such as bus voltages, reactive power estimates, and voltage flow information. When coupled with weather reports, alarms, contingency analysis results, and state estimator calculations, operators tend to become inundated with data, creating severe losses of SA. Additionally, they must search for information across many of these systems just to make a single decision creating the requisite memory trap, as operators must remember and mentally integrate data across the various screens. This lack of data integration provides a real challenge for operators, straining cognitive resources and increasing the risk of operational errors. The inclusion of variable generation data and displays to the control room further threatens to overwhelm even the most experienced of control room operators.

The vast majority of software provided as information aids to the control room have poorly configured user interfaces, which further strains operators' SA. Commonly, falling prey to misplaced salience, each tool uses a different color scheme to display similar information, leading to misunderstanding and interpretation errors. Furthermore, these tools tend to neglect basic human factors requirements (e.g., font sizes, proper color contrast ratios between symbols or text and the background, color cue redundancy), which makes information difficult to read and interpret. Similarly, most software systems fail to make the most critical information also the most salient. Instead, operators must visually fight the allure of multiple flashing lights and a smorgasbord of colors to find the information they really need.

The high number of false alarms that occur in the average control room leads to additional problems for operator SA. These include alarms that occur due to incorrect limits or other data lapses, or they may represent events that operators are working to resolve but remain on the displays for days and weeks on end. This detractor to SA also includes those alarms that occur for a variety of extraneous reasons (e.g., communication interruption), which the operators know does not require a response. These false, or unnecessary, alarms act to reduce the effectiveness of alarm displays, thereby obscuring real problems with the system and increasing the risk that an actual alarm will be neglected in the mix because of complacency or the "cry wolf syndrome." Although creating alarms to direct operators' attention to problems is a well-intentioned approach, poor integration of these tools with the realities of day-to-day operations can significantly undermine their benefit.

Lacking awareness about the state of automation and the systems or entities that automation controls (i.e., becoming out-of-the-loop) is a well-known challenge associated with the use of automated systems. For example, control room operators did not realize that certain diagnostic tools were offline and not updating during the August 2003 blackout, directly creating an inaccurate SA of the state of the system and resultant decision errors.

These highly prevalent problems in current T&D control rooms can be linked directly to the prevalence of technology-centered design (see Figure 2). Engineers have made significant progress in installing sensors (e.g., phasor measurement units and smart meters) and granting access to vast data sources across the system. In addition, different companies have produced software tools that provide portions of needed data and capabilities, but in most cases, the data are not integrated across tools and sensors or transformed into the real information that operators need for making their decisions. This piecemeal acquisition and development approach has left many gaps and has created a situation in which the operator has to seek the needed data across systems and displays, sort through the masses of available data, and mentally integrate or transform it into what is needed. This process tends to be very slow, leading to poor SA, high workload, and multiple opportunities for errors to be made.

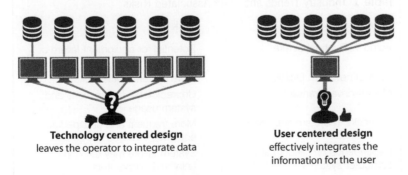

Technology centered design
leaves the operator to integrate data

User centered design
effectively integrates the
information for the user

FIGURE 2

Technology-centered designs tend to inhibit situation awareness and operator performance when compared with user-centered designs.

1.3 Industry trends and SA

As the scope and complexity of grid operations continues to grow, especially with respect to variable energy resources, the electric power industry must focus on understanding and addressing the challenges of integrating today's operations with the smart grid of the future. With the evolution of reliability patterns in power systems caused by new open markets; reduced security margins; and increased wind, solar, and other renewable generation portfolios, the advent of intelligent grids and sophisticated power system algorithms alone will not suffice in preventing human errors in control centers. Operators also must be able to adapt to new system conditions, understand their new role in this much more complex system, and be able to react to system failures and emergency situations. The requisite need to master complex information and make correct decisions depends directly on their ability to derive and maintain SA in this highly complex and less predictable environment.

Several key trends currently are affecting the power T&D industry, which may further deteriorate SA in control centers and other parts of the utility industry. Table 1 lists some of these key trends along with their associated risks. These risks introduce greater uncertainty in day-to-day operations, especially in the face of variable generation concerns. The more uncertainty that operators have to cope with, the more difficult it is to achieve accurate and sufficient SA needed for the reliable operation of the future power grid.

There are generally four modes of operating T&D networks from the perspective of the control room, based on the actions taken by systems operators: (1) reactive, (2) preventive, (3) predictive, and (4) proactive. Most control rooms operate primarily under reactive and preventive modes, replacing components to prevent problems or reacting to a failed breaker, for example. In addition, the current use of contingency analysis tools in an effort to be predictive is limited. Enhanced system reliability will occur when operators have the tools that allow them to predict changes in demand and generation associated with a mix of both conventional and renewable energy sources and to be able to use this information for proactive, rather than just reactive, system operation. Equipping system operators with tools that allow for greater emphasis on predictive information is a requisite to achieving higher levels

Table 1 Industry Trends and Their Associated Risks	
Trend	**Risk**
Aging T&D infrastructure	Potential for equipment failure or malfunction
Lack of new T&D facilities	More system bottlenecks
Market-driven transactions	Unpredictable transactions and system usage
Cutbacks in system maintenance	More frequent and unexpected equipment failures
Dependence on telecom and computer systems	System failure may result in uninformed operators
Aging workforce	Inexperienced operators, system planners, and engineers

T&D, transmission and distribution.

of SA, enabling fast, proactive decisions in the face of dynamic events. Predictive and proactive support tools will be especially critical in an environment that encompasses sources that have more inherent uncertainty.

As an example, consider the SCADA and alarm applications found in classical energy management systems (EMS) currently in use at many T&D control centers. In many cases, these applications are based on nominal power system conditions with scant thought given to how the control staff deals with data generated during disturbances. As a result of technological advances of the past several decades, today's newer protocols and higher bandwidth allow detailed substation changes to be reported to the control center far more quickly and completely than in the past. These advances, however, have created the new challenge of data overload for the system operator. This plethora of data, a symptom of the big-data era, still needs to be organized in a manner that will allow system operators to quickly recognize and respond to disturbances.

In addition, these tools need to enhance the ability of operators to see the patterns and trends that provide for prediction of near-future changes in key system parameters. Effective forecasting models, built to address the operators' SA needs, provide the basis for proactive decision making that will be needed with variable energy resources. As utilities start to operate their systems closer to operational limits, with greater reliance on variable energy resources, more disturbances are likely to occur, resulting in increased threats to the power system and to the customer. Future EMS systems must take on the challenge of reducing and refining the extreme loading of low-level data on the control staff, thus increasing operators' ability to rapidly detect and handle disturbances.

Dealing with the challenges of SCADA, alarm management, and the uncertainty and variability of renewable generation, in addition to the significant SA decrements that have been observed in recent blackouts, requires that the industry directly address SA in the design of T&D management systems. SA is also important for utility-wide responsiveness in cases in which the focus is on achieving operational excellence, and ensuring environmental and regulatory compliance. Typical decision support tools include operational dashboards that display near real-time information, such as reliability metrics, as well as key performance indexes (KPIs), to key decision makers at executive and other senior management levels in the utility. Well-designed dashboards to enhance SA are an important focus for new business intelligence tools being developed for utility companies. Similarly, some EMS industries are examining "Smart Alarm" systems in which repetitive conditions are identified, thus allowing multiple low-level alarms to be grouped intelligently with regard to the underlying problem. This method of analyzing alarms will allow a more organized approach to categorizing alarms and showing their relationships to other alarms, allowing the system operator to more quickly determine the health or state of the grid.

2. Optimizing situation awareness in power system tool designs

Unlike technology-centered system designs, a user-centered design approach starts from the perspective of the T&D operator. The information from all of the various sources and sensors is integrated to fit the goals, tasks, and needs of the users from an operational perspective, greatly simplifying the cognitive work needed to achieve SA and avoiding the many SA demons that undermine it. User-centered design provides a means for better harnessing information technologies to support human work.

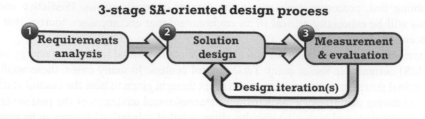

FIGURE 3

Designing for situation awareness (SA) involves three user-centric phases.

Successful system designs must deal with the challenge of combining and presenting the vast amounts of data now available from the many technological systems present to provide true SA: Not just the raw data (i.e., Level 1 SA) but, more critically, integrated information that provides the needed comprehensions and projections. An important key to developing new technologies is in understanding that true SA *only exists in the mind of the human operator*—new data sources, additional displays, and novel tools do not automatically equal SA. Presenting a ton of data will do no good unless it is successfully transmitted, absorbed, and assimilated in a timely manner by the human to form SA. Unfortunately most systems fail in this regard, leaving significant SA problems in their wake.

SA-Oriented Design (SAOD) is a well-established user-centered process for building decision-support and data visualization tools that optimize operator SA based on the scientific research foundations in this field. Described in *Designing for Situation Awareness* [4], SAOD provides a three-phase methodology (see Figure 3) that focuses on creating robust solutions to enhance an operator's awareness of what is happening in a given situation, which can dramatically improve decision making and performance. SAOD is specifically oriented toward designing solutions that enhance people's natural capabilities and overcome the major challenges (i.e., the demons) to SA that exist in twenty-first-century technology-centric control room tool designs.

3. The future of SA in grid operations

The evolution of computer-based applications for analytics and visualization in control centers has focused on the basic tasks performed by grid operators. Initially these tasks were explicitly three-fold—monitoring, information gathering, and control—and improved only Level 1 SA for operators. The second generation of tools targeted the tasks of decision making and taking action. As the industry transforms, the next generation of analytical applications and visualization tools will need to assist operators in coping with exposure (i.e., how vulnerable the system is becoming), and risk mitigation. Achieving these last two goals clearly requires applications that allow operators to look ahead and assess future system conditions before they occur. This ability to project into the future directly addresses the third and highest level of SA, enabling operators to become more proactive in their control of the power grid.

A certain amount of SA currently is achieved in control centers with advanced analytical decision support and variable generation-forecasting tools. EMS applications for higher levels of SA include state estimators, wind forecasting, and contingency analysis tools. Although the current algorithms in these applications have performed well at providing some Level 2 SA, they provide only a limited look into future operating conditions. There is still a strong reliance on the operator to transform the data to make final operational decisions. As shown by previous blackouts, this may not always result in the most efficient utilization of assets.

Achieving the predictive power of Level 3 SA will become increasingly important as power systems become more stressed, the number of expert operators retiring escalates, the presence of wind power and other renewable resources rises, and the number of new regulatory and environmental laws to which utilities must comply grows. These changes will require systems with more predictive tools and faster simulation capabilities to give the operators and other control room personnel ample time to assess and determine the best course of action to maintain system reliability.

Presently most electrical utilities use contingency analysis tools to help monitor their system. These tools look at the present conditions of the system and determine the effect of loss of certain grid elements on the remaining elements for a certain area to only the first contingency. Because the electrical grid is constantly changing because of shifting load peaks, temperatures, and other contributing causes, the system operator also needs a tool that will allow them to better forecast the future. For example, such a tool would take a snapshot of the present day and use models to better predict expected temperatures or loadings. This predictive feature would allow the system operator to better prepare, schedule, and organize the grid and their responses to handle such an upcoming situation.

The design of information displays is a critical system feature that needs to be improved. The operator must have the ability to use a single, integrated user interface to quickly access required information. More and more industries are adding different software to their overall operations, but they are giving little attention to how the operator must interact to access these tools. Effective solutions for improving SA in power network operations involve far more than just adopting trendy new visualizations. Detailed identification of the real SA requirements of operators is fundamental, followed by science-based, principled user-centered display design. These new information displays need to incorporate both effective human factors design principles, as well as SAOD principles.

The electric power industry has received several widely felt wake-up calls in the form of blackouts in Europe, North America, India, and other parts of the world, where loss of SA was a critical factor. On the positive side, a large database of research and established design guidelines exist for directly improving SA that can be adapted readily to this industry. On the downside, it is easy for complacency to set in, reducing the likelihood that utilities and software providers will take effective action until the next blackout occurs. Although it is always easy to say that "It couldn't happen here" or "It was due to failure of the human operator," in reality, all systems are vulnerable unless the industry acknowledges and addresses these inherent weaknesses to enhance overall reliability. The coming integration of renewable power sources and the inherent variability and uncertainties they induce makes this all the more important.

References

[1] Electricity Consumers Resource Council. The economic impacts of the August 2003 blackout. Available: http://www.elcon.org; 2004.
[2] Georgilakis PS. Technical challenges associated with the integration of wind power into power systems. Renew Sust Energ Rev 2008;12:852–63.

[3] Endsley MR. Toward a theory of situation awareness in dynamic systems. Hum Factors 1995;37(1):32–64.

[4] Endsley MR, Jones DG. Designing for situation awareness: an approach to user-centered design. New York: Taylor & Francis; 2012.

[5] Connors ES, Endsley MR, Jones L. Situation awareness in the power transmission and distribution industry. In: Proceedings of the 51st Annual Meeting of the Human Factors and Ergonomics Society. Santa Monica (CA): Human Factors and Ergonomics Society; 2007. pp. 215–9.

[6] Lenox MM, Connors ES, Endsley MR. A baseline evaluation of situation awareness for electric power system operation supervisors. In: Proceedings of the 55th Annual Meeting of the Human Factors and Ergonomics Society. Santa Monica (CA): Human Factors and Ergonomics Society; 2011. pp. 2044–8.

Managing Operational Uncertainty through Improved Visualization Tools in Control Centers with Reference to Renewable Energy Providers

31

Richard Candy

Eskom Transmission, System Operator, South Africa

1. Introduction

The installation of renewable generation on power systems is tantamount to throwing large stones into a quiet pond. The consequences are similar in that both cause waves and surges that disrupt the harmony of the system. To cater to the disruption and the unpredictability of renewable energy providers, additional tools are needed to both visualize and manage the uncertainty and to soften the impact on the control staff. Classical SCADA and alarm systems simply do not have the ability to deal with the situation. A completely different approach is needed, one in which the environmental factors that drive the uncertainty in renewable energy and the SCADA monitoring systems are combined, on a common platform, to provide the control staff with predictability and full situational awareness of these disruptors.

2. Background on SCADA, RTUs, and protocols

Since the first remote terminal units (RTUs) were installed in the 1970s, the instantaneous substation state has been communicated to the master station via individual (indexed) single- and double-bit binary values and multi-bit (12–32 bit) analog values [1].

This philosophy has not changed since then and is still the basis of the second edition of the IEC 60870-5-101 [2] protocol, which is defined as the international standard protocol to move substation data to a SCADA master.

Some additional functionality has been added such as 32 bit bit-strings, file transfer, and the inclusion of GPS time stamping on individual status and analog values. The IEC 60870-5-101 protocol and its supporting documentation is designed for secure, deterministic, and reliable supervisory control and data acquisition for energy management systems around the world. In this role it is robust, structured, and an extremely functional product which does what it sets out to do. However, in the twenty-first century, it does not serve the needs of the control staff, especially during disturbances. In

Renewable Energy Integration. http://dx.doi.org/10.1016/B978-0-12-407910-6.00031-4

fact, using the protocol in its current form actually hinders the rapid understanding and the consequent restoration of the power system following a disturbance.

3. Current IEC 60870-5-101 situation

As things currently stand with IEC 60870, each data element in the substation is assigned a unique address value that has a corresponding record in the SCADA master data base at the control center. Each change in any substation element, be it status or analog, is reported individually to the SCADA master computer, which uses the assigned address value to link the RTU detected state change with the correct record in the SCADA master data base.

4. SCADA alarm processing

In most transmission and distribution systems, the philosophy and design of the alarm processing application are based on a quiescent network state, not a disturbed state. Research has shown that during disturbances, nearly 80% of alarm data is consequential and of virtually no benefit to the control staff [3]. In addition, the design of the SCADA system usually requires that each message be acknowledged by the control staff before it can be deleted. This implies that until a new philosophy is developed, control staff are doomed to suffer from time-based sequential "single dimension" data overloading during disturbances. It is also important to remember that nearly all the reported consequences are reversed once the initiating cause is corrected. This in turn causes a surge in "return-to-normal" messages, which again have to be acknowledged (unless "auto acknowledgment" is implemented for return-to-normal messages), again taking up valuable control staff time.

For example, take a power system disturbance that generates 3200 alarms in an hour with a spike of 1800 in 1 min at the height of the disturbance. The 80% rule means that only 3200 x 0.2 = 620 messages are important and deserve control staff attention, and the remaining 2680 messages just use up resources and annoy the control staff.

A number of factors spring to mind:

- How do we know which alarm messages are important?
- How do we prevent the control staff from having to deal with the non-required alarms in the first place?
- Why do all the unimportant consequential alarms (2680) still have to be read, acknowledged, and deleted? Assuming just 2 s to read and delete each message, it will take just over 1½ h to read 2680 messages, let alone actually understand the content. This is in addition to the 20 min needed to process the important alarms.

At the start of a disturbance, control staff have three problems:

1. What caused the event?
2. What must be done immediately to prevent the first set of consequences causing more problems?
3. Are there other problems either happening or about to happen in other parts of the power system that need to be taken care of?

4.1 Additional problems—environmental and renewable energy producers impacts

Apart from the above, power system control staff are increasingly being forced to deal with and plan for the impact of environmental events affecting the renewable energy producers on their portion of the power system, with no means of either visualizing or reasoning about the relationships that these external events have on the power system. Internationally, the lack of situational awareness of environmental events and their impact on both renewable energy providers and on the power system have been blamed for recent blackouts around the world [4,5].

SCADA systems cannot and do not predict the future either in the short or medium term. However, today's energy management systems are equipped with tools such as contingency analysis and short circuit analysis that can identify the consequences of defined changes to the power system as a direct result of artificial or environmental events. The problem with these tools is that their output is not integrated with rest of the energy management system. The results are usually presented in tabular format, on separate displays, independently of the SCADA one-line displays used by the control staff.

4.2 Ring fencing

One of the most unnoticed aspects of SCADA systems is that they effectively ring fence the control staff from all environmental or weather-related events that are happening to the power system. This means that unless knowledge of the external events is integrated within the SCADA system, they cannot be presented to the staff in the control room. For example, unless grass fires or lightning storms are monitored by the SCADA system, there is no way in which the control staff can know which storms or fires are likely to physically interact with the power system or how weather-related events will impact the output of the renewable energy providers.

Currently there are no tools that are capable of combining SCADA, environmental, and weather data on a common platform that can provide control staff with a comprehensive overview of all threats and risks to the power system in the short or medium term. Nor is there a way in which the different inputs can be visualized within a single comprehensive view that will allow all the different threats and weather impacts to the power system to be seen at the same time. Nor is there a way in which to reason about how the different environmental factors will affect the current power system state and automatically notify the control staff of any potential threats or problems.

The classical power system SCADA and visualization tools do not assist in solving the problems listed above. A major shift in thinking and design is needed. We need to move away from data processing and rather provide control staff with the results of information processing using situational awareness platforms on which all events that can impact or change the health of the power system can be seen at the same time. The architecture of a proposed situation awareness platform is shown in Figure 1.

5. Situational awareness platform

In deploying the situational awareness platform, the existing SCADA infrastructure is left as is with the new functionality wrapped around it.

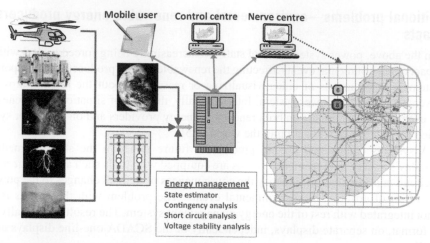

FIGURE 1 Situational awareness visualization architecture.

The fundamental premise behind the design is as follows:

- Combine all SCADA and non-SCADA data sources on a common geographic platform
- Provide information processing rather that data processing
- Design for disturbances first and then add the quiescent functions and alarm responses.
- Notify the user of the existence of the problems and allow them to examine them when they need to.
- Design the platform to allow for the deployment of individual applications whose job it is to evaluate changes and threats to the power system and notify the control staff.

To achieve the desired results the following principles are applied:

- Use object-oriented design methodology to model the power system and all source data.
- Implement a situational awareness visualization platform, based on geographical information system that combines both SCADA and environmental data on the same platform. This will then provide both a current view of the power system as well as a 15 min to 1 h predictive view of the consequences and threats to the power system.
- Develop an "Apps Store" to support user functions. Each application is designed to interact with the bay and substation objects on all platforms. The individual "apps" also have the ability to communicate with one another as well as sending messages to users via the visualization platform.
- Implement a single source definition facility for all power system objects that can be used to update and maintain the data bases across all platforms.
- Install a substation server at each substation that accesses the substation remote terminal (or gateway device) as well as all the secondary plant intelligent electronic devices in the substation. The substation server provides local plant condition monitoring, advanced alarm

processing, and event state determination and has an independent link to the situational awareness server as well as all substations electrically connected to the substation (power station).

- At the substation, provide substation users with an augmented reality tool and applications on their portable devices. The substation server platform provides the web services support for this functionality.

The building blocks listed above are discussed in more detail as follows.

5.1 Object orientation

At the heart of the proposed set of changes, is the need to move away from the classical, SCADA based, single state event reporting and move into object orientated "state" processing. Fortunately the foundation for the functionality is already incorporated in the common information model.

The proposed design uses three object types, which are overlaid on the geographical layout of the power system. The three object types are:

Static: Objects that do not move, e.g., substations, power stations, power lines, transformers depots, load networks, buildings.

Column: These are container objects aligned with a predefined grid layout (30×30 km squares) that spans then entire country. Column objects are used as place holders for future, current and historical time series data related to the ground position of the grid square.

Dynamic: Objects that have nonspecific lifespans and have transitory locations. These include storms, hurricanes, grass fires, fog banks, rainfall, snow, etc.

5.2 Static objects

Static objects have a defined parent/child relationship as is typically found in power systems. For example a substation object would typically have the following child bay objects as indicated in Figure 2:

- Busbars (per voltage level)
- Diameter bay (1, tie, 2)
- Line—(feeder, line, load)
- Transformer
- Coupler
- Section
- Reactor
- Capacitor
- Static Var compensator
- Unit (generator)

Power station objects would have the same bay types as a substation but also include the physical generation production facilities, which are modeled as bays internal to the power station. Where necessary the power station model used three dimensions to define the location of each bay.

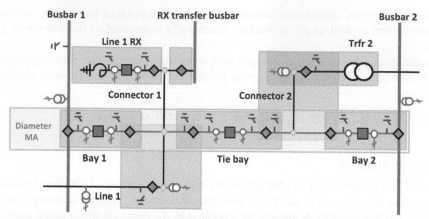

FIGURE 2 A typical 765 kV breaker and a half substation layout showing the different bays.

Power lines are made up of sections demarcated by the bend towers. Each bend section is made up of the towers between the bends, conductor type, and finally the bend section state. Depending on the object type, be it a substation, power station, or building, the source location will use either X and Y or X, Y, and Z offsets to define the locations of the child objects.

5.3 Column objects

At the transmission level, column objects are based on the surface rectangle demarcated by the lines of latitude and longitude, divided by 3, giving roughly nine squares per degree. Each square is approximately 30×30 km, as indicated in the right-hand side of Figure 1 and in more detail in Figure 5.

Each grid square has a ground level origin point defined by its X, Y, and Z geographical location and acts as storage location for, among other things:

- Current weather data
- Short-term weather forecast data (six hourly updated every hour, in 15 min intervals)
- Long-term weather forecast data (24 h and 7 days updated daily in hourly intervals)
- Fire index—current, forecast, and historical
- Ground resistance for geomagnetically induced current analysis as a result of solar flares
- Cloud cover—current, forecast, and historical for renewable energy producers
- Solar radiation—current, forecast, and historical for renewable energy producers
- Rainfall—current, forecast, and historical—also used for flood analysis
- Grass fires—current, forecast, and historical—linked to vegetation and fire risk analysis
- Vegetation—current, forecast, and historical—associated with fire risk
- Magnetometer data—derived and updated every minute as a result of solar flares
- Lighting strike activity—current and historical—used to track storms
- Rain radar—also linked to flood analysis—(not bit maps)

All static and dynamic objects define their reference in relation to the origin of the parent grid square to which they belong. This allows them to obtain their related column data automatically. For distribution, the grid area would vary from 10×10 km to even 1×1 km, depending on the available data, the desired results, the model detail, and computer processing power available. The address of the parent grid square allows child objects to be referenced and updated when needed, without affecting the contents. This is a crucial aspect of the functionality.

5.4 Dynamic objects

Dynamic objects are instantiated only when the parent application driver receives new source data. All dynamic objects have a defined time-to-live counter that gets updated each time new data are sent to it. Each clock cycle results in all dynamic objects receiving a message to count down the time-to-live counter. When the counter reaches zero, the dynamic object is deleted.

5.5 Applications

As indicated previously, the purpose of the visualization platform is to merge SCADA and environmental data on a common geographical platform. However, there is no way in which to derive the relationships between SCADA data items and the environmental values without tools that can link them together. The tools are a combination of independent applications that access the data and information in the different objects and the objects themselves accessing column data and determining abnormalities.

The applications are triggered to run either by the clock or by another object detecting a state change. In the event that an object requests the services of an application, the trigger will usually be linked to the contents of the parent column object or as the result of a message from another object. The relationship between static objects, dynamic objects, and their applications is shown in Figure 3.

The key to the use of customized applications is the standard structure of each of the objects within the visualization space. By defining a standard interface that allows all applications to interact with all the other different objects without problems, the door is opened for developers to design applications for a situational awareness apps store. Based on the fact that most utilities have the same type of problems, an apps store will allow utilities to share applications rather than having to write their own each time. The key is the definition of a set of standard object interfaces and interaction rules for all objects and applications.

5.6 Visualization architecture

All the basic building blocks have been defined, which, when combined, allow for a structure similar to that shown in Figure 4 to be built.

All objects are housed within the pseudo three-dimensional visualization space, and historical data are held in the historian, allowing objects, applications, and users to examine trends and look for anomalies. The applications interact with the objects that are updated via the push server interface, located at the bottom of Figure 4.

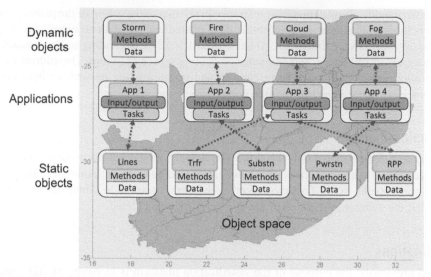

FIGURE 3 Application interaction with both static and dynamic objects.

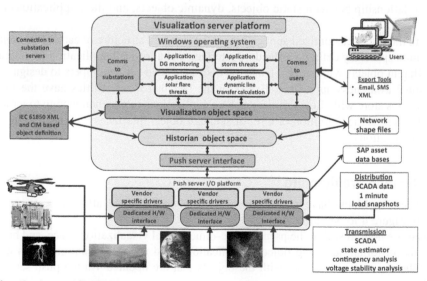

FIGURE 4 Visualization architecture.

The push server provides I/O processing for each data source, error processing, firewall services, and connection problem notifications. Each data source has a dedicated hardware driver and associated input application to process the incoming data or information. On arrival, each new data source packet is validated by the dedicated I/O application before being converted to the correct format and sent to the destination object for processing. The location of all objects is tracked via a "yellow pages" facility, allowing all services or users to contact whatever object they need to.

As can be seen in Figure 4, examples of the different source data include the following: lightning, grass fires, helicopter position, transformer condition data (dissolved gas analysis), storms, magnetometer data, environmental data, energy management SCADA, contingency analysis, and voltage stability data, to list a few. As discussed earlier, each data source is associated with different object types, which are discussed in the next section.

6. Abnormal state notification

In the event that an anomaly is detected by one or more applications or objects, notification of the problem is forwarded to the parent objects for display in one of two ways:

1. The display a notification icon next to all levels, in the parent/child hierarchy.
2. Alarm notification message sent to the alarm window for attention if no action has occurred within the defined response time.

The graphical notification is done using one of four color-coded diamond icons grouped together as shown in Figure 5. At a point definition time, each data element is assigned to one the four icons that,

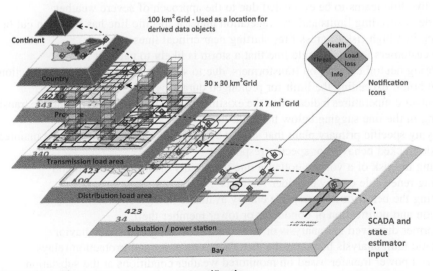

FIGURE 5 Illustration of icon data propagation and notification.

when high, indicates that one of the following events types has occurred on the device and hence the bay and thus in the station:

- Heath
- Threat/warning
- Load loss
- Information

When an application or bay detects a problem, users are notified by the placement of an icon next to the problem device. The device automatically notifies its parent of the problem, which then results in the display of the summated icon at the bay, at the station, and at the parent region. This allows users to see the existence of all the problem types at all levels of the visualization space. To assist the user at run time, filter buttons are provided that allow clearer identification of the problem space.

Once notified of the existence of a problem, via the visualization servers, users can either access the details indicated by the icon, by selecting the icon which then produces a popup window providing a summary of the problem, or they can use the communication links to hyperlink to the source substation server to obtain additional details and information relating to the problem.

Figure 5 shows the detail of how abnormal conditions are propagated up the data pyramid.

7. Benefits of situational awareness visualization platforms

The purpose of the visualization facility is to provide users with an awareness of the current state of the power system. Some examples of the benefits of situational awareness are:

- Notification of dual treats to a power line, e.g., storm at one end and grass fire at the other.
- Notify live-line-teams to be evacuated due to the approach of severe weather.
- Fire index exceeding limits and the vegetation under a specific line has not been cut or cleared, resulting in a high risk of grass fires starting near critical lines.
- Notify customers fed by a single line that a storm is likely to cause it to trip.
- Monitoring neutral currents on transformers, due to solar flares, coronal matter ejection, or coronal holes, exceeding the limit for prolonged durations.
- Ambient air temperatures indicate that the existing load on the line is exceeding the transfer limit, resulting in the line sagging below the legal limit.
- Identifying specific primary plant that should be scheduled for condition-based maintenance based on tracked behavior or specific responses.
- Detecting the risk of a voltage collapse due the rapidly changing Var demand
- Verifying renewable energy provider forecasts
- Predicting the behavior of renewable energy available in a given area
- Detection and notification of conductor or tower member theft
- Transformer dissolved gas analysis or harmonics indicating unusual behavior
- Automatic fault analysis based on the disturbance recorder and protection relays
- Optimized power transfer based on monitored weather conditions at the substation
- Insulator pollution monitoring based on partial discharge signature changes

- Surge arrestor monitoring
- Circuit breaker closing and opening time limit violations
- Wide area monitoring indicating sub synchronous oscillations
- Line fault prediction, location, and diagnostics
- Disturbance recorder trigger interpretation and notification

7.1 Examples

Classical SCADA systems are unable to link environmental events to what is detected and reported to the control staff via the energy management software. In other words, there is no way in which to establish which contingencies are likely to happen as there is no way to "know" which external events will result in a specific contingency being triggered. Nor is there a way to link the potential threats and their impact on the power system since there is no process that combines external threats with the power application software tools. The situational awareness platform is designed explicitly to link contingency results to known threats to the power system.

7.2 Application to renewable energy integration

The situational awareness platform provides the utility with tools for monitoring and tracking renewable energy providers:

- Validating submitted output against actual weather forecast data every hour
- Predicting generation performance based on forecasted weather data hourly
- Predicting generation start-up and drop-off using forecasted weather data
- Determining regional total available renewable energy availability and performance

In addressing the above functionality the following assumptions are made:

1. Forecast weather data and the solar radiation data for each column are updated hourly, for the next 6 h, in 15 min intervals.
2. Each individual renewable energy provider updates its equivalent static object hourly with the available and predicted generation for the next 6 h or next 24 h
3. A regional assessment application exists to compute the results and display anomalies.

With the above in place, it is a simple exercise for each renewable energy provider object, each time it is updated, to access its parent column object and retrieve the forecasted data and compute the generation output and compare it the renewable energy plant's predicted output. By the same token, each time the forecast weather data are updated, the column object requests all renewable energy provider objects within the grid area to validate their data using the latest forecast information.

  Each renewable energy provider object will notify the users of any contradictions or anomalies if there is a difference between what the renewable energy provider thinks it will generate compared to what the calculated output will be using the weather forecast data. This allows the control staff to take the necessary action to protect themselves against either miscalculation, sudden drop-off, or unexpected rises in generation output by the renewable energy providers.

By having an application evaluate all the renewable energy providers in a particular area, the control staff are notified well in advance of potential voltage and demand problems in the area, allowing them to take remedial action in time.

8. Conclusion

The introduction listed the following aspects as problems faced by the control staff which can affect their ability to manage the operational uncertainty:

1. How do we know which alarm messages are important?
2. How do we prevent the control staff from having to deal with the unimportant alarms?
3. How do we know what caused an event?
4. What must be done to prevent the first set of consequences causing more problems?
5. What other problems are about to happen in other parts of the power system?

Commentary on these items is listed below.

Items 1, 2, and 3

The objective is not to send individual alarm messages to the control staff but rather to provide them with complete information messages that define what happened and list any abnormalities in the plant response to an initiating event, i.e., which events should have occurred compared to what did occur [6]. This requires a deterministic knowledge of how the plant should behave under specific conditions.

In addition, substation objects have their object state updated in real time. Each substation object derives its state from the individual bay and busbar connections. Thus the substation "knows" which bays are connected to which busbars and the state of the busbars. This allows the station object to notify the individual bay, in real time, of the busbar state that is used to tailor the alarm messages sent from the individual bays to the control staff.

For example, during a power outage, the busbar voltage drops to zero. The substation object automatically notifies all the bays connected to the zero voltage busbar of the fact. This allows the bays to suppress all alarms related to a zero voltage or mark them as log only. Similarly, if there is a failure common to a number of bays, their bay state is notified and the information is used to either suppress or mark the alarm messages as log only, thus hiding them from the control staff.

During bus strip events, the station object is notified by each bay that it has tripped and the state of the different protection alarms that occurred at the time. This allows the station object to determine which bay caused the trip and which ones were tripped by the busbar protection. The station object derives the root cause of the bus strip, identifies which breakers operated normally and those that did not, and sends the control staff an event summary. To achieve this result, each bay has to be able to "know" which protection scheme operated and use this to update the station object. The station object "knows" which bays are connected to which busbars and thus is able to detect how the busbar protection should have operated to clear the fault. The station object can then cross check the individual bay operations, based on the bay state protection alarm values, and can thus formulate the summary response for the control staff.

Alarm suppression also applies to individually reported line trips. In essence a fault in zone 1 or zone 2 will typically cause the breakers at each end of the line to trip at the same time. Each bay has the ability to know the ID of the bay at the other end the circuit and is thus able to request information from it to confirm the time and trip type. In a typical dual end line trip, the two bays exchange information and combine the protection information to form a single alarm message, which is then sent to the control staff [7].

Items 4 and 5

Prevention of consequential events is really dependent on the ability of the visualization platform applications to determine and notify the users of potential problems in time for them to take corrective action. The aim of the visualization server is to determine what will happen as a result of a specific changes taking part within the visualization space. Applications use the power application results to determine the consequences of a predicted network state and display the results graphically to the users via the visualization framework on each user's workstation or tablet.

References

[1] Candy Richard. Changing the tele-control protocol to support substation based intelligent alarm processing and direct control staff interaction with the substation RTU/gateway. Johannesburg, South Africa: IEEE PES Power Africa 2007 Conference and Exposition; July 2007. 16–20.
[2] International Electrotechnical Commission. International standard IEC 60870-5-101. Telecontrol equipment and systems - part 5-101 transmission protocols - companion standard for basic telecontrol tasks. 2nd ed. July 8, 2003.
[3] Davis Jack. To err is human, to control is crucial. Eng Manag J; August 2002:177.
[4] NERC. Technical analysis of the August 14, 2003, blackout what happened, why, and what did we learn?. Report to the NERC Board of Trustees by the NERC Steering Group; July 13, 2004.
[5] U.S. Department of Energy's Office of Electricity Delivery and Energy Reliability. Economic benefits of increasing electric grid resilience to weather outages. Executive Office of the President; August 2013.
[6] Candy Richard. Reduction in control staff data overloading by including the substation and bay state in reported substation data; 2008. Southern African Power System Protection Conference.
[7] Candy R. Application of object orientation and artificial intelligence to disturbed state alarm processing in an electrical transmission system [Ph.D. thesis]. Johannesburg: University of the Witwatersrand; November 2004.

Dynamic Line Rating (DLR): A Safe, Quick, and Economic Way to Transition Power Networks towards Renewable Energy

32

Peter Schell

General Manager Ampacimon, Liège, Belgium

1. Introduction

Although it has been common since the creation of transmission and distribution networks to assign a fixed rating (i.e. maximum current carrying capacity) to assets and overhead lines in particular, in practice the rating is not a fixed value. Instead it depends on the ambient conditions the asset and the overhead line in particular are experiencing. Indeed, how much current can go through a line is determined by two limits: the maximum temperature the conductor can reach without damage and the minimum clearance to the ground. The conductor becomes longer the warmer it gets and therefore sags, reducing clearance. In practice, overhead lines and, in particular, the tower heights are designed such that minimum clearance will be reached when the conductor reaches its maximum design temperature, typically 85 °C. The current at which the conductor reaches its maximum temperature and minimum clearance is determined by how much the conductor is cooled by the ambient conditions around it. Ambient air temperature, sun radiation, and most important wind speed and direction determine the cooling. Therefore, to determine a fixed rating, a conservative worst case assumption must be made regarding the ambient conditions. Typically this will be an ambient air temperature of 25–35 °C, solar radiation of about 1000 W/m^2, and a perpendicular wind speed of 0.5–0.6 m/s. These conditions do happen, but not very often! This "name plate" rating has been replaced by seasonal ratings in areas with strong differences between seasons, so that nowadays, in regions such as the United States and Europe, lines typically have three ratings: winter, spring/autumn, and summer that assume different ambient temperatures. More details can be found in Cigre Technical Brochure 299 [1].

2. What is dynamic line rating

Dynamic line rating (DLR) is the ambition to rate overhead lines based on the actual and forecasted ambient conditions and use those, much higher dynamic values to operate the network more efficiently.

To make a simple analogy: Operating the power system is like walking along a cliff. Without DLR the operators are blindfolded and must keep a significant distance away from the edge to be safe; i.e.

Renewable Energy Integration. http://dx.doi.org/10.1016/B978-0-12-407910-6.00032-6

they operate their system far below what is possible at a static rating. DLR is like taking the blindfold away! It allows the operators to walk significantly closer to the cliff's edge in absolute safety and enjoy the benefits this brings, i.e. operate the network based on a dynamic rating and take advantage of the significant additional capacity in favorable conditions.

The potential gains are substantial. A temperature difference of 10 °C between the real ambient temperature and the one assumed for the seasonal rating will increase the rating of the line by about 10% (see Figure 1). No solar radiation compared to the assumed 1000 W/m^2 will increase the rating of the line by about 20% (see Figure 2). But the most significant gain is due to the cooling effect of the wind. A perpendicular wind speed of 2.5 m/s is enough to increase the rating by 50% (see Figure 3). By comparison, it takes about 4 m/s of wind for a wind turbine to start producing energy and about 12 m/s to reach 100% output.

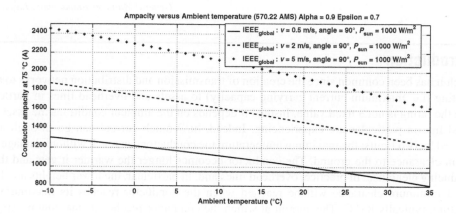

FIGURE 1

Rating versus ambient temperature.

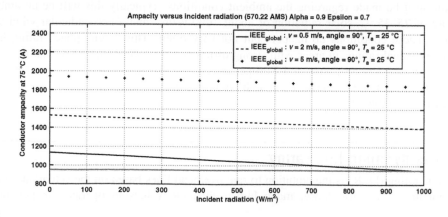

FIGURE 2

Rating versus solar radiation.

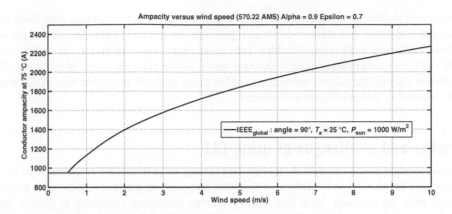

FIGURE 3

Rating versus perpendicular wind speed.

Based on on-site measurements over a 12 month period in a coastal area of Europe, the median gain is between 40% and 50% and a gain of 20% is available 90% of the time (see Figure 4).

It is important to note that using the extra capacity will not shorten the useful lifetime of the conductor. On the contrary, the monitoring system will make sure the conductor remains within its operational limits all the time. Like a car engine, the line is harmed only when operated above its design temperature, not at high speeds/loads as long as it is cooled correctly!

DLR is also possible for other types of assets, for example, underground cables, but based on a different dynamic phenomenon. Obviously, the underground cable is in a much more stable

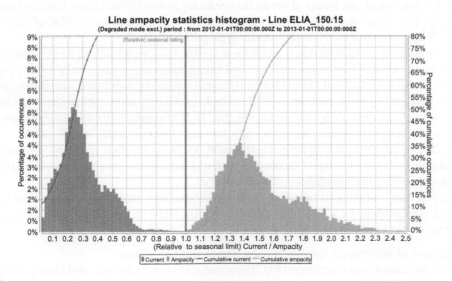

FIGURE 4

Gain distribution for a typical coastal location in Europe.

environment and the cooling effect of the surrounding ground is very constant over time with very small seasonal variations. But there is substantial potential for dynamic gains because of the very long time constants of these underground cables. Indeed, it takes many hours for the cable to reach its maximum operating temperature even at currents much higher then the name plate rating. Based on accurate monitoring of the cable's temperature, it becomes possible to exploit this extra capacity for temporary overloads and contingency situations.

3. Benefits and challenges of using DLR

Given the substantial potential of DLR, it might seem surprising that it has not been used in the past. The main reason for this is the limited value of changing ratings in a power system with very constant power flows over the network. With large power stations providing electricity to nearby towns and industry, the networks were designed to handle peak loads and this needed to be the case year-round (this meant very limited need to use dynamic ratings). Today the situation has changed. Intermittent renewable energy, distributed generation, and the advent of "prosumers" have generated very significant changes in the power flows over the network. The flows are impacted not only by the consumption and network topology, as in the past, but also by the amount of intermittent renewable electricity being injected as well as the behavior of the prosumers.

In this new situation, for a given level of consumption there is an increased need for network capacity to deal with all the different possible flow patterns and to connect the new renewable production sites, which typically are not located in areas with substantial network capacity. More network capacity is needed but not every day, as was the case before. Because of this evolution and the increasing "not in my backyard" mentality, traditional network planning is becoming more and more complicated. It is more complicated to identify where future needs will arise with a sufficient level of certainty. It takes longer and longer to obtain the necessary permits to build new lines and the general variability of the flows means less MWh transported for a given number of MW of line capacity. Suddenly DLR becomes a much more interesting proposition!

This is true for inter-regional flows, which have significantly increased due to the need to balance consumption and demand over larger areas to avoid having uneconomic amounts of reserve production capacity. It is also true for local lines at lower voltages connecting distributed, renewable generation back to the stronger, higher voltage network. If we look specifically at the critical challenge to connect both on- and offshore wind production to the grid, then DLR is a particularly interesting solution due to the positive correlation between wind power injection and dynamic rating (see Figure 5).

Using DLR in conjunction with limited curtailment of wind production allows one to nearly double the available connection capacity of the network, offering a quick and economic solution to one of the key problems slowing the introduction of wind energy.

The key element to taking advantage of DLR and integrating it in the larger operational scheme of the network is to combine DLR with other technologies that allow the operator to have (partial) control of the flows. Several technologies come to mind:

- Curtailment of intermittent renewable energy as mentioned above,
- Flexible AC transmission networks (FACTS)/phase shifting transformers and high voltage direct current lines (HVDC) for "long distance" lines,
- Storage and demand side management for the flows at the distribution level.

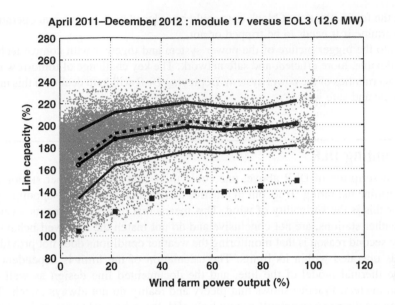

FIGURE 5

Correlation between wind production and DLR. Yellow = mean, Orange = median, black = mean ± standard deviation and green is 98th percentile.

Having control of the flow allows the operator to maximize the use of the varying capacities. More importantly, he can deal in an effective way with the situations where not enough capacity is available. Looking at it another way, having DLR significantly increases the return on investment of the relatively expensive control technologies by providing accurate set points for the control algorithms instead of the very conservative seasonal ratings.

The macroeconomic gains of applying the solutions above is substantial. As part of the EU-funded FP7 project Twenties, these potential gains were evaluated to be about 250 million € annual gain for an investment of about 10 million € if border capacities in central western Europe were increased by 20%, which is certainly feasible using DLR in conjunction with the existing FACTS transformers.

It is important to mention a further aspect that is key to the seamless integration of DLR: the ability to forecast the values both intra-day and day(s) ahead. Even if there is a trend toward a more real-time management of the power system, most of the operational decisions in the power system are taken days or at least hours in advance. Without the ability to forecast the rating and deal with the inherent uncertainties related to forecasting, it becomes impossible to use DLR except for the limited number of decisions that are taken close to real time. Recently, DLR forecasting has been introduced and proven within the aforementioned-mentioned FP7 Twenties Program.

In the not too distant future, integrating DLR with protection devices especially at the distribution level will open the door to a more automated operation of the network. Indeed, it is entirely possible to protect the network based on dynamic ratings instead of the currently used static or seasonal ratings. Past experience has shown that tripping lines too early can lead to cascade reactions and ultimately to black-outs. In the future, the risk of black-outs could be reduced by using the real operational limits of the line to determine if it needs to be tripped or not.

DLR fits into the bigger picture of the power system and together with control technologies will allow a more flexible, more efficient and safe network. The key challenge of tomorrow is to deal with the growing uncertainties of the power system. Deploying DLR will help deal with this uncertainty in a timely, economic and safe way.

4. Implementing DLR

Based on the description of DLR above, one could think that using weather stations is all that is needed to determine the rating. True, but in practice this method provides disappointing results. The first reason for this is the variability of wind. This means that the measurements, even with a large number of weather stations, are not conclusive and do not match the observed behavior of the line very well. The second reason is that monitoring the weather conditions does not provide a guarantee that the line is operated within its limits. The calculation of the limit is dependent on the measurements, the thermal model of the line, and the documented line design as well as the actual installation parameters. Practice shows that theory and reality do not always match! This does not mean that using weather measurements cannot be useful. It can be used to determine the potential for DLR before deciding to implement a direct monitoring solution and to improve the quality of the forecasting.

If we look at the different possible direct monitoring methods, measuring conductor temperature would be the first option that comes to mind. In practice this has proven to be much more difficult then one would think. First, the measurement units alter the temperature locally and this must be corrected. Second, the temperature along the conductor will vary quite significantly, especially when the conductor is highly loaded. It is possible to do a distributed temperature measurement using a fiber optic cable (as is done routinely in underground cables), but this technology is significantly more expensive to implement then all other options available.

Therefore, the most promising technologies today are those that measure the mechanical behavior of the conductor, either the mechanical tension or the sag, via the frequencies of the movements of the conductor. These methods measure the behavior of the conductor directly and avoid the issues associated with the point measurements of conductor temperature. They also have a distinct advantage from a security point of view because in practice, it is not unusual to find spans that reach minimum clearance at a conductor temperature well below design temperature, typically due to slip or excessive creep of the conductor over time.

But having a reliable real-time monitoring system is not enough. To influence the key decisions regarding the power system, one needs to be able to forecast the rating for the coming hours and days. Forecasting the coming hours can be done by extrapolating the real-time measurements using the most appropriate statistical tools and methods. But to forecast the line rating for the next day, weather forecasting data are compulsory. Weather forecasting alone is not sufficient for the same reasons that weather measurements alone do not work. What is needed is the combination of weather forecasting and monitoring results. This combination will generate DLR forecasts that can be tuned to different specific needs. Day-ahead forecasts for markets or maintenance planning need to be the certain minimum transfer capacity, i.e. the lower bound of the probability distribution. On the other hand, a

DLR forecast used to cope with the injected renewable energy for the following day should be closer to the mean of the probability distribution.

Without a reliable real-time monitoring system to guarantee safe operation closer to the operational limits and to validate (and improve) the results of the forecast, it is not feasible to achieve the required level of confidence and trust that is required to start using the extra current carrying capacity uncovered by DLR.

5. Conclusions

In many parts of the world the objective for the coming decade is to make society more sustainable while keeping the economy competitive and improving the standard of living. Translated to the electric power system, this means a more sustainable electricity production with a (very) high level of security of supply at an economically acceptable price. The grid plays a very important role in reaching these difficult objectives. It must become stronger, more flexible, and more efficient. DLR can and should play an important role in tomorrow's grid because it helps increase the capacity of the grid in a flexible agile and very cost-effective way. It is the perfect complement to the technologies that increase the control over the flows, such as FACTS in particular, but also HVDC, storage, and demand side management. All of this will support the integration of renewable generation.

The monitoring hardware, communication infrastructure, and forecasting algorithms are now available to make DLR technically and operationally feasible.

Reference

[1] CIGRE Technical Brochure 299. Guide for the selection of weather parameters for bare overhead conductor ratings www.e-cigre.org; 2006.

DLR forecast used to fill the injected renewable energy for the following day should be closer to the mean of the probability distribution.

Without a reliable real-time monitoring system to guarantee safe operation closer to the operational limits and to validate (and improve) the results of the forecast, it is not possible to achieve the required level of confidence and trust that is required to start using the real-time current carrying capacity afforded by DLR.

Conclusions

In many parts of the world the objective for the coming decade is to make society more sustainable while keeping the economy competitive and improving the standard of living. Translated to the electric power system, this means a new sustainable electricity production with a (very) high level of security of supply at an economically acceptable price. The grid plays a very important role in tackling these difficult objectives, it must become steeper, more flexible, and more efficient. DLR can and should play an important role in tomorrow's grid because it helps increase the capacity of the grid in a flexible, agile and very cost-effective way. It is the (perfect) complement to the technologies that increase the control over the flows, such as FACTS in particular, but also HVDC, storage, and demand side management. All of this will support the distribution of renewable generation.

The monitoring network, communication infrastructure, and forecasting algorithms are now available to make DLR technically and operationally feasible.

Reference

[1] CIGRE Technical Brochure 299, Guide for the selection of weather parameters for bare overhead conductor ratings, www.e-cigre.org, 2006.

Monitoring and Control of Renewable Energy Sources using Synchronized Phasor Measurements

33

Luigi Vanfretti[1], Maxime Baudette[2], Austin White[3]

[1] *Associate Professor and Docent Electric Power Systems Department School of Electrical Engineering KTH Royal Institute of Technology, Stockholm, Sweden luigiv@kth.se; Special Advisor in Strategy and Public Affairs, Statnett SF, Oslo, Norway luigi.vanfretti@statnett.no,* [2] *PhD Student Electric Power Systems Department School of Electrical Engineering KTH Royal Institute of Technology, Stockholm, Sweden,* [3] *Senior Engineer at Oklahoma Gas & Electric Co. Oklahoma, USA*

1. Introduction

Recent environmental concerns with fossil fuel production have stemmed a global trend of increased share of renewable energy production. In several western countries, the full potential of hydro-power has already been exhausted, which has resulted in wind energy being the fastest growing energy technology in the last two decades [1]. Moreover the expansion of the share of electricity production from wind power and solar will continue, especially in Europe [2] and in the USA where 30 states have enforceable renewable portofolio standards or other mandated renewable capacity policies [3].

Unpredictable challenges for power system operation and control continue to emerge as renewable energy sources are brought into commercial operation. With the increased amount of intermittent power sources, one of the observed impacts is an adverse effect on generation power output within a balancing authority's control area. Wind power integration has recently presented challenges in short term operation with regards to system dynamics. Particularly, in the case of Oklahoma Gas & Electric (OG&E), wind power curtailment has been enforced to maintain power supply quality and continuity when the interaction of wind farms with the power grid brings undesirable dynamics into play [4]. In this case, these undesirable dynamics came in the form of sub-synchronous (i.e. fast) oscillations which were first detected due to the impact of the power quality supply at the consumer level. A very recent article in the New York Times [5] shows that these unpredictable challenges are likely to continue to emerge when new plants are brought into operation due to the lack of operation experience and adequate tools for monitoring and control. Photo-voltaic (PV) generation in the German Electric Grid [6] has increased grid stability issues (e.g. "The 50.2 Hz Risk") which may endanger the operation of the entire interconnected grid. It becomes apparent that new means for monitoring and control technologies are needed to cope with the new challenges that the integration of renewable energies bring to the power grid. Real-time monitoring and control tools could help mitigate these undesired

Renewable Energy Integration. http://dx.doi.org/10.1016/B978-0-12-407910-6.00033-8

phenomena, by providing software applications to operate the system with more flexibility, better adapting to phenomena emerging from intermittent generation sources [7].

The effects of increasing amounts of wind power in a power system has been studied, and several problems regarding transient stability have been investigated [8]. However, it is only very recently that some Transmission System Operators (TSOs) have noticed the presence of forced sub-synchronous resonance in the grid due to large scale wind power plants. These oscillations have been observed thanks to the recent adoption of Phasor Measurement Unit (PMU) technology and typically appear in the 3–15 Hz frequency range [4,9–13]. The oscillations can occur during high wind penetration periods, but are typically aggravated by a change in system impedance due to a system fault or transmission line switching. The impact can be substantial and even observable at the consumer level as voltage flicker. The reasons for these forced oscillations are turbine–converter interactions [14] and they differ from the more commonly observed inter-area oscillations below 1 Hz.

As the consequences can be observed at all voltage levels and because these wind farm oscillations have not yet been specifically characterized, it could be supposed that such oscillations could damage equipment on the transmission system or at the customers end. It could also be supposed that the phenomenon could be amplified by the increasing use of wind power. Thus a better ability to monitor the behaviors of wind farms connected to the bulk grid appears necessary. PMU technology allows for means of detection of resonance conditions and it has been found that curtailing the wind power plant output disrupts the oscillatory condition enough to reduce the impact to customers. Such an example is shown in Figure 1, where a forced oscillation is brought under control by curtailing the wind power plant output by 50%.

This chapter gives an overview of how synchrophasor technology can be applied for developing real-time PMU applications, which help in monitoring and control of unwanted dynamics that are a product of renewable energy sources interacting with the power grid. The remainder of this chapter is organized as follows. Section 2 gives an overview of PMU-based monitoring and control systems and describes different environments for developing PMU applications. Section 3 illustrates the development of real-time monitoring tools for the detection of sub-synchronous wind farm oscillations, as an example of how synchrophasor technology can be effectively used for dynamic monitoring of renewable energy sources. Section 4 presents testing and validation experiments performed using historical data and laboratory experiments. Finally, Section 5 gives an outline for the development of new PMU applications that can aid in the monitoring and control of renewable energy sources and their interaction with the grid.

2. Real-time monitoring using synchrophasors

2.1 Wide area monitoring systems

The ability to monitor fast dynamic behaviors such as the sub-synchronous oscillation phenomena is tied to the performance of the measurement equipment used. The sampling rate has a strong impact on the highest frequency observable from the measurements. Traditional equipment and monitoring tools have an asynchronous sampling frequency of typically one sample every few seconds. While this was sufficient to monitor very slow steady state phenomena, it fails to capture faster dynamic behaviors, see Figure 1.

PMUs are devices able to measure the voltage and current phasors of the three phase network with a reporting frequency of typically 30–60 samples per seconds. The frequency of the system is

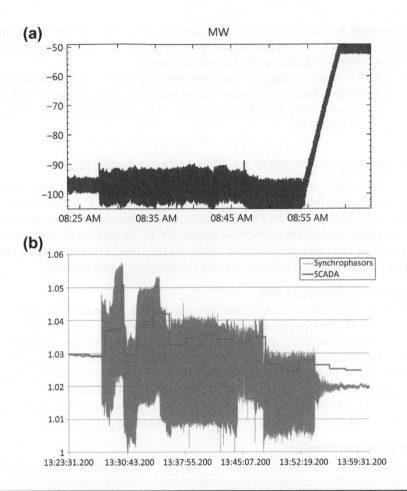

FIGURE 1

Phasor measurement unit (PMU) and traditional equipment measurements Supervisory Control and Data Acquisition (SCADA) of a wind power plant during a fast dynamic event (a) Curtailment of the power output (PMU) (b) Comparison between Phasor measurement unit and SCADA.

internally computed at a higher sampling rate and reported also at 30–60 samples per second.[1] This allows the study of dynamic phenomena occurring up to a frequency of half the reporting rate. The reference for the angle of the phasors is derived from a GPS clock and all the measurements are coupled to a time stamp allowing for the alignment and synchronization of PMUs spanning an entire interconnection.

Phasor Data Concentrators (PDCs) have the role of collecting and forwarding the data of several PMUs. Since all the measurements have a time stamp acquired from the GPS reference, the PDC can

---

[1]Instantaneous active and reactive power can be computed from the voltage and current phasors.

align the measurements in time. PDCs can be configured for several usages, the main being the ability to output a stream of time-aligned measurements from several PMUs. It can also be configured to store and archive data or perform calculations on the measurements.

The usage of PMUs to collect data on wide-area power systems is rather recent. It has enabled system operators to monitor their transmission system with more information as well as to conduct advanced analyses [7], particularly for inter-area modes. This technology offers a great potential for building monitoring systems [15].

2.2 Fast prototyping environment for PMU software applications

The traditional implementation of Wide Area Monitoring Systems (WAMS) relies on a monolithic software architecture. That forces the implementation of applications in a closed software solution (often proprietary) or directly into the PDC. In contrast, the infrastructure deployed at SmarTS Lab uses a modular approach based on application connecting to the PDC to receive measurements. A deeper comparison of both approaches can be found in [16]. This section described the approaches adopted at SmarTS Lab and OG&E for developing synchrophasor applications.

2.2.1 Statnett's synchrophasor SDK

To implement PMU applications using a modular paradigm, a Software Development Kit (SDK) developed by Statnett SF (the Norwegian Transmission System Operator) [17] can be utilized. The aim of the SDK is to facilitate research, fast prototyping and testing of real-time synchrophasor applications.

The chosen development environment is LabView, in which creating graphical user interfaces is simple. However, there is no standard programing interface in LabView to communicate with PMUs or PDC servers using the IEEE C37.118.2 protocol [18] and bring PMU measurements in the programing environment. Statnett's Synchrophasor SDK [17] provides a real-time data mediator, enabling this feature for LabView, as well as a library of functions that allow fast prototyping.

2.2.2 SmarTS Lab environment

The traditional set-up for WAMS is replicated at SmarTS Lab. The adopted approach makes use of Hardware-In-the-Loop (HIL) simulation, replacing the power system by models executed in real-time on a simulator equipped with reconfigurable analog inputs/outputs [19]. The rest of the WAMS architecture is replicated, with PMUs and a PDC server, building a prototyping platform depicted in Figure 2.

The lab is also equipped with a PDC server dedicated to the gathering of PMU measurements from Nordic universities, as a part of the STRONg^{2}rid project [20]. This platform, depicted in Figure 2, allows for the testing of applications using real-time data from the Nordic grid.

2.2.3 Oklahoma Gas and Electric

Oklahoma Gas and Electric Company (OG&E) began using synchrophasor technology in 2008 with a single hardware PDC and eight multifunction PMUs (line protection relays.) The live synchrophasor data was streamed through the PDC to a PC software client for visualization. It was quickly realized that a need existed to further examine data from interesting events which had occurred in the past. Unfortunately, the software purchased at the time did not allow for viewing of historical data. So the utility developed a system to archive the data to a database along with a custom software application named PhasorView that could display the synchrophasor data, both live and archived. With the initial fleet of only eight PMU's on the 345 and 500 kV Extra High Voltage (EHV) system,

(a)

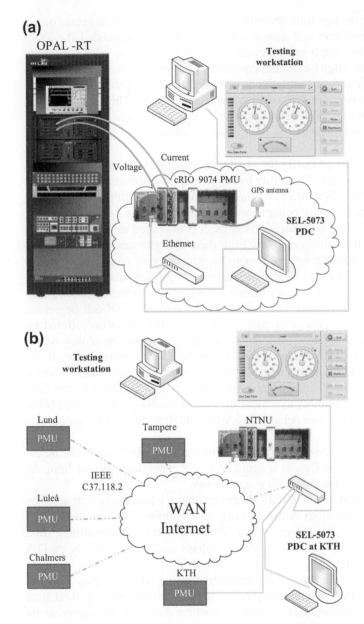

FIGURE 2

Diagram of the set-up at SmarTS Lab for wide area monitoring systems development and testing (a) Hardware-in-the-loop simulation (b) Nordic phasor measurement units (PMUs).

(b)

OG&E made observations at a rate of 30 samples per second and established a baseline for what would be considered normal operating conditions. They also joined the North American Synchrophasor Initiative (NASPI) and began streaming data to the host site at the Tennessee Valley Authority (TVA). This allowed for the utility to contribute their portion of the grid to the system wide view of the U.S. Eastern Interconnection. For the first time, the utility was able to observe how events on the OG&E system affected the interconnection and vice versa. The utility has expanded the PMU coverage of the transmission system by simply adding communications to existing substations with PMU capable devices already installed. High bandwidth communications were added to all EHV and other critical substations as part of a security initiative in 2005. PMUs have gradually been networked over the years, which now cover about 40% of the transmission system. This includes 100% of the EHV system, 100% of wind farms, 90% of the fossil generation fleet, and 34% of the HV system. In total there are now over 200 transmission lines, autotransformers and generators monitored by PMUs.

OG&E has been using synchrophasors primarily for situational awareness, disturbance analysis, and wind power plant integration/monitoring. The utility has a tool for the Supervisory Control and Data Acquisition / Energy Management System (SCADA/EMS) software to bring synchrophasor data into the state estimator, which is currently being tested. The technology is used to assess the stability of the system and proactively find equipment problems. It is also used to monitor how the system responds to faults and assess the voltage recovery from these disturbances. *Synchrophasors have also proved very useful for integrating renewable energy into the grid and monitor power quality.* Application development at OG&E has been an iterative approach spanning several years. The company has maximized the use of open source software (OSS) tools to achieve desired goals. With the release of the Grid Protection Alliance's OpenPDC in October 2009, OG&E began the process to restructure the synchrophasor system to take advantage of the flexibility offered by OpenPDC. One of the major concerns regarding the initial configuration was the ability of the hardware PDC to handle hundreds of PMU devices available on the OG&E system. The single hardware data concentrator could handle approximately 40 PMUs. In order to connect more devices, it was necessary to scale out the system architecture. This means that all data could not be centrally concentrated. OpenPDC allowed for the flexibility to achieve the goal of handling hundreds of PMUs. The company developed a custom VB.net Action Adapter inside OpenPDC to do all of the processing and inserting of data into the SQL database.

The company's first problem back in 2008 was visualization of the data, which led to the development of the PhasorView software. PhasorView is a application that plots data queried from the historian. Figure 3 shows a screenshot of the PhasorView application.

The interactive Geographic Information System (GIS) interface on the left represents an accurate view of the transmission system and substations on which radar and lightning data can also be displayed. The plots on the right show the synchrophasor data for different power system quantities, from the selected PMUs. The bottom right polar plot shows the system voltage angle spread. Above the GIS interface is a legend for the selected PMUs.

The software in live mode serves as a situational awareness tool, using GIS to display the lines and substations within the service territory along with plots of the synchrophasor data. Real time weather radar and lightning strike data is displayed to anticipate where disturbances may occur on the system. With the company's widespread PMU deployment, the technology is used as a system wide fault recorder. Synchrophasor data serves as the top level overview and then the substation level relay and Digital Fault Recorder (DFR) data can be used to further investigate the event. One of the challenges of having over 200 terminals is how to effectively give a good overview of the system. OG&E addressed this by assigning a priority to each terminal that dictates which data is displayed for a given map location and zoom level.

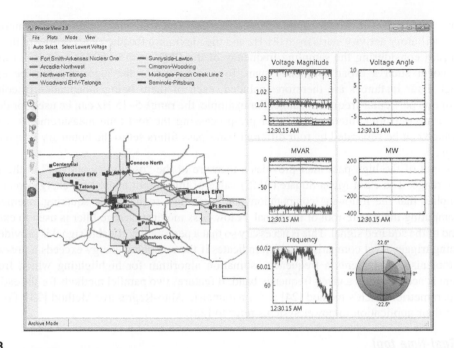

FIGURE 3

Screenshot of PhasorView at Oklahoma Gas & Electric.

3. Detection tools for wind farm oscillation monitoring

The effects of high frequency oscillations presented in the Introduction are undesirable [4], thus new efficient detection tools were developed. These monitoring algorithms allow fast oscillation detection from PMU measurements. The approach developed at SmarTS Lab is presented in Section 3.1 and the in-house tool developed at OG&E in Section 3.2.

3.1 Monitoring tool at SmarTS Lab

Traditional monitoring tools for inter-area oscillations estimate the frequency and the damping for each oscillatory mode with two separate algorithms. A similar strategy is adopted in this case, with one algorithm dedicated to the estimation of the amount of energy in the oscillations and the other dedicated to frequency estimation. The three main algorithms used by the PMU application are described next.

3.1.1 Fast oscillation detection and frequency estimation

The proposed oscillation detection algorithm in this paper builds from work in [21]. As highlighted in [22], it is desirable for a fast oscillation detection tool to provide information about oscillatory behavior at different bands of the spectrum. The frequency spectrum of interest with potential oscillatory activity starts from 0.1 Hz and the maximum frequency according to the Shanon criterion [23] considering the sampling frequency of the PMUs (up to 50 Hz reporting rate in the lab). To cover such a broad frequency span, four instances of the algorithm can be executed in parallel. Four instances are therefore included, each of them being independently configured to monitor one particular frequency range. For example, the range 5–15 Hz can be used for detection of wind farm controller interactions. After pre-processing the real-time measurements, each frequency range can be separated by four different band-pass filters set to the boundary frequencies of each range.

The output of the high-pass filter provides a measure of the "oscillatory activity" at the selected frequency range. The Root Mean Square (RMS) value of this output is used for energy computation, which implies that the following computations are performed sequentially: squaring, averaging and finally computing the square-root of the signal. A low-pass moving average filter is used to extract the main trend of the squared signal. This is necessary so that a persistent and stable signal is provided to the forthcoming trigger level comparison, which indicates if the computed energy exceeds a pre-set level.

The tool also implements a frequency estimation algorithm for highlighting which frequency component is active within a given frequency band. It features two parallel methods for the estimation, the non-parametric Welch's method [24] and a parametric Auto-Regressive Method [23]. For further details on their application to power systems refer to [25].

3.1.2 Real-time tool

The *Monitoring Tool* has been developed to detect oscillations occurring in the power system from real-time PMU measurements. The reporting rate allows to identify different categories of phenomena, which are listed in the previous Section. The *Monitoring Tool* therefore presents an interface with four instances, of the algorithms running in parallel, grouped under the name *Module*.

As mentioned earlier, the *Monitoring Tool* has been developed with the objective of providing an intuitive visual tool. The resulting Graphical User Interface (GUI) is depicted in Figure 4.

The graph on the top-right of the interface is a representation of the buffered input signals received by the tool. the signal displayed can be chosen by the user among the available input signals. This graph is used for tuning the outlier removal algorithm. The top left part of the GUI serves to configure the tool and is divided into tabs to cover the configuration of each *Modules* (highlighted on Figure 4.) In a *Module*, the graph on the right presents the power spectrum density, which is the output of the frequency estimation algorithm. The graphical display on the left presents the results of the energy detection algorithm with LED indicators and a graph. The graph provides a history to easily corroborate the energy computed with the input signal displayed.

3.1.3 Offline replay tool

The original idea was to build a tool consuming real-time data from PMU/PDC streams. At this time no recorded measurements from TSOs were available, but later during the project some recorded measurements were received from OG&E, leading to the development of the *Replay Tool*, in order to exploit them.

In the *Monitoring Tool* presented previously, several configuration options have been highlighted. While some options do not require any knowledge of the power system studied and can therefore be

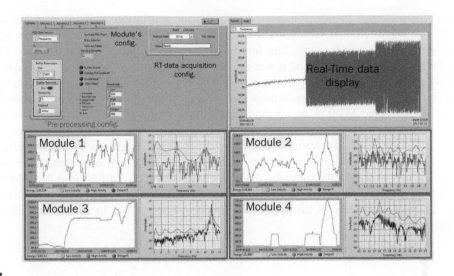

FIGURE 4

Screen shot of the interface of the Monitoring Tool.

tuned directly, others have to take in consideration the properties of the power system, especially for the frequency estimation algorithm parameters. The tool thus needs calibration and archived data was used for this purpose.

The *Replay Tool* has been developed using the same code as the real-time tool. However, additional software has been developed so that it can use archived data instead of using live PMU streams. The interface is almost identical to the *Monitoring Tool*. The processing algorithms are identical, however this tool has some specific features. For example it allows to scroll along the replayed data, which can be a useful feature to get a quick overview of the content of the selected file.

3.2 Oklahoma Gas & Electric

OG&E, along with many other utilities in the U.S. Great Plains region, has a large wind generation resource potential. Many large scale wind farm facilities varying in size from 100 to 300 MW have been brought online, with many more under development. Currently the Southwest Power Pool's generation interconnection queue is approaching 30 GW of wind resources, making it one of the nation's most prominent sources of renewable energy. Determining how these vast resources will be integrated into the regional power grid proves to be a challenge and synchrophasor technology is able to provide the tools necessary to do so in a reliable manner. Each new wind farm facility brought online in OG&E's service territory is accompanied by PMU measurements at the point of interconnection. In December 2010, the utility began observing sub-synchronous oscillations on the transmission system in a concentrated portion of the grid in northwestern Oklahoma as shown in Figure 1.

These oscillations were found during periods of high wind generation, above 80% of the nameplate capacity. The voltage oscillations observed were as high as 5% fluctuation at an oscillatory frequency of around 14 Hz as shown in Figure 1. This level of voltage fluctuation exceeded the IEEE 141-1993 standards for objectionable flicker, and it was confirmed that the impact was observable to the area's distribution customers. The problem has been localized to specific wind farms and the utility is undergoing efforts with the turbine manufactures to resolve the problem. This phenomenon could not be observed with traditional SCADA monitoring and without synchrophasor technology, the problem would have taken much longer to identify and resolve. The benefit of having PMU measurements at the point of wind farm interconnection is to ensure that customers receive clean power while maintaining the level of system stability necessary for reliable power system operation. After observing these instances of voltage oscillations, the company implemented a Fast Fourier Transform (FFT) based detection program to detect the oscillations and send email notifications when it requires corrective action. Figure 5 shows the application running on the server in real-time to detect these oscillations.

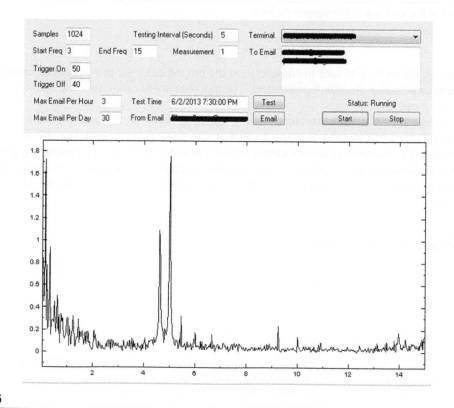

FIGURE 5

Fast Fourier transform real-time application developed by Oklahoma Gas & Electric.

4. Testing and validation

4.1 SmarTS Lab

For testing the monitoring tool, Hardware-In-the-Loop (HIL) simulation was used with the development of a power system model capable of recreating the event described in Section 1 and additional perturbations [26]. The power system model is equipped with two variable loads introducing random variation and sinusoidal variation that will excite low frequency dynamics in the power system. The model performs the simulation of wind farms interaction by injecting oscillations at 10.83 Hz, which was chosen as the equivalent of the 13 Hz oscillations in a 50 Hz system (the original case occurred in the USA, where the nominal frequency is 60 Hz).

The validation of the algorithms is performed with a scenario including several perturbations according to the experiment protocol summarized in Table 1, enabling the testing of the tool's performance in different situations. The experiment has been performed several times to ensure that the configuration of the different parameters of the processing algorithms was appropriate. From the experiment scenario presented in Table 1, only the oscillation injection is reported is this Chapter. The full testing experiment of the tool is reported in [27].

The simulation of the power system model is started with both wind farms receiving an average wind speed of 12 m/s with 10% turbulence. The loads are also configured to have a sinusoidal profile at different frequencies.

The processing algorithms of the *Monitoring Tool* detect the slow dynamic activity resulting from both load variations and wind turbulence, see *Module 1* on Figure 4, where the spectral estimator highlights both frequency components at 0.4 and 0.8 Hz.

Forced oscillations were then injected at first with 0.05 p.u. amplitude at the point of common coupling of the first wind farm. They can be observed in the frequency graph of the tool, see Figure 6(a). It can be noticed that the frequency range containing 10.83 Hz is active with the flag *Danger!!!*, see Figure 6(b), while the other frequency ranges remain inactive or with a low activity. This shows the fast reaction of the tool, its selectivity and its ability to estimate in real-time the level of energy in these oscillations. The frequency estimation algorithm does not update as quickly as the energy detection algorithm, the frequency of the injected oscillations is thus not detected as quickly. However, it can be noticed that the parametric method starts to show distinctively a peak at the right frequency, see Figure 6(b).

After the beginning of the injection with 0.05 p.u. amplitude the injection is increased up to 0.07 p.u. amplitude. The resulting oscillations have a larger amplitude, as shown on the real-time data display on Figure 4, and the energy detection algorithm identifies an increase in the energy level in the oscillations. The frequency estimation algorithm also detects very precisely the frequency at which the oscillations are occurring as shown in *Module 3* on Figure 4.

Table 1 Experimental Testing Protocol

Start	Oscillation Injection	Perturbations	Major Fault	End
Random load variation	Set perturbation at 10.83 Hz	Generating minor faults	Three phase fault and line opening	End of the oscillation injection

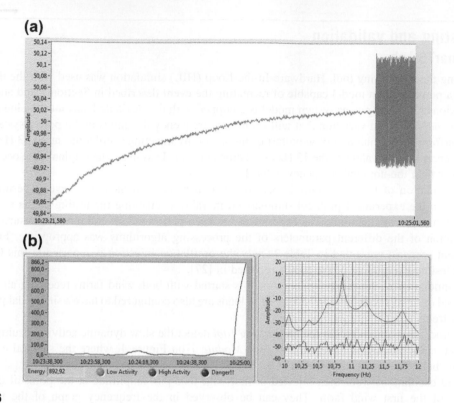

FIGURE 6

Partial screen shot of the monitoring tool during the oscillation injection at 10.83 Hz (a) Frequency of the system (b) Module (10–12 Hz).

Additional information on functionalities and testing results are available in [27].

4.2 Validation of the OG&E FFT detection program

The OG&E FFT detection program utilizes the open source library Exocortex[2]. To validate that the program correctly calculates the FFT, the output is compared to MathCad's FFT algorithm which is taken as reference[3]. Figure 7(a) shows the FFT analysis from both tools on a parcel of PMU measurements when oscillatory activity was present on the grid.

The same data set was queried from the database and placed into a .csv file. Using MathCad to read the .csv file, the built in FFT algorithm is then used to generate an identical plot as shown in Figure 6.

---

[2]http://www.exocortex.org/dsp/.
[3]www.ptc.com/product/mathcad.

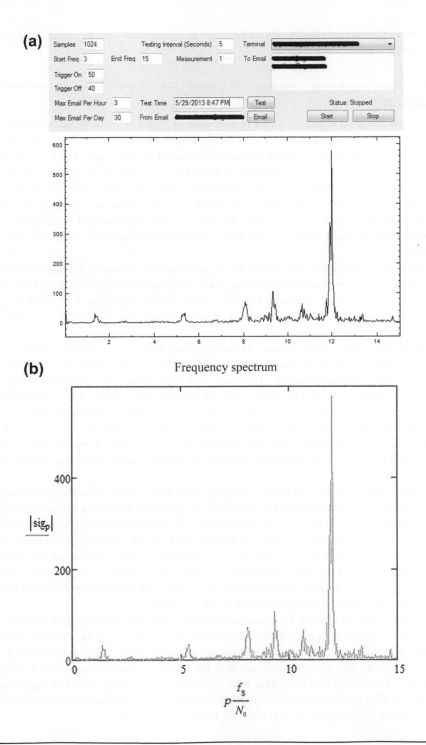

FIGURE 7

Screenshots for validation during an oscillatory event (a) Oklahoma Gas & Electric fast Fourier transform detection program (b) Offline validation with MathCad.

Using these two different calculation algorithms, the company was able to validate the proper functionality of the FFT detection program.

5. Conclusions

This chapter described how synchronized phasor data applications can be developed to help grid operators in monitoring and control of renewable energy sources when unpredictable dynamic interactions arise. These type of challenges were illustrated for the case of sub-synchronous wind farm oscillations which have been captured in OG&E's grid. The two applications described in this chapter illustrate how PMU data can offer added value providing real-time data to advanced algorithms that go beyond monitoring and allow fast-detection of dynamic interactions of renewable sources in the grid. The accuracy and value of the implemented applications was tested and validated through both real-time hardware-in-the-loop simulation and through replay of historical data.

This chapter also highlights the flexibility of independent software development of two sub-synchronous wind farm oscillation monitoring tools implemented under different paradigms and software environments. These different software environments illustrate how new synchrophasor applications can be conceived and implemented without relying on a monolithic and closed software development system.

At this time the only control action available to operators is to enforce wind farm output curtailment. Another means for mitigation is network reinforcement, which is costly and will often take a relatively long time to complete. However, sub-synchronous oscillations arising from wind farm interactions can be mitigated through the use of controllable devices [10]. One important aspect to further investigate is how the use of synchrophasor data can help in providing rich measurements containing information on sub-synchronous oscillations into the controls of Static Var Compensators, and other controllable devices. Recent applications to damping of low-frequency oscillations suggest that PMU-based sub-synchronous oscillation compensation could be effectively applied [28].

The prediction of unwanted dynamics emerging from interactions of renewable energy sources with the grid will ultimately depend on the ability to properly represent these dynamics instudy models [29]. As grid conditions change continuously, the representation of wind farmclusters into computer simulation programs for prediction will have to be updated. This task canbe time consuming, and one important use of phasor data could be the accurate estimation ofaggregated wind farm models from near-real time measurements. The representation of theseaggregated models could be updated using phasor data and supplied to dynamic securityassessment tools which can help pin-point undesirable dynamics during stressed network operation.

Indeed, the flexibility offered by non-conventional software development systems for PMU applications offer unlimited opportunities to conceive new software tools. A clear example is the development of mobile monitoring applications presented in [20]. This example brings real-time synchrophasors outside the control room and into the hands of analysts and persons with different roles in a utility. Liberating real-time data from the control room may offer the opportunity to ask questions and develop understanding which could be used for developing new software applications that could facilitate the integration of renewable energy into the grid.

Acknowledgments

The economical support of the institutions and funding bodies listed below is sincerely acknowledged: L. Vanfretti was supported by Statnett SF, the Norwegian Transmission System Operator, the STandUP *for* Energy collaboration initiative and Nordic Energy Research through the STRONg^{2}rid project. M. Baudette was supported by EIT KIC InnoEnergy through Action 2.6 of the Smart Power project and by Statnett SF, the Norwegian Transmission System Operator.

References

[1] Manwell JF, McGowan JG, Rogers AL. Wind energy explained: theory, design and application. John Wiley & Sons; 2010.

[2] Milborrow D. Europe 2020 wind energy targets, EU wind power now and the 2020 vision. Windpower Monthly Special Report; March 2011.

[3] U.S. Energy Information Administration (EIA). Most states have renewable portofolio standards [Online]. Available: http://www.eia.gov/todayinenergy/detail.cfm?id=4850; February 2012.

[4] White A, Chisholm S. Relays become problem solvers. Transm Distrib World; November 2011.

[5] Cardwell D. Intermittent nature of green power is challenge for utilities. New York Times [Online]. Available: http://www.nytimes.com/2013/08/15/business/energy-environment/intermittent-nature-of-green-power-is-challenge-for-utilities.html?smid=pl-share; August 2013.

[6] von Appen J, Braun M, Stetz T, Diwold K, Geibel D. Time in the sun: the challenge of high PV penetration in the German electric grid. IEEE Power Energy Mag 2013;11(2):55–64.

[7] (RAPIR Chair) Patel M. Real-time application of synchrophasors improving reliability. Princeton (NJ): North American Electricity Reliability Corporation, Tech. Rep; 2010.

[8] Wiik J, Gjerde JO, Gjengedal T. Impacts from large scale integration of wind energy farms into weak power systems. In: Proceedings of the international conference on power system technology; 2000. pp. 49–54.

[9] Larsen E. Wind generators and series-compensated AC transmission lines. In: IEEE power and energy society general meeting; 2012. pp. 1–4.

[10] Suriyaarachchi DHR, Annakkage U, Karawita C, Kell D, Mendis R, Chopra R. Application of an SVC to damp sub-synchronous interaction between wind farms and series compensated transmission lines. In: IEEE power and energy society general meeting; 2012. pp. 1–6.

[11] Sahni M, Badrzadeh B, Muthumuni D, Cheng Y, Yin H, Huang S, et al. Sub-synchronous interaction in wind power plants-Part II: an ERCOT case study. In: IEEE power and energy society general meeting; 2012. pp. 1–9.

[12] Adams J, Pappu V, Dixit A. ERCOT experience screening for sub-synchronous control interaction in the vicinity of series capacitor banks. In: IEEE power and energy society general meeting; 2012. pp. 1–5.

[13] Ma HT, Brogan PB, Jensen KH, Nelson RJ. Sub-synchronous control interaction studies between full-converter wind turbines and series-compensated ac transmission lines. In: IEEE power and energy society general meeting; 2012. pp. 1–5.

[14] White (OGE) A, Chisholm (OGE) S, Khalilinia (WSU) H, Tashman (WSU) Z, Venkatasubramanian (WSU) M. Analysis of subsynchronous oscillations at OG&E. In: NASPI-NREL synchrophasor technology and renewables integration workshop–Denver, CO; June 7, 2012.

[15] Vanfretti L, Chenine M, Almas MS, Leelaruji R, Angquist L, Nordstrom L. SmarTS Lab; a laboratory for developing applications for WAMPAC systems. In: IEEE power and energy society general meeting; July 2012. pp. 1–8.

[16] Vanfretti L, Van Hertem D, Gjerde J. Smart transmission grids vision for Europe. In: Hills P, Mah D, Li VOK, Balme R, editors. Smart grids and sustainable energy transformation. Springer; 2014.

[17] Vanfretti L, Aarstrand VH, Shoaib Almas M, Peric V, Gjerde JO. A software development toolkit for real-time synchrophasor applications. In: Proceedings of the IEEE PowerTech; 2013.

[18] IEEE Std C37.118.2-2011 (Revision of IEEE Std C37.118-2005) IEEE standard for synchrophasor data transfer for power systems; 2011. pp. 1–53.

[19] eMEGAsim PowerGrid real-time digital hardware in the loop simulator–OPAL RT. [Online]. Available: http://www.opal-rt.com/.

[20] Almas MS, Baudette M, Vanfretti L, Løvlund S, Gjerde JO. Synchrophasor network, laboratory and software applications developed in the STRONg2rid project. IEEE PES general meeting, submitted for publication.

[21] Hauer J, Vakili F. An oscillation detector used in the BPA power system disturbance monitor. IEEE Trans Power Syst February 1990;5(1):74–9.

[22] Trudnowski D. Fast real-time oscillation detection. In: Proc. North American SynchroPhasor initiative (NASPI) work group meeting, Orlando, FL; February 29–March 1, 2012.

[23] Hayes MH. Statistical digital signal processing and modeling. John Wiley & Son; 1996.

[24] Welch P. The use of fast Fourier transform for the estimation of power spectra: a method based on time averaging over short, modified periodograms. IEEE Trans Audio Electroacoust June 1967;15(2):70–3.

[25] Sanchez-Gasca J. Identification of electromechanical modes in power systems. Special Publication, TP462. IEEE Power & Energy Society; July 2012.

[26] Baudette M. Fast real-time detection of sub-synchronous oscillations in power systems using synchrophasors [Master's thesis]. KTH, Electric Power Systems; 2013.

[27] Vanfretti L, Baudette M, Al-Khatib I, Almas MS, Gjerde JO. Testing and validation of a fast real-time oscillation detection PMU-based application for wind-farm monitoring. In: 2013 first international Black Sea conference on communications and networking (BlackSeaCom), Batumi, Georgia; July 2013.

[28] Uhlen K, Vanfretti L, De Oliveira MM, Leirbukt A, Aarstrand VH, Gjerde JO. Wide-area power oscillation damper implementation and testing in the Norwegian transmission network. In: IEEE power and energy society general meeting; 2012. pp. 1–7.

[29] Zavadil R, Miller N, Ellis A, Muljadi E, Pourbeik P, Saylors S, et al. Models for change. IEEE Power Energy Mag 2011;9(6):86–96.

Every Moment Counts: Synchrophasors for Distribution Networks with Variable Resources

34

Alexandra von Meier, Reza Arghandeh

California Institute for Energy and Environment, University of California, Berkeley

1. Introduction

Historically, power distribution systems did not require elaborate monitoring schemes. With radial topology and one-way power flow, it was only necessary to evaluate the envelope of design conditions (i.e., peak loads or fault currents), rather than continually observe the operating state. But the growth of distributed energy resources, such as renewable generation, electric vehicles, and demand response programs, introduces more short-term and unpredicted fluctuations and disturbances [1]. This suggests a need for more refined measurement, given both the challenge of managing increased variability and uncertainty and the opportunity of recruiting diverse resources for services in a more flexible grid. This chapter addresses how the direct measurement of voltage-phase angle might enable new strategies for managing distribution networks with diverse, active components. Specifically, it discusses high-precision micro-synchrophasors, or phasor measurement units (μPMUs), that are tailored to the particular requirements of power distribution to support a range of diagnostic and control applications, from solving known problems to opening as yet unexplored possibilities.

2. Variability, uncertainty, and flexibility in distribution networks

Electric transmission and distribution systems are formally distinguished by voltage level, but harbor profound differences in design and operation. These differences explain the diverse sets of challenges encountered in the context of renewables integration, as well as the historical lag of distribution behind transmission systems in terms of observability and sophistication of measurement. Broadly speaking, distribution systems tend to be low-tech, aging, and due for upgrades [2–4].

Architecture: For economy and simplicity of protection, distribution systems are generally laid out radially, with legacy equipment, such as protective relays and voltage regulation devices, designed on the assumption of one-directional power flow from the substation toward loads. Although easier to operate in principle, radial design also presents liabilities: When distributed generation introduces reverse power flow, some older controls may malfunction. In addition, radial design makes the mathematical estimation of the operating state more difficult, by removing the redundancy afforded by Kirchhoff's laws: in other words, the estimate for voltages and currents at one node cannot be corroborated by those at neighboring nodes. Distribution systems also have many more nodes or

Renewable Energy Integration. http://dx.doi.org/10.1016/B978-0-12-407910-6.00034-X

429

branch points than transmission networks, because each secondary transformer represents a load bus. With many more nodes than measuring points (and without smart meter data available in near real-time), it becomes even more difficult to perform state estimation. Traditional distribution operations never required this level of analysis, but the uncertainties introduced by diverse distributed resources make it increasingly important to assess the actual operating state of the system.

Variation: At the local scale, we lose the statistical effects of aggregating many customers that is assumed in transmission-level analysis. Consequently, there is more variation in both time and space. The load duration curve is "peakier" for an individual distribution feeder than for an entire service territory, and ramp rates as a percentage of load can be much steeper. In addition, phase imbalances are much more important, often ranging in the tens of percent, making it necessary to consider all three phases individually. Variation also means that local idiosyncrasies, such as load types and topography, are more important: no two distribution feeders are exactly alike, and it can be difficult to extrapolate analytical findings from one area to another. This underscores the need for carefully monitoring individual distribution circuits as penetration levels of active components increase.

Exposure: Closer proximity to many types of hazards (flora, fauna, and human activities) means more exposure and vulnerability for distribution circuits: unsurprisingly, most customer outages originate in the distribution system. With any number of local factors affecting distribution, but without redundant supply paths, distribution operations often revolve around switching procedures, such as isolating sections or restoring service to customers as safely and quickly as possible. One implication is that improved distribution reliability is a likely area for early benefits from advanced monitoring. Exposure to the elements and the actions of many individuals also means a higher degree of uncertainty in distribution operations, whereas operating errors and malfunctions pose an immediate and physical risk.

Opacity: Historically, distribution operators have relied on field crews as their eyes and ears to report on system status. Despite increasing prevalence of supervisory control and data acquisition (SCADA), it is still often necessary to send someone in a truck to verify, for example, whether a switch is open or closed, or to pinpoint the location of a downed line. This has important practical implications, but also poses an analytical challenge: although the power flow calculation in transmission networks assumes that the topology of the network and the physical characteristics of all branches are known exactly, such information tends not to be reliably available for distribution circuits.

Various tools have been developed and implemented to provide distribution operations and planning with more detailed and timely information. Even so, creating situational awareness out of disjointed data streams remains a challenge. Circuit models, where available, may be based on unreliable input data and questionable assumptions. Physical measurements from the field remain a limiting factor for analysis, human operators, and automated control systems alike.

The lack of visibility on distribution systems follows from simple economics: there has never been a pressing need to justify extensive investment in sensing equipment and communications. Even with the growing need for monitoring capabilities, the costs must be far lower to make a business case for measurement devices on a distribution circuit compared with the transmission setting.

Arguably, the time for such a business case is fast approaching. Distributed energy resources are beginning to pose both challenges and opportunities for actively managing distribution circuits. As discussed elsewhere in this volume, renewable and other non-traditional resources create a need for coordination at higher resolution in both space and time, from protection to voltage regulation and other power quality issues. They also represent a new menu of options for grid support functions, such as volt-VAR optimization, energy storage on time scales anywhere from cycles to hours, or even intentional islanding. From both the perspective of avoiding adverse customer impacts and taking optimal advantage of new resources, increased flexibility in operating distribution circuits is called for. But given the particular data richness of distribution circuits, this means that much more, better, and faster information from far behind the substation will be needed to make intelligent and economical decisions. This chapter suggests the possibility that distribution system monitoring and control might skip a generation of SCADA developed for transmission systems and proceed straight into the twenty-first century, bringing us to the state of the art in alternating current (a.c.) measurement: synchrophasors.

3. **Microsynchrophasor (μPMU) technology**

The essence of synchrophasors, or PMUs, is the precise time stamping of voltage measurements to compare the phase angle among different locations [5]. This technology became feasible in the 1980s with readily accessible GPS time signals. Real (active) power flow between two points on an a.c. network varies mainly with the voltage angle difference δ. When the line impedance is mainly inductive, real power flow P_{12} can be approximated by the following scalar equation:

$$P_{12} \approx \frac{V_2 V_1}{X} \sin \delta \tag{1}$$

where X is the line reactance and V_1 and V_2 are the voltage magnitudes. Voltage magnitude and phase angle at each node in a network are considered the *state variables*, because they uniquely determine a.c. power flow throughout the network.

Direct measurement of the state variable δ offers some basic advantages. Voltage angle can serve as a proxy for local current measurements, where installation of current sensors is inconvenient. If both voltage and current magnitude and angle are measured, this provides maximal information about the system state from just one instrumented node. Because voltage angle varies across an a.c. network in a continuous profile, power flow patterns can be inferred from angle gradients, without explicitly measuring branch currents. Beyond steady-state analysis, the key benefit of synchrophasor measurements lies in observing dynamic behavior, including rapid changes on the scale of cycles rather than seconds.

Today, PMUs are used almost exclusively on transmission systems. Some of the most notable benefits have come from observing subsynchronous oscillations across wide areas, such as the Western Interconnect in the United States, that threaten a.c. system stability. By directly identifying oscillation modes and their associated damping, operators can take specific appropriate actions (e.g., derating transmission lines) as necessary, thus supporting both reliability and asset use.

Although "distribution PMUs" may already be deployed at distribution substations (e.g., embedded in protective relays), the use of their measurements is mainly in reference against phase angles elsewhere on the transmission grid, not the distribution feeder. By contrast, the purpose of microsynchrophasors is specifically to compare voltage angles at different points on distribution circuits, behind the substation. According to Eqn (1), such angle differences on distribution systems will tend to be much smaller, owing to smaller power flows and shorter distances. For example, a typical voltage phase angle difference for a distribution feeder at full load might be 0.1°/mile, compared with tens of degrees between transmission nodes.[1] Consequently, transmission PMUs with typical errors near $\pm 1°$ may not provide enough precision for meaningful distribution measurements.

In addition to requiring greater angular resolution, distribution synchrophasor measurements will likely have to contend with more noise, including harmonics and small transients associated with nearby devices or switching operations on a circuit. For this reason, we expect that it will prove useful to combine PMUs with power quality measurements, so as to analyze and interpret angle data in proper context.

Figure 1 illustrates the capabilities of a new μPMU device developed and manufactured by Power Standards Laboratory, based on a commercially available power quality recorder, the PQube (www. powerstandards.com). A key capability of the μPMU is to combine high-resolution angle measurements with detailed characterization of waveforms, including harmonics and transients. The figure shows relevant quantities are approximately situated on a logarithmic time scale for visual comparison [6]. An absolute floor of attainable angular resolution is set by timekeeping accuracy, while the high sampling rate serves to characterize harmonics. Useful rates of data recording and communication will vary depending on what practical applications in distribution systems are to be supported, which will differ in the quantity, quality, and timeliness of measurement data they require.

The authors are using this μPMU in a project funded by the US Department of Energy's Advanced Research Projects Agency-Energy (ARPA-E) program.[2] We envision from several up to tens of μPMUs installed at multiple locations throughout a distribution feeder (e.g., at the substation, end of feeder, or laterals) and any key distributed generation facilities. The μPMU connects at the secondary voltage level (with voltage inputs from 100 to 690 V_{LL}) or through potential transformers at substations or line devices. Each μPMU may upload its precisely time-stamped measurements through a suitable communication layer (e.g., 4G wireless) to a flexible local network termed μPnet, where angle measurements will be compared. The μPnet concept will build on the simple Measurement and Actuation Profile developed by the University of California, Berkeley, as a foundation for managing both real-time and archival data from a wide variety of physical sources [7,8]. Initial μPMU deployment schemes and networking will be tested in several field installations under the ARPA-E project [6].

4. Applications for μPMU measurements

A broad spectrum of potential distribution system applications could hypothetically be supported by synchrophasor data, as has been noted in the literature [9–11]. This section aims to characterize some selected applications and speculate about the potential advantage afforded by voltage angle measurement to support them.

Topology status verification means detecting or confirming the open or closed status of operable switches or breakers by comparing voltage angle directly across them or validating power flow solutions under different network topologies. Fast and reliable empirical topology identification could

---

[1]It should be noted that the approximate Eqn (1) is less apt when applied to distribution lines, where resistance is significant compared to inductance (greater R/X ratio). However, the general statement holds that phase angle differences in distribution are one to two orders of magnitude smaller than in transmission.

[2]This three-year award, DE-AR0000340, commenced in 2013 and includes research partners CIEE, UC Berkeley, Lawrence Berkeley National Laboratory, and Power Standards Lab.

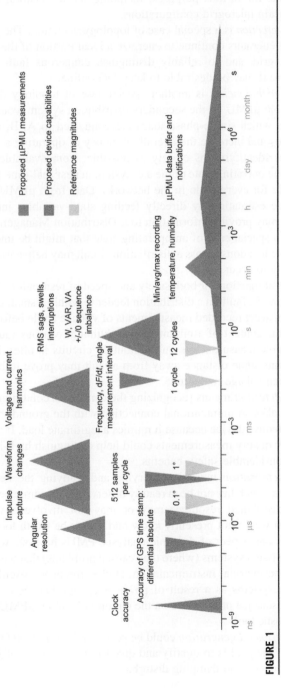

FIGURE 1

Time scale of microphasor measurement unit (μPMU) functionalities.

enhance safety where remote indicators are unavailable or considered unreliable. It could also support circuit switching operations for various purposes, including service restoration after an outage, or safely accommodating certain microgrid configurations.

Unintentional Island Detection is a special case of topology detection. The goal is to immediately identify when distributed generators continue to energize a local portion of the network that has been separated from the main grid, and to reliably distinguish dangerous fault situations from other abnormal conditions, where it may be desirable to keep DG online.

Phase identification and balancing is another special case of topology detection. Direct angle measurement with a portable μPMU on the secondary distribution system would be a uniquely quick and easy way to ascertain which single-phase loads are connected to A, B, or C. Improved phase balancing can reduce losses and increase the overall efficiency of operating a three-phase system.

State estimation means identifying as closely as possible, from available network models and empirical measurements, the operating state of the a.c. system in near real-time (i.e., to specify voltage magnitude and phase angle for every node in the network). Data from μPMUs could ease the difficulties of distribution state estimation by directly feeding state variables into a Distributed State Estimator, which, in turn, may provide information to a Distribution Management System.

Reverse power flow is a special case of an operating state that might be undesirable. Phase angle measurements from suitable locations across a distribution circuit may help anticipate when and where reverse power flow is likely to occur.

Fault location is a critical function for both safety and speed of restoration. The goal is to infer the actual geographical location of a fault on a distribution feeder to within a small circuit section between protective devices, by comparing recorded measurements of voltage angle before and during the fault and interpreting these in the context of a circuit model. Although various fault location algorithms exist, the quality of available measurements on distribution circuits is often insufficient to support them. Voltage angle at points some distance away from a fault may prove to be a more sensitive indicator of fault location than voltage magnitude.

High-impedance fault detection means recognizing the dangerous condition where an object such as a downed power line makes an unintentional connection with the ground, but does not draw sufficient current to trip a protective device because it mimics a legitimate load. If a combination of high-resolution angle and power quality measurements could help distinguish high-impedance faults from loads, this would afford considerable safety benefits.

Dynamic circuit characterization involves observing and studying the behavior of distribution circuits on short time scales, which has not been readily observable. Dynamic behavior in the presence of high penetrations of distributed resources may or may not involve anything problematic or actionable; the point is that we do not presently know and it may be worth looking.

Oscillation detection is one aspect of dynamic analysis. PMUs have shown subsynchronous oscillations to exist on transmission systems (where they caused problems) that were neither predicted by models nor observed by conventional instrumentation. Higher-frequency oscillations could conceivably occur on distribution systems as a result of power exchange between and among distributed energy resources, or any resonance phenomena on the circuit [12]. If so, μPMU measurements would be an ideally suited diagnostic tool.

Characterization of distributed generation could be performed with μPMU measurements at small time scales below a cycle. The goal is to qualify and quantify the behavior of inverters in relation to stabilizing system a.c. frequency and damping disturbances in power angle or frequency. This could help exclude adverse grid impacts, such as resonance or simultaneous trips of distributed generators, while facilitating their recruitment for advanced ancillary services, such as inertia or transient mitigation, in the future.

Unmasking loads from net metered DG would involve inferring the amount of load being offset by DG behind a net meter through measurements and correlated data obtained outside the customer's premises. Estimating the real-time levels of renewable generation versus loads would allow for better anticipation of changes in the net load, by separately forecasting the load and generation, and for assessing the system's risk exposure to sudden generation loss. At the aggregate level, this information is of interest to system operators for evaluating stability margins and damping levels in the system.

Fault-induced delayed voltage recovery (FIDVR) is an unstable operating condition that results from the interaction of stalled air conditioners with capacitor bank controls [13]. μPMU data might help anticipate FIDVR before it occurs, by identifying in near real-time the varying contribution to total customer load from devices such as single-phase induction motors in residential and small commercial air conditioners that pose an increased risk.

Table 1 summarizes prospective advantages of high-resolution voltage angle measurements compared with conventional techniques to support the listed diagnostic applications.

The described applications focus on diagnostic capabilities. However, synchrophasor data may enable more refined management and active control of distribution systems. Possible control applications include the following:

Protective relaying. Reverse power flow was noted previously as a condition that can be important to diagnose and avoid, but another approach is to use protection schemes that safely accommodate reverse flow. Without requiring a costly replacement of protective devices, it may be feasible to develop supervisory differential relaying schemes based on μPMU data that recommend settings to individual devices according to overall system conditions [14].

Microgrid coordination. Microgrid monitoring and control has to satisfy several objectives: load sharing among DG units, voltage and frequency regulation in islanded and grid connected modes, island detection and resynchronization, supply and demand optimization, and real-time monitoring of disturbances and harmonics. As an accurate measure of system state, μPMU data could support control algorithms in all of the previously described aspects.

Generation and load within a power island can be balanced through conventional frequency regulation techniques, but explicit phase angle measurement may prove to be a more versatile indicator. In particular, angle data may provide for more robust and flexible islanding and resynchronization of microgrids. A convenient property of PMU data for matching frequency and phase angle is that the measurements on either side need not be at the identical location as the physical switch between the island and the grid. A self-synchronizing island that matches its voltage phase angle to the core grid could be arbitrarily disconnected or paralleled, without interruption of load. Initial tests of such a strategy with angle-based control of a single generator enabled smooth transitions under continuous load with minimal discernible transient effects [15].

Comparison of angle difference between a microgrid or local resource cluster and a suitably chosen point on the core grid could enable the cluster to provide ancillary services as needed, and as determined by direct, physical measurement of system stress rather than a price signal (e.g., by adjusting power imports or exports to keep the phase angle difference within a predetermined limit). A variation of this approach, known as angle-constrained active management, has been demonstrated in a limited setting with two wind generators on a radial distribution circuit [16].

Table 1 Potential Diagnostic Applications with Microphasor Measurement Units Measurements

Diagnostic Application	Competing Conventional Strategies	Likely Advantage of Voltage Angle	Likely Technical Challenges
Topology detection	Direct SCADA on switches	Possibly fewer measurement points, independent validation	Algorithm using minimal placement
Unintentional island detection	Various	Possibly faster, greater sensitivity and selectivity, possibly less expensive	Speed
State estimation	Computation based on V mag (voltage magnitude) measurements	Possibly fewer measurement points, better accuracy, faster convergence	Algorithm using minimal placement
Reverse power flow detection	Detect with PQ sensor (V mag, I mag (current magnitude), and angle)	May extrapolate to locations not directly monitored	Algorithm using minimal placement
Fault location	Various	Possibly better accuracy (i.e., locate fault more closely with δ than V)	Need high resolution, fast data
High-impedance fault detection	Various, difficult	Possibly better sensitivity and selectivity with δ	Unknown
Dynamic circuit monitoring	High-resolution PQ instruments, none for δ	Uniquely capture oscillations, damping	Data mining for relevant phenomena
Oscillation detection	None	Unique	Unknown
Load and DG characterization	Limited observation with PQ instruments	Uniquely capture dynamic behaviors	Data mining, proximity to subject
Unmasking load/DG	None	May be unique	Unknown
FIDVR detection	Detected with V mag	Possibly less expensive, faster	Easy
FIDVR prediction	None	May be unique	Unknown

SCADA, supervisory control and data acquisition; FIDVR, Fault-induced delayed voltage recovery.

Volt-VAR optimization. Voltage angle measurement would not afford an inherent advantage over magnitude for feeder voltage optimization, but the capability to support this important function alongside other applications could add significantly to the business case for µPnet deployment.

5. Moving forward

This chapter proposed high-resolution voltage phase angle measurement as a new option for accurate and flexible monitoring and control of distribution networks in the presence of variability and uncertainty. The central hypothesis is that a broad range of specific diagnostic and control applications will depend on improved visibility and transparency of the distribution system, meaning better knowledge of the system state in real-time. To effectively manage distribution networks with high renewable penetration, demand response and distributed control, high-precision monitoring systems will be needed to provide clear, accurate, and complete observation of varying system behavior. The authors believe that μPMUs are a strong candidate for creating this functionality in an economical manner.

Before any of the specific applications discussed herein can be evaluated in practice, it will be necessary to simply observe what phenomena can, in fact, be detected at the resolution of the μPMU, and what can be reliably concluded from those observations. A key challenge will be to distill raw phase angle measurements into tools that support situational awareness and ultimately produce actionable operational intelligence, without the clutter of excess data. Also, to advance opportunities for active coordinated control based on μPMU measurements, the requirements for hierarchical, layered, distributed control of aggregated distributed resources and loads (in particular, islandable clusters) need to be studied in relation to voltage phase angle.

Active management of transparent distribution systems at high granularity in space and time may become a necessity simply to accommodate new resources without adverse impacts on power quality and reliability. It also implies exciting possibilities for new operating strategies (e.g., distributed resources might smoothly transition between connected and islanded states), and be capable of providing local power quality and reliability services on the one hand and support services to the core grid on the other hand, as desired at any given time. This type of flexibility is essentially a form of redundancy, without a clear business case at present. However, considerations of security and infrastructure resiliency may support the development of such strategies in the future. Whatever course the evolution of distribution systems takes, increased visibility and precise measurement are sure to be critical aspects of the new infrastructure.

References

[1] Jung J, Cho Y, Cheng D, Onen A, Arghandeh R, Dilek M, et al. Monte Carlo analysis of plug-in hybrid vehicles and distributed energy resource growth with residential energy storage in Michigan. Appl Energy August 2013;108:218–35.

[2] Li Z, Guo J. Wisdom about age [aging electricity infrastructure]. Power Energy Mag IEEE 2006;4:44–51.

[3] Brown RE, Humphrey BG. Asset management for transmission and distribution. Power Energy Mag IEEE 2005;3:39–45.

[4] Research ED, Group I. Failure to act, the economic impact of current investment in electricity infrastructure. American Society of Civil Engineers; 2011.

[5] Phadke AG, Thorp JS. Synchronized phasor measurements and their applications. Springer; 2008.

[6] von Meier A, Culler D, McEachern A, Arghandeh R. Micro-synchrophasors for distribution systems. In: IEEE 5th innovative smart grid technologies conference, Washington D.C; 2014.

[7] Katz RH, Culler DE, Sanders S, Alspaugh S, Chen Y, Dawson-Haggerty S, et al. An information-centric energy infrastructure: the berkeley view. Sustainable Comput Inf Syst 2011;1:7–22.

[8] Dawson-Haggerty S, Jiang X, Tolle G, Ortiz J, Culler D. sMAP: a simple measurement and actuation profile for physical information. In: Proceedings of the 8th ACM conference on embedded networked sensor systems; 2010. pp. 197–210.

[9] Schweitzer E, Whitehead D, Zweigle G, Ravikumar KG, et al. Synchrophasor-based power system protection and control applications. In: International Symposium Modern Electric Power Systems (MEPS); 2010. pp. 1–10.

[10] Schweitzer E, Whitehead DE. Real-world synchrophasor solutions. In: The 62nd annual conference for Protective relay engineers; 2009. pp. 536–47.

[11] Wache M, Murray D. Application of synchrophasor measurements for distribution networks. In: Power and energy society general meeting, 2011. IEEE; 2011. pp. 1–4.

[12] Arghandeh R, Onen A, Jung J, Broadwater RP. Harmonic interactions of multiple distributed energy resources in power distribution networks. Electr Power Syst Res 2013;105:124–33.

[13] Lu N, Yang B, Huang Z, Bravo R. The system impact of air conditioner under-voltage protection schemes. In: Power systems conference and exposition, 2009. PSCE'09. IEEE/PES; 2009. pp. 1–8.

[14] Centeno V, King R, Cybulka L. Application of advanced wide area early warning systems with adaptive protection—phase 2. US Department of Energy, USA; 2011.

[15] Borghetti A, Nucci CA, Paolone M, Ciappi G, Solari A. Synchronized phasors monitoring during the islanding maneuver of an active distribution network. Smart Grid IEEE Trans 2011;2:82–91.

[16] Ochoa LF, Wilson DH. Angle constraint active management of distribution networks with wind power. In: Innovative smart grid technologies conference Europe (ISGT Europe), 2010. IEEE PES; 2010. pp. 1–5.

Big Data, Data Mining, and Predictive Analytics and High Performance Computing

35

Phillippe Mack

1. Introduction

The performance metrics of available storage, computing power, telecommunication bandwidth, and speed have reached numbers that can easily make our head spin. These metrics have evolved from the early needs of mankind to store, process, and communicate information.

In 1937, Herman Hollerith, who would later form IBM, designed a paper-punch machine to tabulate census data. It had taken the U.S. Census Bureau eight years to complete the 1880 census, but thanks to Hollerith's invention, that time was reduced to just one year. In 2013, for the same volume of matter, holographic technologies stored an average of more than 1 TB of data in compact disc format (Figure 1).

In the fourth century BC, Ancient Greeks were using hydraulic telegraphs to send messages. In 2013, a direct high-speed trading link between Chicago and New York cut the round-trip time by more than half a millisecond to just under 15 ms; it is capable of transferring several hundred billion information bytes per second.

Mechanical and electrical computing machines have been around since the nineteenth century. In 1941, Konrad Zuse built the Z3, which is regarded as the first working programmable, fully automatic modern computing machine. It was able to process 20 floating point operations per second. In 2013, the world's fastest supercomputer, National University of Defense Technology Tianhe in Guangzhou, China, was able to process 233.86 peta floating point operations per second.

In the last decade, the Internet has triggered changes in society as great as the industrial revolution in the eighteenth century, significantly affecting economic trends and dramatically changing people's personal and professional lives.

Information technology has evolved from localized hardware systems to privately connected networks and now to ambient and seamlessly interconnected devices. Memory, processing systems, and storage are now abstractions shared between billions of devices. The Internet is the real backbone of this cloud of devices that support and supply services that can be accessed from anywhere and are always available.

The amount of information generated by the systems that drive finance, manufacturing processes, health care, insurance, and utilities is reaching volumes that are becoming impossible to manage by humans without the support of computers and advanced algorithms.

Renewable Energy Integration. http://dx.doi.org/10.1016/B978-0-12-407910-6.00035-1

FIGURE 1

The first 5-MB IBM hard disk drive in 1956.

Computers were initially designed to compute faster and more reliably than humans. Hardware development has led to a dramatic increase in processing speeds, storage capacity, and data transmission speed. Computers have now become responsible for the huge and ever-increasing flow of information that, more and more often, can only be processed by smart algorithms to extract nuggets of valuable knowledge.

Sociologists, doctors, and marketers long ago understood that computers could help them understand the complex behavior of humans. In 1662, John Graunt published "Natural and Political Observations Made upon the Bills of Mortality," an analysis of mortality data to try to predict the outbreak of the bubonic plague in London. This is one of the first known and reported uses of data to detect patterns. Since then, exploring data has led to various research activities in statistics, neural networks, decision theory, and most recently machine learning.

Today, data mining is no longer confined to laboratories at universities. It has become an essential business tool. A lack of knowledge on how to leverage the constant flow of data is probably a competitive disadvantage. Data mining is at the heart of the success of leading companies, such as Amazon and Google. It is gradually being introduced in traditional industries as the key to better understanding customer behavior and as a layer of intelligence for their own operations and products, such as predictive maintenance of cars, power plants, trains, aircraft maintenance, voice recognition in phones, and patient health.

2. Sources of data in utilities

2.1 Renewable energy data

Renewable sources of energy, particularly wind and solar, have grown very fast in recent years. This is good news considering the huge challenges faced in limiting the impact of climate change and preserving the planet for future generations. However, this remarkable growth in renewables also comes with some major technical and economic challenges.

Compared to traditional sources of electrical energy, renewable sources are generally smaller capacity units scattered throughout the grid at the transmission and distribution levels. They are dependent on weather conditions, so their available capacity is uncertain. In a system where the balance between consumption and production is critical, increased uncertainties threaten the integrity of system performance. It is important to understand these uncertainties and to try to reduce their impact on grid performance by not integrating renewables indiscriminately.

Fortunately, electrical grid operators are investing a lot in smart grid technologies to monitor their operations. At the same time, renewable energy providers have installed systems and devices that provide advanced monitoring and control. Most of these systems generate huge volumes of information that needs to be managed and processed to their full potential to be useful for the various market players.

2.2 Grid operation data

For a transmission system operator (TSO), a supervisory control and data acquisition (SCADA) system is critical for properly monitoring grid operations. SCADA systems gather a lot of measurements, such as voltages, currents, power injection, and power consumption at each node of the grid. Operators can have a fairly accurate real-time view of the steady state of the transmission grid. This constant stream of information is usually processed by various power-system computer software applications, such as state estimators, optimal power flow, or static and dynamic security analyzers. These tools give operators a better insight about the current state of the grid, but they also generate another layer of information that operators need to process.

Experienced operators have been able to use this flow of information to build up expertise and adopt ad-hoc good practices. However, the opening of the electricity markets a few years ago has created a real challenge for even the most experienced operators. Operators have been forced to change their integrated production-transmission view of the system. Also, the numerous incentives to integrate more renewable sources combined with the exit of nuclear generation in some countries in the EU continue to challenge familiar views of system behavior.

Uncertainty on both the demand and generation sides of the market is increasingly the new normal. Continuous changes coupled with vexing complexities are coming at a pace faster than operators can absorb to allow them to develop new mental models of the system.

To try to overcome these challenges, new monitoring systems have to be installed on the grid. Phasor management units (PMUs), for example, should help operators diagnose problems earlier and prevent events that would threaten the dynamic security of the system.

Sources of data not directly connected with the grid's technical operations (e.g., weather conditions or market conditions) obviously can affect how the grid is used by market players and how energy flows on the grid.

Grid control centers are flooded with an amount of data that is almost impossible for operators to handle. To use this data efficiently, operators need tools to help them detect patterns in data that are important for decision making.

2.3 Renewable production data

The most common sources of renewable energy are biomass, wind, solar, and hydro. The main issue with renewable sources of energy, apart from biomass, is their strong dependency on local weather conditions. The uncertainties they inject can continuously unbalance a system. As the uncertainty inherent in weather forecasting is very difficult to eliminate totally, the grid and the electricity market will have to learn to deal more and more with increased operational uncertainty. In addition to the uncertainty induced by renewables, there are other sources of uncertainty that are discussed in other chapters of this book. The market players that can successfully use available data to forecast and reduce operational uncertainties will gain a significant competitive advantage.

2.4 Wind farm data

Typically, modern windmills have more than 300 data points that are used for control purposes and to monitor degradation of mechanical, electrical, and structure components. Vibration sensors are commonly available in wind turbines, mainly to monitor the health of particular components. For example, shock pulse methods are used to diagnose the degradation of ball bearings. More and more sophisticated and dedicated sensors are available on the market. With current information and communication technologies, there are no technical or economic impediments to collecting and storing data in a data center. However, the value of the information brought by the sensors is limited if the data are not correlated with other measurements and the potential high value in the data cannot be used to optimize these assets.

Operation and maintenance of wind turbines constitutes a significant share of the annual costs of a wind turbine, up to 20–25% of the cost per kilowatt supplied. Diagnostic centers are being built to store data and monitor operation of wind farms. Centralized operation and control systems help to reduce operation and maintenance costs significantly. However, a diagnostics center is more like a data center than a real intelligence center, where data scientists mine data and improve and predict performance of wind farms.

2.5 Solar panel data

Solar panels are less complex to operate and maintain than wind turbines, but their large-scale integration can also have significant impacts at the distribution level. Electrical distribution networks have been designed to supply flows of electricity downward from the transmission grid to small loads, such as home users or small- and medium-size enterprises. Initially, the grid was not designed to support upward flows of energy from solar panels. When the sun shines in an area where a lot of photovoltaic (PV) panels are installed, it can cause thermic and overvoltage issues for the grid.

Ideally, solar panels managed at a domestic level would be part of a smart metering infrastructure that delivers consumption and production information at a higher sampling rate (i.e., several times a day). Also, solar panels are bundled with inverters that could be used by the distribution system operator (DSO) or TSO to provide distributed ancillary services for voltage or even frequency control. Together with smart appliances such as ovens, refrigerators, air conditioning, and heating systems, PV panels would be able to react to signals sent by the energy supplier or the DSO. In addition, DSO could manage information, such as power from inverters, to analyze and centrally control problems related to inadequate local voltage.

Output from solar panels can be greatly reduced because of airborne dust, bird droppings, or even snow. Monitoring and diagnosing the yield performance properly is also something that needs to be considered when collecting and processing solar panel data.

2.6 Load metering data

Advanced metering infrastructure (AMI) is deployed in many countries. AMI enables two-way communication meters reading electrical consumption at a higher frequency. This information can be processed in real-time and signals sent to manage demand. Demand-side management requires a lot of data processing to understand the load patterns and to design proper signals that enable optimal use of the distribution grid and manage reserve and frequency response.

Collecting data at each meter is critical for energy suppliers and distribution system operators to better understand the patterns of electrical energy users so they can elaborate market signals and understand customer behavior.

2.7 Market data

The rules that govern liberalized, competitive electricity markets play an important role because external factors, such as oil prices and market subsidies, can have a significant impact in determining the supply-demand equilibrium. Intraday electricity markets, ancillary services, balancing markets, and capacity markets provide a continuous flow of information about market conditions and indirectly about electrical system conditions, adding to the data tsunami in the control center.

2.8 Simulation data

Besides real-time data collected from grid operations, simulation tools used in system and operational planning generate large amounts of data to support decision-making and planning for the short term and long term.

Specific local weather simulators can also be used to forecast capacity of wind, hydro, and solar generation.

3. The big data era

As described, data can be generated by many sources in electric power systems and market operations. As showned in Figure 2, the hardware cost to store all this data is getting cheaper and cheaper. The intelligence hidden in these piles of data can certainly be extracted to better integrate renewable energy.

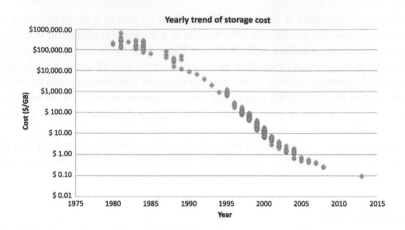

FIGURE 2

Trend of data storage cost.

We can certainly speak in terms of Big Data when we look at the multiple velocity sources and volumes of data involved in renewable energy integration. Standard data management systems would have difficulty in managing such high volumes and constant flows of information efficiently.

After the blackouts that occurred in the United States in 2003 and 2006, the Department of Energy and the Federal Energy Regulatory Commission recommended that utilities and grid operators install synchro-phasor-based transmission monitoring systems to collect the real-time data. PMUs increase data significantly, up to 6.2 billion data points per day at a size of up to 60 gigabytes using 100 synchrophasors. This is Big Data!

3.1 Tools to manage big data

Scientific applications, such as the Square Kilometer Array radio telescope, can produce 50 teraoctets of analyzed data from a stream of 7000 teraoctets of raw data per day. Electrical power systems and market operations do not generate such enormously high volumes of data but can still benefit from the technologies that are now mature and used by Google, Amazon, and Facebook and in commercial applications in banking, finance, and the telecom industries.

Standard data management systems cannot manage such Big Data. However, in the past few years, many initiatives have been launched to develop new technologies that can store such high volumes of information. New technologies such as MongoDB, Cassandra, CouchDB, and Dynamo deliver cloud-based, open-source data tools for industries from e-commerce to telecommunication. In addition, NoSQL databases are an important building block for the solutions that would enable smart grids and better integration of renewable energy.

Some of these technologies are not yet mature and may perform differently depending on the context. Even in simple benchmark tests, significant performance gaps can be observed between different NoSQL (Figure 3) and SQL technologies. This means that technologies must be assessed carefully to match the requirements.

3.2 Data mining

Data mining is a set of processes intended to transform raw data into new and useful knowledge. Standard guidelines have been defined for implementing a data mining process. When implementing a data mining project, it is important to follow a standard and proven methodology to avoid common pitfalls and reduce risks. One of the most widely used is the cross industry standard process for data mining (CRISP) [2] methodology (Figure 4).

3.2.1 Data mining techniques

There are many methods available to mine large data sets for valuable information. From visualization to the most advanced machine learning algorithms, there are a plethora of tools to extract knowledge from data. Common machine learning algorithms can be applied to typical problems in power systems. All algorithms use a data table as an input where the rows are called the objects (a snapshot of the transmission grid state identified by a timestamp for example) and the columns are called the attributes (e.g., the voltages levels, productions, and currents in lines of the transmission grid state).

Machine learning algorithms are typically divided into two main types of algorithms: supervised and unsupervised. Unsupervised algorithms are useful when you do not have a specific output variable to explain or predict. Typically, this category of algorithm is used for:

1. Link analysis problems to discover correlations or associations between attributes. Link analysis would be used to discover a group of measurements that are strongly correlated, such as voltages of the nodes located in an area.
2. Clustering problems to detect clusters of objects that have similar behavior, such as states of the power grid that are similar.

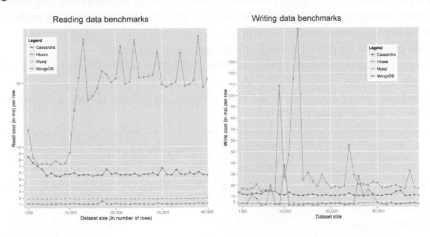

FIGURE 3

NoSQL read/write database performance.

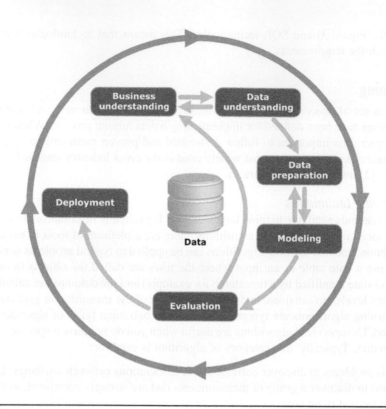

FIGURE 4

CRISP methodology.

Supervised algorithms are useful when you have a value (an output attribute) that you want to explain, predict, or model with other attributes (input attributes). Typically, this category of algorithm is used for:

1. Classification problems to predict an attribute having discrete and symbolic values.
2. Regression problems to predict an attribute having numerical values.
3. Optimization problems to discover the conditions that lead to a higher value or lower value of the output.

The following table shows examples of the most common algorithms in each category and also the type of problem where it can be used [3].

Algorithms	Type of Problem				
	Link Analysis	Clustering	Optimization	Classification	Regression
Association rule	x				
Bayesian networks	x			x	x
K-Means		x			
Hierarchical clustering	x	x			
PRIM analysis			X		
K nearest neighbor				x	x

	Type of Problem				
—cont'd					
Algorithms	**Link Analysis**	**Clustering**	**Optimization**	**Classification**	**Regression**
Artificial neural networks				X	X
Decision tree				X	X
Ensemble trees				X	X
Support vector machine				X	X
Multilinear regression					X

3.2.1.1 Association rules
Association rules are commonly used for basket analysis. These kinds of algorithms are used by companies, such as Amazon, to discover products that are bought together. They can be used in power systems to discover patterns in the status of breakers.

3.2.1.2 Bayesian networks
Bayesian networks and influence diagrams are models that can be partially or completely built by an expert. The parameters interact in the same way as in a decision tree and can be explicitly visualized, which makes it a great tool to model and store the knowledge extracted from historical data and/or from experts.

3.2.1.3 K-Means
K-Means is one of the most popular clustering techniques based on Euclidian distance between objects. K is the number of clusters you want to extract from the data set.

3.2.1.4 Hierarchical clustering
Hierarchical clustering is a model that is viewed as a dendrogram (Figure 5). A dendrogram is a tree diagram frequently used to illustrate the arrangement produced by hierarchical clustering.

3.2.1.5 PRIM analysis
PRIM (Patient Rule Induction Method) is an algorithm that was introduced by Friedman and Fisher in 1999. Its objective is to find subregions in the input space with relatively high or low values for the target variable.

3.2.1.6 K nearest neighbor
Case-based reasoning algorithms are a very simple approach. They use neighboring cases to infer the properties of new objects. K nearest neighbor (K-NN) is an example of a popular case-based reasoning algorithm.

3.2.1.7 Artificial neural networks
Neural networks are effective machine learning algorithm sets that are able to model complex nonlinear problems. Neural networks have been used to predict electricity prices using various stochastic inputs, such as weather forecasts and assets availability. However, they are black-box models that are tricky to tune and maintain.

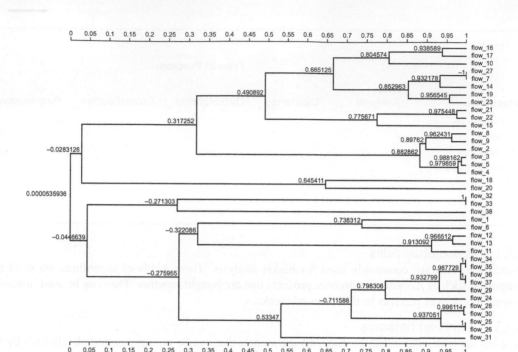

FIGURE 5

Example of a dendrogram computed on power lines flow.

3.2.1.8 Decision trees

Decision trees (Figure 6) are commonly used because they predict the output as well as explain how the value of the output is decided. The set of "if-then" rules of a decision tree can be easily understood and validated by domain experts. Decision trees can be used on historical and simulated data to build security rules for operators or directly for automatic control systems.

3.2.1.9 Ensemble trees

Ensemble trees are effective machine learning techniques that aggregate several models (e.g., several decision trees) to obtain better predictive performance and robustness. Ensemble trees are also useful for ranking and selecting the most important attributes that influence output.

3.2.1.10 Support vector machines

Support vector machines are the next generation of machine learning algorithms. They can be used to model highly complex nonlinear problems, but they are much more complex than neural networks and usually require strong expertise in the area.

3.2.1.11 Multilinear regression

Linear regression models are used to model linear relationships between the output attributes and the input attribute. Linear regression is a very popular technique because many problems can be modeled accurately enough with linear relationships.

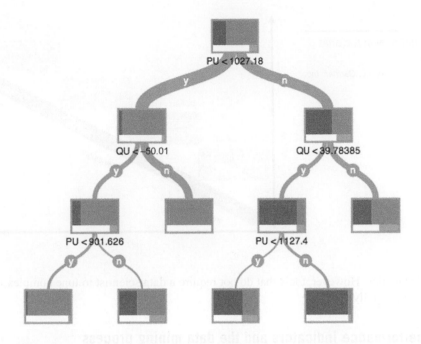

FIGURE 6

Example of a decision tree.

3.3 Predictive analytics

Predictive analytics is a subset of data mining dedicated to the construction of predictive models. The goal of predictive analytics is to identify useful correlated patterns and parameters to predict the occurrence of events based on past data. In business analytics terminology, predictive analytics usually follows the descriptive analytics step, which is more focused on understanding and describing past events. Prescriptive analytics would then use the outputs of the descriptive and predictive analytics steps to recommend actions that would improve business performances. Figure 7 shows a commonly accepted methodology.

Predictive analytics and prescriptive analytics would typically input content to a situation awareness system so that operators can take action to prevent the impact of forecast events. For example, based on one-day forecasts of available renewable energy, it would recommend some steps to maximize the security of the transmission grid.

Predictive analytics is a class of technology that, if applied regularly in the area of power grid operations, would make the grid smarter and more responsive in a near future. It is also a very attractive set of technologies for managing and reducing uncertainties induced by renewable energy integration and demand-side management systems.

The main barriers to a broader adoption of predictive analytics are the lack of knowledge about the tools and the unavailability of tools that are easy to use by the nonexperts in the field of machine

Value versus difficulty in the analytics journey.

Source: Gartner, Inc.

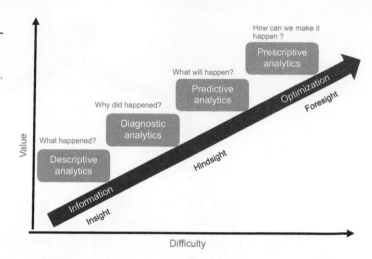

learning and statistics. However, tools that do not require a data scientist to tune complex models are becoming commercially available.

3.4 Key performance indicators and the data mining process

Key performance indicators (KPIs) are metrics usually used by managers to support their decision-making process. They provide synthetic information on how well technical or organizational processes are performing. A KPI must be clearly and well defined—how to compute it, who will use it, when to provide it, and how to provide it—so it is used and understood by the right people: For example, power imbalance is a KPI that is monitored in real-time on SCADA screens by operators in the control room of an electrical transmission system. The cost of imbalance is a KPI monitored on a weekly or monthly basis in the reporting system of managers of transmission system operators. The links between the two indicators have to be taken into account when aligning objectives for operators and managers.

However, merely reporting an indicator is not enough. What can the manager do if the KPI is drifting? Which decision should be taken to improve performance safely and how can it be sure that when progress is reported, the reason is well understood, under control, and sustainable? How should the manager avoid unexpected drifts in the future and mitigate them?

Here, Big Data and data mining can play a significant role. The KPI might be just the visible part of the iceberg. Data mining can sort through the oceans of collected data very efficiently and select the specific data that are key to diagnosing variability and drift in performance, predicting performance trends, and prescribing and planning proper decisions to sustain and improve performance.

In general, KPIs and continuous improvement programs, such as Six Sigma, are critical to the successful integration of Big Data and analytics paradigms.

3.5 High performance computing and big data analytics

Big Data would have little value if computing power were not adequate to extract valuable knowledge from huge volumes of information. A few years ago, when we talked about high-performance

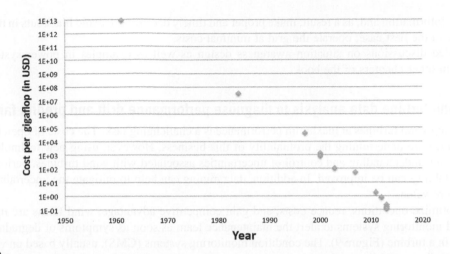

FIGURE 8

Trend of cost per gigaflop.

computing, we thought of supercomputers. Although hardware supercomputers, such as quantum computers, are still a hot topic, techniques such as virtualization and in-memory computing offer competitive solutions. Virtualization is dramatically democratizing high-performance computing as it becomes available at a much lower cost (Figure 8).

Infrastructure such as AMAZON EC2, which offers computing power as a service in their elastic cloud, is probably the best known service available today. The Hadoop Map/Reduce paradigm is a good example of algorithms that take advantage of computing resources scattered in the cloud and aggregating them in a smart way.

Thanks to distributed architecture, not only data can be stored and retrieved efficiently, but computing power can also be mutualized and distributed in a cloud of services. For example, the Raspberry PI foundation has developed a cheap PC-like board the size of a credit card for less than $25. Using open-source software, very large and compact clusters of several thousand nodes can be easily built for a very competitive price and offering. Facebook is another interesting example; they build large clusters (farms) of Mac Minis as part of the backbone of their social network.

4. Examples of applications

4.1 Visualization and grid situational awareness

The need to provide operators with a real-time and accurate view of the current state of the power system has recently focused development in power systems applications on situational awareness (SA). SA tools include dashboards of KPIs that synthesize the actual performance of the grid on multiple criteria such as losses, security margins, and other indicators.

SA tools allow operators to quickly focus their attention where and when it is required most. Modern visualization technologies can greatly assist operators to find the needle in the haystack of available information and, as a result, make proper and timely decisions to avoid blackouts in the worst case and, in the best case, operate the grid at minimal costs.

Detailed discussions on situation awareness design as well as examples in power systems are detailed in other chapters of the book.

4.2 Wind turbine data analysis to diagnose performance drift and predict failure

Maintaining wind turbines at maximum performance is a challenging task. The costs of operation and maintenance are undermining the profitability of this business. However, using new technologies to gain a better understanding and control of uncertainties associated with wind turbines, performance, and profitability can be improved. In addition, data mining can help to mitigate the uncertainties and create more value.

To optimize operations, reduce costs, and gain competitive advantage, wind farms are installing advanced monitoring systems to alert the maintenance team as soon as symptoms of degradation are detected in a turbine (Figure 9). The condition monitoring systems (CMS), usually based on vibration analysis to diagnose failures of rotating equipment such as wind turbines, are quite limited and require fine tuning of thresholds. They can be difficult to manage and generate many false alarms, so when the maintenance team gets on site there is often nothing to do.

Data mining has proven it can deliver reliable solutions to extract critical parameters and patterns to support the design of smart agents that screen the whole set of measurements, including historical data combined with standard CMS indicators and the measurements used for turbine operations. If multivariate symptoms of failure are detected, it automatically sends a work order or a request to the asset management system. With the support of advanced analytics, experts in the diagnostics center decide on what action to take to get the wind turbine back to optimal performance.

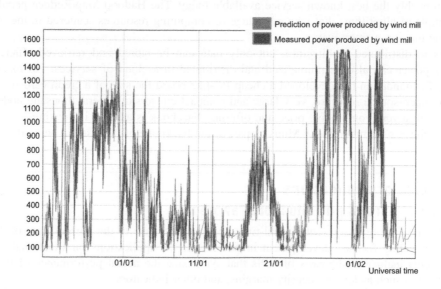

FIGURE 9

Predicted windmill power production versus actual production.

4.3 Predictive analytics to support dynamic security assessment in a control center

The major challenge faced by European TSOs is the increasing demand to integrate intermittent generation. In particular, uncertainties significantly complicate the traditional deterministic paradigm of dynamic security assessment. Data mining, used in the both offline and online applications, can help deal with the uncertainties so that traditional deterministic tools can move from a common deterministic approach to one that is a probabilistic.

In offline applications, data mining helps to identify the distribution density of various parameters such as injection of renewable energy at nodes in the transmission grid, distribution of transients in power lines and how these variables are correlated (e.g., the probability that two wind farms inject correlated amounts of power). This helps to create a set of realistic and representative starting points and trajectories of the grid states for the next two days. Based on the Monte Carlo method for realistic simulations, an offline dynamic security assessment can be used to qualify the security level of a particular state of the gird [1].

These simulations require strong computing power to be performed on a daily basis. For example, as soon as a database of possible scenarios is available from the simulated datasets, data mining can, in real-time, help predict dynamic security margins, taking into account uncertainties of renewable energy injections.

4.4 Predictive analytics to support industrial demand-side management

Integration of renewable energy, such as wind and solar, also means opportunities for other market players. Many energy-intensive industries have the opportunity to use flexibility in their energy consumption to compensate for imbalances due to the uncertainties of wind and solar generation.

Analytics on historical operations data can help these industries better understand their energy patterns and identify real-time flexibility in their energy production. With the support of predictive analytics, manufacturing and process industries can more accurately forecast their energy consumption and eventually deliver the value of their flexibility as a service to the TSO to compensate for the uncertainties of renewable energy.

---

5. The future is now

In this chapter, we have seen that utilities faced with the many challenges of massive renewable integration could greatly benefit from data mining and Big Data technologies. In the smart grid era, Big Data and data mining technologies are essential components to leverage the value of each market player and to optimize the use of assets.

However, Big Data and analytics are still a long way from being considered as a core component in the development of smart grid projects. This may explain why, despite the huge investments made by utilities in smart meters, most of them still struggle to get significant business value from the continuous flow of information they collect. The trend is clear that in tomorrow's economy analytics will be the cornerstone of business applications (Figure 10).

Integrating Big Data and analytics is not a trivial task. We have seen that there are numerous mature technologies available. It is very important to assess the right mix of technologies to design a Big Data

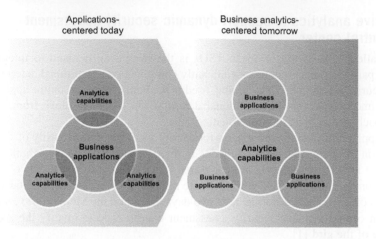

Applications-centered today

Business analytics-centered tomorrow

FIGURE 10

Analytics at the core of business and IT.

Gartner.

solution that matches business needs. Big Data and data mining projects have to be driven by the business, not by technologists.

It is also clear that there is still a significant gap between what the energy business understands about Big Data and what Big Data can actually provide as value for the energy business. Training and communications are key to aligning energy business expertise and Big Data science. Opportunity exists to develop new solutions based on cross-thinking of data scientists and energy experts.

References

[1] Wehenkel L. Automatic learning techniques in power systems. Boston: Kluwer Academic; 1998.
[2] Shearer C. The CRISP-DM model: the new blueprint for data mining. J Data Warehousing 2000;5:13–22.
[3] PEPITE SA, http://docs.mydatamaestro.com/.

Epilogue

J. Charles Smith

In the words of Yogi Berra, that great baseball icon, "It is tough to make predictions, especially about the future". Recognizing his words of caution, I will go ahead and make a prediction that we will see a future of ever increasing amounts of variable generation in the form of wind power and solar photovoltaics on the power system. And along with that will come all of the variability and uncertainty that they bring, as well as an ever-increasing fraction of asynchronous generation to deal with. How much variable generation? Would you believe 40% of annual energy produced for load? How about 50%? Those levels are being studied seriously right now. And at those levels of energy penetration, you can be sure that there will be many hours when the capacity penetration will be at or above 100%, depending on export and storage opportunities. Do those numbers sound high to you? Ten years ago, 10% sounded high. Only 5 years ago, 20% sounded high. Today, 30–40% sounds high. But if the past is any guide, we will be talking routinely of 40–50% in another 5–10 years, and higher numbers after that. There is nothing that a power systems engineer worth his salt enjoys likes a good challenge, and integration of high levels of variable generation like wind and solar energy are just the challenge!

The rate of growth and the exact path that the growth follows will depend in large part on energy policy that is yet to be made in the US, and perhaps less so in Europe and China. But several requirements are clear. The growth cannot take place, and the markets cannot operate efficiently without a corresponding growth in the transmission system. Transmission is the glue that holds the system together and allows for system expansion, and without it, the system will be inherently constrained. Most jurisdictions around the world have now recognized this, and you hardly hear a discussion of big wind anymore without hearing a corresponding discussion about transmission. Ten years ago in the US, many state legislatures passed Renewable Portfolio Standards (RPS) without considering transmission, but now even state legislatures are talking about transmission as they reexamine their RPS legislation. There is a lot of discussion today about the need for a Smart Grid, which incorporates increased levels of monitoring, communication, and control, especially at the distribution level. The term is also applied to power electronic controllers on the transmission system, but the fact remains that a significant amount of wire in the air will be required to meet aggressive renewable energy targets in the future. The advent of voltage source converters for High-voltage direct current lines will allow more innovative and customized transmission solutions to be developed in the future.

The requirement for increased sources of flexibility to manage the inherent variability in the wind and solar generation to maintain system reliability is growingly increasingly visible as well. What do we mean by flexibility? In the past, capacity adequacy, in units of MW, was the primary metric used to determine system reliability. Today, it is not only MW, but the ability to move the MW rapidly (MW/min) that needs to be provided to manage the system and maintain reliable operation, and that's what we mean by flexibility. Flexibility comes from many sources that have been well explained in the preceding chapters; the supply side; the demand side; energy storage, market rules; and operating policies and procedures. In the future, I think we will see a greater emphasis placed on demand side sources of flexibility: Think aggregated demand side resources, plug hybrid electric vehicles, integration of heat systems with electric systems. The opportunities are huge.

Variable generation forecasting is a third requirement for the reliable integration of large shares of variable generation into the electricity system of the future. We have also heard about this in previous

455

chapters; day ahead forecasting for improved economics of operation under a unit commitment and economic dispatch paradigm; short-term ramp forecasts for more secure and reliable operation; near real-time forecasts for incorporation into the 5-min dispatch and for improved situational awareness; and nodal forecasts for minimizing transmission congestion. Great strides have been made in forecasting in the past 10 years, but there is still more to be done. Ensemble forecasting, artificial intelligence learning systems, and probabilistic forecasts all have continued contributions to make.

Market design and operation remain an evolving arena where significant gains can be achieved. There is now a near-universal recognition of the large operational and economic benefits to be gained from the aggregation of VG over large geographic areas, corresponding to a large market footprint. This allows a reduction in the per unit variability of the wind generation fleet due to the decreasing correlation in wind plant output with increasing distance, a reduction in the wind plant output forecast error, and an increased generation fleet with which to balance the system. In addition to large markets, fast markets also offer significant benefit to the ability to integrate large volumes of VG. With an hourly dispatch, balancing within the hour is done with regulation resources, which are the most expensive form of generation. If the generation participates in a 5-min economic dispatch, tremendous flexibility can be extracted from the generation stack, and regulation only need be supplied for the deviations within the 5-min interval. The ability to update the wind plant forecast close to real time is also important, because the accuracy of the forecast improves significantly as you get closer to real time. Trying to operate the system in real time with yesterday's forecast no longer makes sense; we can do much better with today's forecasting tools.

The exact nature of the markets to be operated is a continuing area of discussion, and sometimes heated discussion! Perhaps one of the largest discussions in that regard is over the need or lack thereof for capacity markets. In some regions, there is a co-optimized energy and ancillary service market, while other regions have the energy and ancillary service market in addition to a capacity market. The discussion comes about due to the fact that zero marginal cost wind energy can depress the prices in energy markets, and with the regulatory caps on energy market prices, there is growing concern that conventional generators will not be able to generate sufficient revenue to cover the amortization of their capital costs in an energy-only market. Some regions have capacity markets, which have been in operation for 5–10 years, and have been doing a good job of providing incentives for new generation capacity and demand resources to maintain desired levels of system reliability. And some regions still have a regulatory paradigm to ensure capacity adequacy. This discussion is going on in Europe in a vigorous fashion as we enter 2014, and its outcome will in large part determine the ultimate success of the electricity market integration effort in Europe. The discussion also continues in the US, and remains to be settled. It is likely that both types of market structures may continue to coexist for a while.

In addition to energy and capacity markets, there are also ancillary service markets and price responsive load markets. Price responsive load tends to be thought of as an energy product, while demand response tends to be thought of as a capacity product. Ancillary service markets are slowly evolving to allow wind generation to participate, but it is a work in progress. Another current topic is the provision of market signals to incentivize flexibility. There is currently a vigorous debate over whether flexibility should be incentivized through a separate market product, or whether strong price signals in an energy market will be sufficient, given the inherent flexibility in the generation stack. This discussion is also ongoing, and will likely continue for the foreseeable future.

In addition to the challenges mentioned above, there is a grand challenge which faces the next generation of power engineers and that is what to do when the last synchronous machine comes off the system. This is a question, which ought to keep the best minds at the universities occupied for the foreseeable future, and system operators awake at night; how to design and operate a power system with no synchronous machines. What then does synchronism mean? Where is the synchronizing torque? What does stability mean? What about low short-circuit ratio considerations? Right now there are limits imposed on VG penetration on some systems due to its asynchronous characteristics. How will they be overcome? Answering these questions will require us to look at system dynamics in a whole new way. But it will certainly provide a new challenge for power engineers and faculty for a long time to come.

Another challenge in the flexibility arena is to think of the possibilities offered if we dispatch load to meet available generation. It may seem a little bit of a stretch now, but if retail customers become exposed to real-time prices, and if those prices are low or negative for many hours, a whole new paradigm could develop. It may also drive more integration of electric and thermal systems, as is already happening in Denmark with the electric system and district heating system with electric boilers. Many people also talk about the grand challenge of energy storage systems. While early applications are being found for short-duration storage systems, price is still a major barrier, especially when you realize that we really need seasonal storage to really take advantage of the seasonal renewable energy cycles, and that is a long ways off, but a laudable goal.

Wind power plants are a very capable source of energy when looked at in terms of the services they can currently supply to the grid. They can provide very fast real and reactive power control during both normal and abnormal conditions, down regulation whenever they are on line and up regulation if they are spilling wind, frequency response with a droop characteristic, and inertial response. Wind plants already perform as a dispatchable resource within limits in some markets. Strong grid codes are essential to secure this performance, and it is in everyone's best interest to require strong grid codes. The market will determine how the owners are compensated for providing these features.

Society is calling for a decarbonized energy future, and wind and solar energy are well positioned to supply a significant part of that future. The interesting thing is that we don't need to wait for all of the challenges above to be addressed, we are already well along the road. But we cannot rest on our laurels; we must continue to make progress down the road to meet all of the challenges of the future. In the words of the singer/songwriter Bob Dylan, "The answer, my friend, is blowin' in the wind", and as we look over our shoulder, what do we see but "here comes the sun"!

Index

Printed and bound by CPI Group (UK) Ltd, Croydon, CR0 4YY

03/10/2024

01040323-0006